Lévy Processes and Stochastic Calculus

CAMBRIDGE STUDIES IN ADVANCED MATHEMATICS

Editorial Board

Bela Bollobas, *University of Memphis*
William Fulton, *University of Michigan*
Frances Kirwan, *Mathematical Institute, University of Oxford*
Peter Sarnak, *Princeton University*
Barry Simon, *California Institute of Technology*

All the titles listed below can be obtained from good booksellers or from Cambridge University Press. For a complete series listing visit http://publishing.cambridge.org/stm/mathematics/csam

75 T. Sheil-Small *Complex polynomials*
76 C. Voisin *Hodge theory and complex algebraic geometry, I*
77 C. Voisin *Hodge theory and complex algebraic geometry, II*
78 V. Paulsen *Completely bounded maps and operator algebras*
82 G. Tourlakis *Lectures in logic and set theory I*
83 G. Tourlakis *Lectures in logic and set theory II*
84 R. Bailey *Association schemes*
85 J. Carlson *et al. Period mappings and period domains*
89 M. C. Golumbic & A. N. Trenk *Tolerance graphy*
90 L. H. Harper *Global methods for combinatorial isoperimetric problems*
91 I. Moerdijk & J. Mrčun *Introduction to foliations and Lie groupoids*

Lévy Processes and Stochastic Calculus

DAVID APPLEBAUM
University of Sheffield

PUBLISHED BY THE PRESS SYNDICATE OF THE UNIVERSITY OF CAMBRIDGE
The Pitt Building, Trumpington Street, Cambridge, United Kingdom

CAMBRIDGE UNIVERSITY PRESS
The Edinburgh Building, Cambridge CB2 2RU, UK
40 West 20th Street, New York, NY 10011–4211, USA
477 Williamstown Road, Port Melbourne, VIC 3207, Australia
Ruiz de Alarcón 13, 28014 Madrid, Spain
Dock House, The Waterfront, Cape Town 8001, South Africa

http://www.cambridge.org

© Cambridge University Press 2004

This book is in copyright. Subject to statutory exception
and to the provisions of relevant collective licensing agreements,
no reproduction of any part may take place without
the written permission of Cambridge University Press.

First published 2004
Reprinted 2005

Printed in the United Kingdom at the University Press, Cambridge

Typeface Times 10/13 pt.   *System* LATEX 2_ε [TB]

A catalogue record for this book is available from the British Library

Library of Congress Cataloguing in Publication data
Applebaum, David, 1956–
Lévy processes and stochastic calculus / David Applebaum.
p.   cm. – (Cambridge studies in advanced mathematics; 93)
Includes bibliographical references and index.
ISBN 0 521 83263 2
1. Lévy processes.   2. Stochastic analysis.   I. Title.   II. Series.
QA274.73.A67 2004
519.2′2–dc22   2003063882

ISBN 0 521 83263 2 hardback

To Jill

And lest I should be exalted above measure through the abundance of revelations, there was given to me a thorn in the flesh, a messenger of Satan to buffet me, lest I should be exalted above measure.

Second Epistle of St Paul to the Corinthians, Chapter 12

The more we jump – the more we get – if not more quality, then at least more variety.
James Gleick *Faster*

Contents

	Preface	*page* ix
	Overview	xv
	Notation	xxiii

1	**Lévy processes**	**1**
1.1	Review of measure and probability	1
1.2	Infinite divisibility	20
1.3	Lévy processes	39
1.4	Convolution semigroups of probability measures	58
1.5	Some further directions in Lévy processes	63
1.6	Notes and further reading	67
1.7	Appendix: An exercise in calculus	69

2	**Martingales, stopping times and random measures**	**70**
2.1	Martingales	70
2.2	Stopping times	78
2.3	The jumps of a Lévy process – Poisson random measures	86
2.4	The Lévy–Itô decomposition	96
2.5	The interlacing construction	111
2.6	Semimartingales	115
2.7	Notes and further reading	116
2.8	Appendix: càdlàg functions	117

3	**Markov processes, semigroups and generators**	**120**
3.1	Markov processes, evolutions and semigroups	120
3.2	Semigroups and their generators	129
3.3	Semigroups and generators of Lévy processes	136
3.4	L^p-Markov semigroups	148

3.5	Lévy-type operators and the positive maximum principle	156
3.6	Dirichlet forms	164
3.7	Notes and further reading	176
3.8	Appendix: Unbounded operators in Banach spaces	176

4	**Stochastic integration**	**190**
4.1	Integrators and integrands	190
4.2	Stochastic integration	197
4.3	Stochastic integrals based on Lévy processes	205
4.4	Itô's formula	218
4.5	Notes and further reading	245

5	**Exponential martingales, change of measure and financial applications**	**246**
5.1	Stochastic exponentials	247
5.2	Exponential martingales	250
5.3	Martingale representation theorems	262
5.4	Stochastic calculus and mathematical finance	267
5.5	Notes and further reading	288
5.6	Appendix: Bessel functions	289

6	**Stochastic differential equations**	**292**
6.1	Differential equations and flows	293
6.2	Stochastic differential equations – existence and uniqueness	301
6.3	Examples of SDEs	314
6.4	Stochastic flows, cocycle and Markov properties of SDEs	319
6.5	Interlacing for solutions of SDEs	328
6.6	Continuity of solution flows to SDEs	331
6.7	Solutions of SDEs as Feller processes, the Feynman–Kac formula and martingale problems	337
6.8	Marcus canonical equations	348
6.9	Notes and further reading	357

	References	360
	Index of notation	375
	Subject index	379

Preface

The aim of this book is to provide a straightforward and accessible introduction to stochastic integrals and stochastic differential equations driven by Lévy processes.

Lévy processes are essentially stochastic processes with stationary and independent increments. Their importance in probability theory stems from the following facts:

- they are analogues of random walks in continuous time;
- they form special subclasses of both semimartingales and Markov processes for which the analysis is on the one hand much simpler and on the other hand provides valuable guidance for the general case;
- they are the simplest examples of random motion whose sample paths are right-continuous and have a number (at most countable) of random jump discontinuities occurring at random times, on each finite time interval.
- they include a number of very important processes as special cases, including Brownian motion, the Poisson process, stable and self-decomposable processes and subordinators.

Although much of the basic theory was established in the 1930s, recent years have seen a great deal of new theoretical development as well as novel applications in such diverse areas as mathematical finance and quantum field theory. Recent texts that have given systematic expositions of the theory have been Bertoin [36] and Sato [274]. Samorodnitsky and Taqqu [271] is a bible for stable processes and related ideas of self-similarity, while a more applications-oriented view of the stable world can be found in Uchaikin and Zolotarev [297]. Analytic features of Lévy processes are emphasised in Jacob [148, 149]. A number of new developments in both theory and applications are surveyed in the volume [23].

Stochastic calculus is motivated by the attempt to understand the behaviour of systems whose evolution in time $X = (X(t), t \geq 0)$ contains both deterministic and random noise components. If X were purely deterministic then three centuries of calculus have taught us that we should seek an infinitesimal description of the way X changes in time by means of a differential equation

$$\frac{dX(t)}{dt} = F(t, X(t))dt.$$

If randomness is also present then the natural generalisation of this is a stochastic differential equation:

$$dX(t) = F(t, X(t))dt + G(t, X(t))dN(t),$$

where $(N(t), t \geq 0)$ is a 'driving noise'.

There are many texts that deal with the situation where $N(t)$ is a Brownian motion or, more generally, a continuous semimartingale (see e.g. Karatzas and Shreve [167], Revuz and Yor [260], Kunita [182]). The only volumes that deal systematically with the case of general (not necessarily continuous) semimartingales are Protter [255], Jacod and Shiryaev [151], Métivier [221] and, more recently, Bichteler [43]; however, all these make heavy demands on the reader in terms of mathematical sophistication. The approach of the current volume is to take $N(t)$ to be a Lévy process (or a process that can be built from a Lévy process in a natural way). This has two distinct advantages.

- The mathematical sophistication required is much less than for general semimartingales; nonetheless, anyone wanting to learn the general case will find this a useful first step in which all the key features appear within a simpler framework.
- Greater access is given to the theory for those who are only interested in applications involving Lévy processes.

The organisation of the book is as follows. Chapter 1 begins with a brief review of measure and probability. We then meet the key notions of infinite divisibility and Lévy processes. The main aim here is to get acquainted with the concepts, so proofs are kept to a minimum. The chapter also serves to provide orientation towards a number of interesting theoretical developments in the subject that are not essential for stochastic calculus.

In Chapter 2, we begin by presenting some of the basic ideas behind stochastic calculus, such as filtrations, adapted processes and martingales. The main aim is to give a martingale-based proof of the Lévy–Itô decomposition of an arbitrary Lévy process into Brownian and Poisson parts. We then meet the important idea of interlacing, whereby the path of a Lévy process is obtained as the almost-sure limit of a sequence of Brownian motions with drift interspersed with jumps of random size appearing at random times.

Chapter 3 aims to move beyond Lévy processes to study more general Markov processes and their associated semigroups of linear mappings. We emphasise, however, that the structure of Lévy processes is the paradigm case and this is exhibited both through the Courrège formula for the infinitesimal generator of Feller processes and the Beurling–Deny formula for symmetric Dirichlet forms. This chapter is more analytical in flavour than the rest of the book and makes extensive use of the theory of linear operators, particularly those of pseudo-differential type. Readers who lack background in this area can find most of what they need in the chapter appendix.

Stochastic integration is developed in Chapter 4. A novel aspect of our approach is that Brownian and Poisson integration are unified using the idea of a martingale-valued measure. At first sight this may strike the reader as technically complicated but, in fact, the assumptions that are imposed ensure that the development remains accessible and straightforward. A highlight of this chapter is the proof of Itô's formula for Lévy-type stochastic integrals.

The first part of Chapter 5 deals with a number of useful spin-offs from stochastic integration. Specifically, we study the Doléans-Dade stochastic exponential, Girsanov's theorem and its application to change of measure, the Cameron–Martin formula and the beginnings of analysis in Wiener space and martingale representation theorems. Most of these are important tools in mathematical finance and the latter part of the chapter is devoted to surveying the application of Lévy processes to option pricing, with an emphasis on the specific goal of finding an improvement to the celebrated but flawed Black–Scholes formula generated by Brownian motion. At the time of writing, this area is evolving at a rapid pace and we have been content to concentrate on one approach using hyperbolic Lévy processes that has been rather well developed. We have included, however, a large number of references to alternative models.

Finally, in Chapter 6 we study stochastic differential equations driven by Lévy processes. Under general conditions, the solutions of these are Feller processes and so we gain a concrete class of examples of the theory developed in Chapter 3. Solutions also give rise to stochastic flows and hence generate random dynamical systems.

The book naturally falls into two parts. The first three chapters develop the fundamentals of Lévy processes with an emphasis on those that are useful in stochastic calculus. The final three chapters develop the stochastic calculus of Lévy processes.

Each chapter closes with some brief historical remarks and suggestions for further reading. I emphasise that these notes are only indicative; no attempt has been made at a thorough historical account, and in this respect I apologise to any readers who feel that their contribution is unjustly omitted. More thorough historical notes in relation to Lévy processes can be found in the chapter notes to Sato [274], and for stochastic calculus with jumps see those in Protter [255].

This book requires background knowledge of probability and measure theory (such as might be obtained in a final-year undergraduate mathematics honours programme), some facility with real analysis and a smattering of functional analysis (particularly Hilbert spaces). Knowledge of basic complex variable theory and some general topology would also be an advantage, but readers who lack this should be able to read on without too much loss. The book is designed to be suitable for underpinning a taught masters level course or for independent study by first-year graduate students in mathematics and related programmes. Indeed, the two parts would make a nice pair of linked half-year modules. Alternatively, a course could also be built from the core of the book, Chapters 1, 2, 4 and 6. Readers with a specific interest in finance can safely omit Chapter 3 and Section 6.4 onwards, while analysts who wish to deepen their understanding of stochastic representations of semigroups might leave out Chapter 5.

A number of exercises of varying difficulty are scattered throughout the text. I have resisted the temptation to include worked solutions, since I believe that the absence of these provides better research training for graduate students. However, anyone having persistent difficulty in solving a problem may contact me by e-mail or otherwise.

I began my research career as a mathematical physicist and learned modern probability as part of my education in quantum theory. I would like to express my deepest thanks to my teachers Robin Hudson, K.R. Parthasarathy and Luigi Accardi for helping me to develop the foundations on which later studies have been built. My fascination with Lévy processes began with my attempt to understand their wonderful role in implementing cocycles by means of annihilation, creation and conservation processes associated with the free quantum field, and this can be regarded as the starting point for quantum stochastic calculus. Unfortunately, this topic lies outside the scope of this volume but interested readers can consult Parthasarathy [249], pp. 152–61 or Meyer [226], pp. 120–1.

My understanding of the probabilistic properties of Lévy processes has deepened as a result of work in stochastic differential equations with jumps over the past 10 years, and it's a great pleasure to thank my collaborators Hiroshi Kunita, Serge Cohen, Anne Estrade, Jiang-Lun Wu and my student

Fuchang Tang for many joyful and enlightening discussions. I would also like to thank René Schilling for many valuable conversations concerning topics related to this book. It was he who taught me about the beautiful relationship with pseudo-differential operators, which is described in Chapter 3. Thanks are also due to Jean Jacod for clarifying my understanding of the concept of predictability and to my colleague Tony Sackfield for advice about Bessel functions.

Earlier versions of this book were full of errors and misunderstandings and I am enormously indebted to Nick Bingham, Tsukasa Fujiwara, Fehmi Özkan and René Schilling, all of whom devoted the time and energy to read extensively and criticize early drafts. Some very helpful comments were also made by Krishna Athreya, Ole Barndorff-Nielsen, Uwe Franz, Vassili Kolokoltsov, Hiroshi Kunita, Martin Lindsay, Nikolai Leonenko, Carlo Marinelli (particularly with regard to LaTeX) and Ray Streater. Nick Bingham also deserves a special thanks for providing me with a valuable tutorial on English grammar. Many thanks are also due to two anonymous referees employed by Cambridge University Press. The book is greatly enriched thanks to their perceptive observations and insights.

In March 2003, I had the pleasure of giving a course, partially based on this book, at the University of Greifswald, as part of a graduate school on quantum independent increment processes. My thanks go to the organisers, Michael Schürmann and Uwe Franz, and all the participants for a number of observations that have improved the manuscript.

Many thanks are also due to David Tranah and the staff at Cambridge University Press for their highly professional yet sensitive management of this project.

Despite all this invaluable assistance, some errors surely still remain and the author would be grateful to be e-mailed about these at dba@maths.ntu.ac.uk. Corrections received after publication will be posted on his website http://www.scm.ntu.ac.uk/dba/.

Overview

It can be very useful to gain an intuitive feel for the behaviour of Lévy processes and the purpose of this short introduction is to try to develop this. Of necessity, our mathematical approach here is somewhat naive and informal – the structured, rigorous development begins in Chapter 1.

Suppose that we are given a probability space $(\Omega, \mathcal{F}, P)$. A Lévy process $X = (X(t), t \geq 0)$ taking values in $\mathbb{R}^d$ is essentially a stochastic process having stationary and independent increments; we always assume that $X(0) = 0$ with probability 1. So:

- each $X(t) : \Omega \to \mathbb{R}^d$;
- given any selection of distinct time-points $0 \leq t_1 < t_2 < \cdots < t_n$, the random vectors $X(t_1), X(t_2) - X(t_1), X(t_3) - X(t_2), \ldots, X(t_n) - X(t_{n-1})$ are all independent;
- given any two distinct times $0 \leq s < t < \infty$, the probability distribution of $X(t) - X(s)$ coincides with that of $X(t-s)$.

The key formula in this book from which so much else flows, is the magnificent *Lévy–Khintchine formula*, which says that any Lévy process has a specific form for its characteristic function. More precisely, for all $t \geq 0, u \in \mathbb{R}^d$,

$$\mathbb{E}(e^{i(u,X(t))}) = e^{t\eta(u)} \tag{0.1}$$

where

$$\eta(u) = i(b,u) - \tfrac{1}{2}(u,au) + \int_{\mathbb{R}^d - \{0\}} \left[e^{i(u,y)} - 1 - i(u,y)\chi_{0<|y|<1}(y) \right] \nu(dy). \tag{0.2}$$

In this formula $b \in \mathbb{R}^d$, a is a positive definite symmetric $d \times d$ matrix and ν is a Lévy measure on $\mathbb{R}^d - \{0\}$, so that $\int_{\mathbb{R}^d - \{0\}} \min\{1, |y|^2\} \nu(dy) < \infty$. If you have not seen it before, (0.2) will look quite mysterious to you, so we need to try to extract its meaning.

First suppose that $a = \nu = 0$; then (0.1), just becomes $\mathbb{E}(e^{i(u,X(t))}) = e^{it(u,b)}$, so that $X(t) = bt$ is simply deterministic motion in a straight line. The vector b determines the velocity of this motion and is usually called the *drift*.

Now suppose that we also have $a \neq 0$, so that (0.1) takes the form $\mathbb{E}(e^{i(u,X(t))}) = \exp\{t[i(b,u) - \frac{1}{2}(u,au)]\}$. We can recognise this as the characteristic function of a Gaussian random variable $X(t)$ having mean vector tb and covariant matrix ta. In fact we can say more about this case: the process $(X(t), t \geq 0)$ is a *Brownian motion with drift*, and such processes have been extensively studied for over 100 years. In particular, the sample paths $t \to X(t)(\omega)$ are continuous (albeit nowhere differentiable) for almost all $\omega \in \Omega$. The case $b = 0, a = I$ is usually called *standard Brownian motion*.

Now consider the case where we also have $\nu \neq 0$. If ν is a finite measure we can rewrite (0.2) as

$$\eta(u) = i(b', u) - \tfrac{1}{2}(u, au) + \int_{\mathbb{R}^d - \{0\}} (e^{i(u,y)} - 1)\nu(dy)$$

where $b' = b - \int_{0<|y|<1} y\nu(dy)$. We will take the simplest possible form for ν, i.e. $\nu = \lambda \delta_h$ where $\lambda > 0$ and δ_h is a Dirac mass concentrated at $h \in \mathbb{R}^d - \{0\}$.

In this case we can set $X(t) = b't + \sqrt{a}B(t) + N(t)$, where $B = (B(t), t \geq 0)$ is a standard Brownian motion and $N = (N(t), t \geq 0)$ is an independent process for which

$$\mathbb{E}(e^{i(u,N(t))}) = \exp\left[\lambda t(e^{i(u,h)} - 1)\right].$$

We can now recognise N as a Poisson process of intensity λ taking values in the set $\{nh, n \in \mathbb{N}\}$, so that $P(N(t) = nh) = e^{-\lambda t}[(\lambda t)^n/n!]$ and $N(t)$ counts discrete events that occur at the random times $(T_n, n \in \mathbb{N})$. Our interpretation of the paths of X in this case is now as follows. X follows the path of a Brownian motion with drift from time zero until the random time T_1. At time T_1 the path has a jump discontinuity of size $|h|$. Between T_1 and T_2 we again see Brownian motion with drift, and there is another jump discontinuity of size $|h|$ at time T_2. We can continue to build the path in this manner indefinitely.

The next stage is to take $\nu = \sum_{i=1}^{m} \lambda_i \delta_{h_i}$, where $m \in \mathbb{N}$, $\lambda_i > 0$ and $h_i \in \mathbb{R}^d - \{0\}$, for $1 \leq i \leq m$. We can then write

$$X(t) = b't + \sqrt{a}B(t) + N_1(t) + \cdots + N_m(t),$$

where $N_1, \ldots, N_m$ are independent Poisson processes (which are also independent of B); each N_i has intensity λ_i and takes values in the set $\{nh_i, n \in \mathbb{N}\}$

where $1 \leq i \leq m$. In this case, the path of X is again a Brownian motion with drift, interspersed with jumps taking place at random times. This time, though, each jump size may be any of the m numbers $|h_1|, \ldots, |h_m|$.

In the general case where ν is finite, we can see that we have passed to the limit in which jump sizes take values in the full continuum of possibilities, corresponding to a continuum of Poisson processes. So a Lévy process of this type is a Brownian motion with drift interspersed with jumps of arbitrary size. Even when ν fails to be finite, if we have $\int_{0<|x|<1} |x|\nu(dx) < \infty$ a simple exercise in using the mean value theorem shows that we can still make this interpretation.

The most subtle case of the Lévy–Khintchine formula (0.2) is when $\int_{0<|x|<1} |x|\nu(dx) = \infty$ but $\int_{0<|x|<1} |x|^2 \nu(dx) < \infty$. Thinking analytically, $e^{i(u,y)} - 1$ may no longer be ν-integrable but

$$e^{i(u,y)} - 1 - i(u, y)\chi_{0<|y|<1}(y)$$

always is. Intuitively, we may argue that the measure ν has become so fine that it is no longer capable of distinguishing small jumps from drift. Consequently it is necessary to amalgamate them together under the integral term. Despite this subtlety, it is still possible to interpret the general Lévy process as a Brownian motion with drift b interspersed with 'jumps' of arbitrary size, provided we recognise that at the microscopic level tiny jumps and short bursts of drift are treated as one. A more subtle discussion of this, and an account of the phenomenon of 'creep', can be found at the end of Section 2.4. We will see in Chapter 2 that the path can always be constructed as the limit of a sequence of terms, each of which is a Brownian motion with drift interspersed with bona fide jumps.

When $\nu < \infty$, we can write the sample-path decomposition directly as

$$X(t) = bt + \sqrt{a}B(t) + \sum_{0 \leq s \leq t} \Delta X(s), \quad (0.3)$$

where $\Delta X(s)$ is the jump at time s (e.g. if $\nu = \lambda \delta_h$ then $\Delta X(s) = 0$ or h). Instead of dealing directly with the jumps it is more convenient to count the times at which the jumps occur, so for each Borel set A in $\mathbb{R}^d - \{0\}$ and for each $t \geq 0$ we define

$$N(t, A) = \#\{0 \leq s \leq t; \Delta X(s) \in A\}.$$

This is an interesting object: if we fix t and A then $N(t, A)$ is a random variable; however, if we fix $\omega \in \Omega$ and $t \geq 0$ then $N(t, \cdot)(\omega)$ is a measure. Finally, if we fix A with $\nu(A) < \infty$ then $(N(t, A), t \geq 0)$ is a Poisson process with intensity $\nu(A)$.

When $\nu < \infty$, we can write

$$\sum_{0 \leq s \leq t} \Delta X(s) = \int_{\mathbb{R}-\{0\}} x N(t, dx).$$

(Readers might find it helpful to consider first the simple case where $\nu = \sum_{i=1}^{m} \lambda_i \delta_{h_i}$.)

In the case of general ν, the delicate analysis whereby small jumps and drift become amalgamated leads to the celebrated *Lévy–Itô decomposition*,

$$X(t) = bt + \sqrt{a}B(t) + \int_{0<|x|<1} x[N(t,dx) - t\nu(dx)] + \int_{|x|\geq 1} xN(t,dx).$$

Full proofs of the Lévy–Khintchine formula and the Lévy–Itô decomposition are given in Chapters 1 and 2.

Let us return to the consideration of standard Brownian motion $B = (B(t), t \geq 0)$. Each $B(t)$ has a Gaussian density

$$p_t(x) = \frac{1}{(2\pi t)^{d/2}} \exp\left(-\frac{|x|^2}{2t}\right)$$

and, as was first pointed out by Einstein [92], this satisfies the diffusion equation

$$\frac{\partial p_t(x)}{\partial t} = \frac{1}{2}\Delta p_t(x),$$

where Δ is the usual Laplacian in $\mathbb{R}^d$. More generally, suppose that we want to build a solution $u = (u(t,x), t \geq 0, x \in \mathbb{R}^d)$ to the diffusion equation that has a fixed initial condition $u(0,x) = f(x)$ for all $x \in \mathbb{R}^d$, where f is a bounded continuous function on $\mathbb{R}^d$. We then have

$$u(t,x) = \int_{\mathbb{R}^d} f(x+y) p_t(y) dy = \mathbb{E}(f(x+B(t))). \tag{0.4}$$

The modern way of thinking about this utilises the powerful machinery of operator theory. We define $(T_t f)(x) = u(t,x)$; then $(T_t, t \geq 0)$ is a one-parameter semigroup of linear operators on the Banach space of bounded continuous functions. The semigroup is completely determined by its infinitesimal generator Δ, so that we may formally write $T_t = e^{t\Delta}$ and note that, from the diffusion equation,

$$\Delta f = \frac{d}{dt}(T_t f)\Big|_{t=0}$$

for all f where this makes sense.

This circle of ideas has a nice physical interpretation. The semigroup or, equivalently, its infinitesimal version – the diffusion equation – gives a deterministic macroscopic description of the effects of Brownian motion. We see from (0.4) that to obtain this we must average over all possible paths of the particle that is executing Brownian motion. We can, of course, get a microscopic description by forgetting about the semigroup and just concentrating on the process $(B(t), t \geq 0)$. The price we have to pay for this is that we can no longer describe the dynamics deterministically. Each $B(t)$ is a random variable, and any statement we make about it can only be expressed as a probability. More generally, as we will see in Chapter 6, we have a dichotomy between solutions of stochastic differential equations, which are microscopic and random, and their averages, which solve partial differential equations and are macroscopic and deterministic.

The first stage in generalising this interplay of concepts is to replace Brownian motion by a general Lévy process $X = (X(t), t \geq 0)$. Although X may not in general have a density, we may still obtain the semigroup by $(T(t)f)(x) = \mathbb{E}(f(X(t) + x))$, and the infinitesimal generator then takes the more general form

$$(Af)(x) = b^i (\partial_i f)(x) + \tfrac{1}{2} a^{ij} (\partial_i \partial_j f)(x)$$
$$+ \int_{\mathbb{R}^d - \{0\}} [f(x+y) - f(x) - y^i (\partial_i f)(x) \chi_{0 < |y| < 1}(y)] \nu(dy). \quad (0.5)$$

In fact this structure is completely determined by the Lévy–Khinchine formula, and we have the following important correspondences:

- drift $\longleftrightarrow$ first-order differential operator
- diffusion $\longleftrightarrow$ second-order differential operator
- jumps $\longleftrightarrow$ superposition of difference operators

This enables us to read off our intuitive description of the path from the form of the generator, and this is very useful in more general situations where we no longer have a Lévy–Khinchine formula. The formula (0.5) is established in Chapter 3, and we will also derive an alternative representation using pseudo-differential operators.

More generally, the relationship between stochastic processes and semigroups extends to a wider class of Markov processes $Y = (Y(t), t \geq 0)$, and here the semigroup is given by conditioning:

$$(T_t f)(x) = \mathbb{E}\big(f(Y(t)) | Y(0) = x\big).$$

Under certain general conditions that we will describe in Chapter 3, the generator is of the Courrège form

$$(Af)(x) = c(x)f(x) + b^i(x)(\partial_i f)(x) + a^{ij}(x)(\partial_i \partial_j f)(x)$$
$$+ \int_{\mathbb{R}^d - \{x\}} [f(y) - f(x) - \phi(x,y)(y^i - x^i)(\partial_i f)(x)]\mu(x,dy). \tag{0.6}$$

Note the similarities between equations (0.5) and (0.6). Once again there are drift, diffusion and jump terms, however, these are no longer fixed in space but change from point to point. There is an additional term, controlled by the function c, that corresponds to killing (we could also have included this in the Lévy case), and the function ϕ is simply a smoothed version of the indicator function that effects the cut-off between large and small jumps.

Under certain conditions, we can generalise the Lévy–Itô decomposition and describe the process Y as the solution of a stochastic differential equation

$$dY(t) = b(Y(t-))dt + \sqrt{a(Y(t-))}dB(t)$$
$$+ \int_{|x|<1} F(Y(t-), x)[N(dt, dx) - dt\nu(dx)]$$
$$+ \int_{|x|\geq 1} G(Y(t-), x)N(dt, dx). \tag{0.7}$$

The kernel $\mu(x, \cdot)$ appearing in (0.6) can be expressed in terms of the Lévy measure ν and the coefficients F and G. This is described in detail in Chapter 6.

To make sense of the stochastic differential equation (0.7), we must rewrite it as an integral equation, which means that we must give meaning to *stochastic integrals* such as

$$\int_0^t U(s)dB(s) \quad \text{and} \quad \int_0^t \int_{0<|x|<1} V(s,x)(N(ds,dx) - ds\nu(dx))$$

for suitable U and V. The usual Riemann–Stieltjes or Lebesgue–Stieltjes approach no longer works for these objects, and we need to introduce some extra structure. To model the flow of information with time, we introduce a filtration $(\mathcal{F}_t, t \geq 0)$ that is an increasing family of sub-σ-algebras of $\mathcal{F}$, and we say that a process U is adapted if each $U(t)$ is $\mathcal{F}_t$-measurable for each $t \geq 0$. We then define

$$\int_0^t U(s)dB(s) = \lim_{n\to\infty} \sum_{j=1}^{m_n} U(t_j^{(n)})\left[B(t_{j+1}^{(n)}) - B(t_j^{(n)})\right]$$

where $0 = t_0^{(n)} < t_1^{(n)} < \cdots < t_{m_n}^{(n)} = t$ is a sequence of partitions of $[0, t]$ whose mesh tends to zero as $n \to \infty$. The key point in the definition is that for each term in the summand, $U(t_j^{(n)})$ is fixed in the past while the increment $B(t_{j+1}^{(n)}) - B(t_j^{(n)})$ extends into the future. If a Riemann–Stieltjes theory were possible, we could evaluate $U(x_j^{(n)})$ at an arbitrary point for which $t_j^{(n)} < x_j^{(n)} < t_{j+1}^{(n)}$. The other integral,

$$\int_0^t \int_{0<|x|<1} V(s, x)[N(ds, dx) - ds\nu(dx)],$$

is defined similarly.

This definition of a stochastic integral has profound implications. In Chapter 4 we will explore the properties of a class of Lévy-type stochastic integrals that take the form

$$Y(t) = \int_0^t G(s)ds + \int_0^t F(s)dB(s) + \int_0^t \int_{0<|x|<1} H(s, x)[N(ds, dx) - ds\nu(dx)] + \int_0^t K(s, x)N(ds, dx)$$

and, for convenience, we will take $d = 1$ for now. In the case where F, H and K are identically zero and f is a differentiable function, the chain rule from differential calculus gives $f(Y(t)) = \int_0^t f'(Y(s))G(s)ds$, which we can write more succinctly as $df(Y(t)) = f'(Y(t))G(t)dt$. This formula breaks down for Lévy-type stochastic integrals, and in its place we get the famous *Itô formula*,

$$df(Y(t))$$
$$= f'(Y(t))G(t)dt + f'(Y(t))F(t)dB(t) + \tfrac{1}{2}f''(Y(t))F(t)^2 dt$$
$$+ \int_{|x|\geq 1} [f(Y(t-) + K(t, x)) - f(Y(t-))]N(dt, dx)$$
$$+ \int_{0<|x|<1} [f(Y(t-) + H(t, x)) - f(Y(t-))](N(dt, dx) - \nu(dx)dt)$$
$$+ \int_{0<|x|<1} \big[f(Y(t-) + H(t, x)) - f(Y(t-)) - H(t, x)f'(Y(t-))\big]\nu(dx)dt.$$

If you have not seen this before, think of a Taylor series expansion in which $dB(t)^2$ behaves like dt and $N(dt, dx)^2$ behaves like $N(dt, dx)$. Alternatively, you can wait for the full development in Chapter 4. Itô's formula is the key to the wonderful world of stochastic calculus. It lies behind the extraction of the Courrège generator (0.6) from equation (0.7). It also has many important applications including option pricing, the Black–Scholes formula and attempts to replace the latter using more realistic models based on Lévy processes. This is all revealed in Chapter 5, but now the preview is at an end and it is time to begin the journey . . .

Notation

Throughout this book we will deal extensively with random variables taking values in the Euclidean space $\mathbb{R}^d$, where $d \in \mathbb{N}$. We recall that elements of $\mathbb{R}^d$ are vectors $x = (x_1, x_2, \ldots, x_d)$ with each $x_i \in \mathbb{R}$ for $1 \leq i \leq d$. The inner product in $\mathbb{R}^d$ is denoted by (x, y) where $x, y \in \mathbb{R}^d$, so that

$$(x, y) = \sum_{i=1}^{d} x_i y_i.$$

This induces the Euclidean norm $|x| = (x, x)^{1/2} = \left(\sum_{i=1}^{d} x_i^2\right)^{1/2}$. We will use the Einstein summation convention throughout this book, wherein summation is understood with respect to repeated upper and lower indices, so for example if $x, y \in \mathbb{R}^d$ and $A = (A^i_j)$ is a $d \times d$ matrix then

$$A^i_j x_i y^j = \sum_{i,j=1}^{d} A^i_j x_i y^j = (x, Ay).$$

We say that such a matrix is *positive definite* if $(x, Ax) \geq 0$ for all $x \in \mathbb{R}^d$ and *strictly positive definite* if the inequality can be strengthened to $(x, Ax) > 0$ for all $x \in \mathbb{R}^d$, with $x \neq 0$ (note that some authors call these 'non-negative definite' and 'positive definite', respectively). The transpose of a matrix A will always be denoted A^T. The determinant of a square matrix is written as $\det(A)$ and its trace as $\text{tr}(A)$. The identity matrix will always be denoted I.

The set of all $d \times d$ real-valued matrices is denoted $M_d(\mathbb{R})$.

If $S \subseteq \mathbb{R}^d$ then its orthogonal complement is $S^\perp = \{x \in \mathbb{R}^d; (x, y) = 0 \text{ for all } y \in S\}$.

The open ball of radius r centred at x in $\mathbb{R}^d$ is denoted $B_r(x) = \{y \in \mathbb{R}^d; |y - x| < r\}$ and we will always write $\hat{B} = B_1(0)$. The sphere in $\mathbb{R}^d$ is the $(d-1)$-dimensional submanifold, denoted S^{d-1}, defined by $S^{d-1} = \{x \in \mathbb{R}^d; |x| = 1\}$.

We sometimes write $\mathbb{R}^+ = [0, \infty)$.

The sign of $u \in \mathbb{R}$ is denoted $\operatorname{sgn}(u)$ so that $\operatorname{sgn}(u) = (u/|u|)$ if $u \neq 0$, with $\operatorname{sgn}(0) = 0$.

For $z \in \mathbb{C}$, $\Re(z)$ and $\Im(z)$ denote the real and imaginary parts of z, respectively.

The complement of a set A will always be denoted A^c and $\bar{A}$ will mean closure in some topology. If f is a mapping between two sets A and B, we denote its range as $\operatorname{Ran}(f) = \{y \in B; y = f(x) \text{ for some } x \in A\}$.

For $1 \leq n \leq \infty$, we write $C^n(\mathbb{R}^d)$ to denote the set of all n-times differentiable functions from $\mathbb{R}^d$ to $\mathbb{R}$, all of whose derivatives are continuous. The jth first-order partial derivative of $f \in C^1(\mathbb{R}^d)$ at $x \in \mathbb{R}^d$ will sometimes be written $(\partial_j f)(x)$. Similarly, if $f \in C^2(\mathbb{R}^d)$, we write

$$(\partial_i \partial_j f)(x) \quad \text{for} \quad \frac{\partial^2 f}{\partial x_i \partial x_j}(x).$$

When $d = 1$ and $f \in C^n(\mathbb{R})$, we sometimes write

$$f^{(r)}(x) \quad \text{for} \quad \frac{d^r f}{dx^r}(x),$$

where $1 \leq r \leq n$.

Let $\mathcal{H}$ be a real inner product space, equipped with the inner product $\langle \cdot, \cdot \rangle$ and associated norm $||x|| = \langle x, x \rangle^{1/2}$, for each $x \in \mathcal{H}$. We will frequently have occasion to use the *polarisation identity*

$$\langle x, y \rangle = \tfrac{1}{4}(||x+y||^2 - ||x-y||^2),$$

for each $x, y \in \mathcal{H}$.

For $a, b \in \mathbb{R}$, we will use $a \wedge b = \min\{a, b\}$ and $a \vee b = \max\{a, b\}$.

We will occasionally use Landau notation, according to which $(o(n), n \in \mathbb{N})$ is any real-valued sequence for which $\lim_{n \to \infty} (o(n)/n) = 0$ and $(O(n), n \in \mathbb{N})$ is any real-valued sequence for which $\limsup_{n \to \infty} (O(n)/n) < \infty$. Functions $o(t)$ and $O(t)$ are defined similarly. If $f, g : \mathbb{R} \to \mathbb{R}$ and $a \in \mathbb{R} \cup \{\infty\}$, then by $f \sim g$ as $x \to a$ we mean $\lim_{x \to a} [f(x)/g(x)] = 1$.

If $f : \mathbb{R}^d \to \mathbb{R}$ then by $\lim_{s \uparrow t} f(s) = l$ we mean $\lim_{s \to t, s < t} f(s) = l$. Similarly, $\lim_{s \downarrow t} f(s) = l$ means $\lim_{s \to t, s > t} f(s) = l$.

1
Lévy processes

Summary Section 1.1 is a review of basic measure and probability theory. In Section 1.2 we meet the key concepts of the infinite divisibility of random variables and of probability distributions, which underly the whole subject. Important examples are the Gaussian, Poisson and stable distributions. The celebrated Lévy–Khintchine formula classifies the set of all infinitely divisible probability distributions by means of a canonical form for the characteristic function. Lévy processes are introduced in Section 1.3. These are essentially stochastic processes with stationary and independent increments. Each random variable within the process is infinitely divisible, and hence its distribution is determined by the Lévy–Khintchine formula. Important examples are Brownian motion, Poisson and compound Poisson processes, stable processes and subordinators. Section 1.4 clarifies the relationship between Lévy processes, infinite divisibility and weakly continuous convolution semigroups of probability measures. Finally, in Section 1.5, we briefly survey recurrence and transience, Wiener–Hopf factorisation and local times for Lévy processes.

1.1 Review of measure and probability

The aim of this section is to give a brief resumé of key notions of measure theory and probability that will be used extensively throughout the book and to fix some notation and terminology once and for all. I emphasise that reading this section is no substitute for a systematic study of the fundamentals from books such as Billingsley [44], Itô [146], Ash and Doléans-Dade [14], Rosenthal [264], Dudley [84] or, for measure theory without probability, Cohn [73]. Knowledgeable readers are encouraged to skip this section altogether or to use it as a quick reference when the need arises.

1.1.1 Measure and probability spaces

Let S be a non-empty set and $\mathcal{F}$ a collection of subsets of S. We call $\mathcal{F}$ a *σ-algebra* if the following hold.

(1) $S \in \mathcal{F}$.
(2) $A \in \mathcal{F} \Rightarrow A^c \in \mathcal{F}$.
(3) If $(A_n, n \in \mathbb{N})$ is a sequence of subsets in $\mathcal{F}$ then $\bigcup_{n=1}^{\infty} A_n \in \mathcal{F}$.

The pair $(S, \mathcal{F})$ is called a *measurable space*. A *measure* on $(S, \mathcal{F})$ is a mapping $\mu : \mathcal{F} \to [0, \infty]$ that satisfies

(1) $\mu(\emptyset) = 0$,
(2)
$$\mu\left(\bigcup_{n=1}^{\infty} A_n\right) = \sum_{n=1}^{\infty} \mu(A_n)$$

for every sequence $(A_n, n \in \mathbb{N})$ of mutually disjoint sets in $\mathcal{F}$.

The triple $(S, \mathcal{F}, \mu)$ is called a *measure space*.

The quantity $\mu(S)$ is called the *total mass* of μ and μ is said to be *finite* if $\mu(S) < \infty$. More generally, a measure μ is *σ-finite* if we can find a sequence $(A_n, n \in \mathbb{N})$ in $\mathcal{F}$ such that $S = \bigcup_{n=1}^{\infty} A_n$ and each $\mu(A_n) < \infty$.

For the purposes of this book, there will be two cases of interest. The first comprises

- **Borel measures** Let S be a subset of $\mathbb{R}^d$. We equip S with the relative topology induced from $\mathbb{R}^d$, so that $U \subseteq S$ is open in S if $U \cap S$ is open in $\mathbb{R}^d$. Let $\mathcal{B}(S)$ denote the smallest σ-algebra of subsets of S that contains every open set in S. We call $\mathcal{B}(S)$ the *Borel σ-algebra of S*. Elements of $\mathcal{B}(S)$ are called *Borel sets* and any measure on $(S, \mathcal{B}(S))$ is called a *Borel measure*.

One of the best known examples of a Borel measure is given by the *Lebesgue measure* on $S = \mathbb{R}^d$. This takes the following explicit form on sets in the shape of boxes $A = (a_1, b_1) \times (a_2, b_2) \times \cdots \times (a_d, b_d)$ where each $-\infty < a_i < b_i < \infty$:

$$\mu(A) = \prod_{i=1}^{d} (b_i - a_i).$$

Lebesgue measure is clearly σ-finite but not finite.

Of course, Borel measures make sense in arbitrary topological spaces, but we will not have need of this degree of generality here.

1.1 Review of measure and probability

The second case comprises

- **Probability measures** Here we usually write $S = \Omega$ and take Ω to represent the set of outcomes of some random experiment. Elements of $\mathcal{F}$ are called *events* and any measure on $(\Omega, \mathcal{F})$ of total mass 1 is called a *probability measure* and denoted P. The triple $(\Omega, \mathcal{F}, P)$ is then called a *probability space*.

Occasionally we will also need *counting measures*, which are those that take values in $\mathbb{N} \cup \{0\}$.

A proposition p about the elements of S is said to hold *almost everywhere* (usually shortened to *a.e.*) with respect to a measure μ if $\mathcal{N} = \{s \in S; p(s)$ is false$\} \in \mathcal{F}$ and $\mu(\mathcal{N}) = 0$. In the case of probability measures, we use the terminology '*almost surely*' (shortened to *a.s.*) instead of 'almost everywhere', or alternatively '*with probability* 1'. Similarly, we say that '*almost all*' the elements of a set A have a certain property if the subset of A for which the property fails has measure zero.

Continuity of measures Let $(A(n), n \in \mathbb{N})$ be a sequence of sets in $\mathcal{F}$ with $A(n) \subseteq A(n+1)$ for each $n \in \mathbb{N}$. We then write $A(n) \uparrow A$ where $A = \bigcup_{n=1}^{\infty} A(n)$, and we have

$$\mu(A) = \lim_{n \to \infty} \mu(A(n)).$$

When μ is a probability measure, this is usually called *continuity of probability*.

Let G be a group whose members act as measurable transformations of $(S, \mathcal{F})$, so that $g : S \to S$ for each $g \in G$ and $gA \in \mathcal{F}$ for all $A \in \mathcal{F}$, $g \in G$, where $gA = \{ga, a \in A\}$. We say that a measure μ on $(S, \mathcal{F})$ is G-*invariant* if

$$\mu(gA) = \mu(A)$$

for each $g \in G$, $A \in \mathcal{F}$.

A (finite) *measurable partition* of a set $A \in \mathcal{F}$ is a family of sets $B_1, B_2, \ldots, B_n \in \mathcal{F}$ for which $B_i \cap B_j = \emptyset$ whenever $i \neq j$ and $\bigcup_{i=1}^{n} B_i = A$. We use the term *Borel partition* when $\mathcal{F}$ is a Borel σ-algebra.

If $\{\mathcal{G}_i, i \in I\}$ is a (not necessarily countable) family of sub-σ-algebras of $\mathcal{F}$ then $\bigcap_{i \in I} \mathcal{G}_i$ is the largest sub-σ-algebra contained in each $\mathcal{G}_i$ and $\bigvee_{i \in I} \mathcal{G}_i$ denotes the smallest sub-σ-algebra that contains each $\mathcal{G}_i$.

If P is a probability measure and $A, B \in \mathcal{F}$, it is sometimes notationally convenient to write $P(A, B) = P(A \cap B)$.

Completion of a measure Let $(S, \mathcal{F}, \mu)$ be a measure space. Define

$$\mathcal{N} = \{A \subseteq S; \exists N \in \mathcal{F} \text{ with } \mu(N) = 0 \text{ and } A \subseteq N\}$$

and

$$\overline{\mathcal{F}} = \{A \cup B; A \in \mathcal{F}, B \in \mathcal{N}\}.$$

Then $\overline{\mathcal{F}}$ is a σ-algebra and the *completion* of the measure μ on $(S, \mathcal{F})$ is the measure $\overline{\mu}$ on $(S, \overline{\mathcal{F}})$ defined by

$$\overline{\mu}(A \cup B) = \mu(A), \qquad A \in \mathcal{F}, \; B \in \mathcal{N}.$$

In particular, $\overline{\mathcal{B}(S)}$ is called the σ-algebra of *Lebesgue measurable sets* in S.

1.1.2 Random variables, integration and expectation

For $i = 1, 2$, let $(S_i, \mathcal{F}_i)$ be measurable spaces. A mapping $f : S_1 \to S_2$ is said to be $(\mathcal{F}_1, \mathcal{F}_2)$-*measurable* if $f^{-1}(A) \in \mathcal{F}_1$ for all $A \in \mathcal{F}_2$. If each $S_1 \subseteq \mathbb{R}^d$, $S_2 \subseteq \mathbb{R}^m$ and $\mathcal{F}_i = \mathcal{B}(S_i)$, f is said to be *Borel measurable*. In the case $d = 1$, we sometimes find it useful to write each Borel measurable f as $f^+ - f^-$ where, for each $x \in S_1$, $f^+(x) = \max\{f(x), 0\}$ and $f^-(x) = -\min\{f(x), 0\}$. If $f = (f_1, f_2, \ldots, f_d)$ is a measurable mapping from S_1 to $\mathbb{R}^d$, we write $f^+ = (f_1^+, f_2^+, \ldots, f_d^+)$ and $f^- = (f_1^-, f_2^-, \ldots, f_d^-)$.

In what follows, whenever we speak of measurable mappings taking values in a subset of $\mathbb{R}^d$, we always take it for granted that the latter is equipped with its Borel σ-algebra.

When we are given a probability space $(\Omega, \mathcal{F}, P)$ then measurable mappings from Ω into $\mathbb{R}^d$ are called *random variables*. Random variables are usually denoted $X, Y, \ldots$. Their values should be thought of as the results of quantitative observations on the set Ω. Note that if X is a random variable then so is $f(X) = f \circ X$, where f is a Borel measurable mapping from $\mathbb{R}^d$ to $\mathbb{R}^m$. A measurable mapping $Z = X + iY$ from Ω into $\mathbb{C}$ (equipped with the natural Borel structure inherited from $\mathbb{R}^2$) is called a *complex random variable*. Note that Z is measurable if and only if both X and Y are measurable.

If X is a random variable, its *law* (or *distribution*) is the Borel probability measure p_X on $\mathbb{R}^d$ defined by

$$p_X = P \circ X^{-1}.$$

We say that X is *symmetric* if $p_X(A) = p_X(-A)$ for all $A \in \mathcal{B}(\mathbb{R}^d)$.

Two random variables X and Y that have the same probability law are said to be *identically distributed*, and we sometimes denote this as $X \stackrel{\mathrm{d}}{=} Y$. For

1.1 Review of measure and probability

a one-dimensional random variable X, its *distribution function* is the right-continuous increasing function defined by $F_X(x) = p_X((-\infty, x])$ for each $x \in \mathbb{R}$.

If $W = (X, Y)$ is a random variable taking values in $\mathbb{R}^{2d}$, the probability law of W is sometimes called the *joint distribution* of X and Y. The quantities p_X and p_Y are then called the *marginal distributions* of W, where $p_X(A) = p_W(A, \mathbb{R}^d)$ and $p_Y(A) = p_W(\mathbb{R}^d, A)$ for each $A \in \mathcal{B}(\mathbb{R}^d)$.

Suppose that we are given a collection of random variables $(X_i, i \in I)$ in a fixed probability space; then we denote by $\sigma(X_i, i \in I)$ the smallest σ-algebra contained in $\mathcal{F}$ with respect to which all the X_i are measurable. When there is only a single random variable X in the collection, we denote this σ-algebra as $\sigma(X)$.

Let S be a Borel subset of $\mathbb{R}^d$ that is locally compact in the relative topology. We denote as $B_b(S)$ the linear space of all bounded Borel measurable functions from S to $\mathbb{R}$. This becomes a normed space (in fact, a Banach space) with respect to $||f|| = \sup_{x \in S} |f(x)|$ for each $f \in B_b(S)$. Let $C_b(S)$ be the subspace of $B_b(S)$ comprising continuous functions, $C_0(S)$ be the subspace comprising continuous functions that vanish at infinity and $C_c(S)$ be the subspace comprising continuous functions with compact support, so that

$$C_c(S) \subseteq C_0(S) \subseteq C_b(S).$$

$C_b(S)$ and $C_0(S)$ are both Banach spaces under $||\cdot||$ and $C_c(S)$ is norm dense in $C_0(S)$. When S is compact, all three spaces coincide. For each $n \in \mathbb{N}$, $C_b^n(\mathbb{R}^d)$ is the space of all $f \in C_b(\mathbb{R}^d) \cap C^n(\mathbb{R}^d)$ such that all the partial derivatives of f, of order up to and including n, are in $C_b(\mathbb{R}^d)$. We further define $C_b^\infty(\mathbb{R}^d) = \bigcap_{n \in \mathbb{N}} C_b^n(\mathbb{R}^d)$. We define $C_c^n(\mathbb{R}^d)$ and $C_0^n(\mathbb{R}^d)$ analogously, for each $1 \leq n \leq \infty$.

Let $(S, \mathcal{F})$ be a measurable space. A measurable function, $f : S \to \mathbb{R}^d$, is said to be *simple* if

$$f = \sum_{j=1}^{n} c_j \chi_{A_j}$$

for some $n \in \mathbb{N}$, where $c_j \in \mathbb{R}^d$ and $A_j \in \mathcal{F}$ for $1 \leq j \leq n$. We call χ_A the *indicator function*, defined for any $A \in \mathcal{F}$ by

$$\chi_A(x) = 1 \quad \text{whenever } x \in A; \qquad \chi_A(x) = 0 \quad \text{whenever } x \notin A.$$

Let $\Sigma(S)$ denote the linear space of all simple functions on S and let μ be a measure on $(S, \mathcal{F})$. The *integral* with respect to μ is the linear mapping from

$\Sigma(S)$ into $\mathbb{R}^d$ defined by

$$I_\mu(f) = \sum_{j=1}^n c_j \mu(A_j)$$

for each $f \in \Sigma(S)$. The integral is extended to measurable functions $f = (f_1, f_2, \ldots, f_d)$, where each $f_i \geq 0$, by the prescription for $1 \leq i \leq d$

$$I_\mu(f_i) = \sup\{I_\mu(g_i), \quad g = (g_1, \ldots, g_d) \in \Sigma(S), g_i \leq f_i\}$$

and to arbitrary measurable functions f by

$$I_\mu(f) = I_\mu(f^+) - I_\mu(f^-).$$

We write $I_\mu(f) = \int f(x)\mu(dx)$ or, alternatively, $I_\mu(f) = \int f d\mu$. Note that at this stage there is no guarantee that any of the $I_\mu(f_i)$ is finite.

We say that f is *integrable* if $|I_\mu(f^+)| < \infty$ and $|I_\mu(f^-)| < \infty$. For arbitrary $A \in \mathcal{F}$, we define

$$\int_A f(x)\mu(dx) = I_\mu(f\chi_A).$$

It is worth pointing out that the key estimate

$$\left|\int_A f(x)\mu(dx)\right| \leq \int_A |f(x)|\mu(dx)$$

holds in this vector-valued framework (see e.g. Cohn [73], pp. 352–3).

In the case where we have a probability space $(\Omega, \mathcal{F}, P)$, the linear mapping I_P is called the *expectation* and written simply as $\mathbb{E}$ so, for a random variable X and Borel measurable mapping $f : \mathbb{R}^d \to \mathbb{R}^m$, we have

$$\mathbb{E}(f(X)) = \int_\Omega f(X(\omega))P(d\omega) = \int_{\mathbb{R}^m} f(x)p_X(dx).$$

If $A \in \mathcal{F}$, we sometimes write $\mathbb{E}(X; A) = \mathbb{E}(X\chi_A)$.

In the case $d = m = 1$ we have *Jensen's inequality*,

$$f(\mathbb{E}(X)) \leq \mathbb{E}(f(X)),$$

whenever $f : \mathbb{R} \to \mathbb{R}$ is a convex function and X and $f(X)$ are both integrable.

The *mean* of X is the vector $\mathbb{E}(X)$ and this is sometimes denoted μ (if there is no measure called μ already in the vicinity) or μ_X, if we want to emphasise the underlying random variable. If $X = (X_1, X_2, \ldots, X_d)$ and $Y = (Y_1, Y_2, \ldots, Y_d)$ are two random variables then the $d \times d$ matrix with (i, j)th entry $\mathbb{E}[(X_i - \mu_{X_i})(Y_j - \mu_{Y_j})]$ is called the *covariance* of X and Y and denoted $\text{Cov}(X, Y)$. In the case $X = Y$ and $d = 1$, we write $\text{Var}(X) = \text{Cov}(X, Y)$ and call this quantity the *variance* of X. It is sometimes denoted σ^2 or σ_X^2. When $d = 1$ the quantity $\mathbb{E}(X^n)$, where $n \in \mathbb{N}$, is called the *nth moment* of X.

The *Chebyshev–Markov inequality* for a random variable X is

$$P(|X - \alpha\mu| \geq C) \leq \frac{\mathbb{E}(|X - \alpha\mu|^n)}{C^n},$$

where $C > 0, \alpha \in \mathbb{R}, n \in \mathbb{N}$. The commonest forms of this are the Chebyshev inequality ($n = 2, \alpha = 1$) and the Markov inequality ($n = 1, \alpha = 0$).

We return to a general measure space $(S, \mathcal{F}, \mu)$ and list some key theorems for establishing the integrability of functions from S to $\mathbb{R}^d$. For the first two of these we require $d = 1$.

Theorem 1.1.1 (Monotone convergence theorem) *If $(f_n, n \in \mathbb{N})$ is a sequence of non-negative measurable functions on S that is (a.e.) monotone increasing and converging pointwise to f (a.e.), then*

$$\lim_{n \to \infty} \int_S f_n(x)\mu(dx) = \int_S f(x)\mu(dx).$$

From this we easily deduce

Corollary 1.1.2 (Fatou's lemma) *If $(f_n, n \in \mathbb{N})$ is a sequence of non-negative measurable functions on S, then*

$$\liminf_{n \to \infty} \int_S f_n(x)\mu(dx) \geq \int_S \liminf_{n \to \infty} f_n(x)\mu(dx),$$

which is itself then applied to establish

Theorem 1.1.3 (Lebesgue's dominated convergence theorem) *If $(f_n, n \in \mathbb{N})$ is a sequence of measurable functions from S to $\mathbb{R}^d$ converging pointwise to f (a.e.) and $g \geq 0$ is an integrable function such that $|f_n(x)| \leq g(x)$ (a.e.) for all $n \in \mathbb{N}$, then*

$$\lim_{n \to \infty} \int_S f_n(x)\mu(dx) = \int_S f(x)\mu(dx).$$

We close this section by recalling function spaces of integrable mappings. Let $1 \leq p < \infty$ and denote by $L^p(S, \mathcal{F}, \mu; \mathbb{R}^d)$ the Banach space of all equivalence classes of mappings $f : S \to \mathbb{R}^d$ which agree a.e. (with respect to μ) and for which $||f||_p < \infty$, where $||\cdot||_p$ denotes the norm

$$||f||_p = \left[\int_S |f(x)|^p \mu(dx)\right]^{1/p}.$$

In particular, when $p = 2$ we obtain a Hilbert space with respect to the inner product

$$\langle f, g \rangle = \int_S (f(x), g(x))\mu(dx),$$

for each $f, g \in L^2(S, \mathcal{F}, \mu; \mathbb{R}^d)$. If $\langle f, g \rangle = 0$, we say that f and g are *orthogonal*. A linear subspace V of $L^2(S, \mathcal{F}, \mu; \mathbb{R}^d)$ is called a *closed subspace* if it is closed with respect to the topology induced by $||\cdot||_2$, i.e. if $(f_n; n \in \mathbb{N})$ is a sequence in V that converges to f in $L^2(S, \mathcal{F}, \mu; \mathbb{R}^d)$ then $f \in V$.

When there can be no room for doubt, we will use the notation $L^p(S)$ or $L^p(S, \mu)$ for $L^p(S, \mathcal{F}, \mu; \mathbb{R}^d)$.

Hölder's inequality is extremely useful. Let $p, q > 1$ be such that

$$1/p + 1/q = 1.$$

Let $f \in L^p(S)$ and $g \in L^q(S)$ and define $(f, g) : S \to \mathbb{R}$ by $(f, g)(x) = (f(x), g(x))$ for all $x \in S$. Then $(f, g) \in L^1(S)$ and we have

$$||(f, g)||_1 \leq ||f||_p ||g||_q.$$

When $p = 2$, this is called the *Cauchy–Schwarz inequality*.

Another useful fact is that $\Sigma(S)$ is dense in each $L^p(S)$, i.e. given any $f \in L^p(S)$ we can find a sequence $(f_n, n \in \mathbb{N})$ in $\Sigma(S)$ such that $\lim_{n\to\infty} ||f - f_n||_p = 0$.

The space $L^p(S, \mathcal{F}, \mu)$ is said to be *separable* if it has a countable dense subset. A sufficient condition for this is that the σ-algebra $\mathcal{F}$ is *countably generated*, i.e. there exists a countable set $\mathcal{C}$ such that $\mathcal{F}$ is the smallest σ-algebra containing $\mathcal{C}$. If $S \in \mathcal{B}(\mathbb{R}^d)$ then $\mathcal{B}(S)$ is countably generated.

1.1.3 Conditional expectation

Let $(S, \mathcal{F}, \mu)$ be an arbitrary measure space. A measure ν on $(S, \mathcal{F})$ is said to be *absolutely continuous* with respect to μ if $A \in \mathcal{F}$ and $\mu(A) = 0 \Rightarrow \nu(A) = 0$. We then write $\nu \ll \mu$. Two measures μ and ν are said to be *equivalent* if they are mutually absolutely continuous. The key result on absolutely continuous measures is

1.1 Review of measure and probability

Theorem 1.1.4 (Radon–Nikodým) *If μ is σ-finite and ν is finite with $\nu \ll \mu$, then there exists a measurable function $g : S \to \mathbb{R}^+$ such that, for each $A \in \mathcal{F}$,*

$$\nu(A) = \int_A g(x)\mu(dx).$$

The function g is unique up to μ-almost-everywhere equality.

The functions g appearing in this theorem are sometimes denoted $d\nu/d\mu$ and called (versions of) the *Radon–Nikodým derivative* of ν with respect to μ. For example, if X is a random variable with law p_X that is absolutely continuous with respect to Lebesgue measure on $\mathbb{R}^d$, we usually write $f_X = dp_X/dx$ and call f_X a *probability density function* (or sometimes a density or a pdf for short).

Now let $(\Omega, \mathcal{F}, P)$ be a probability space and $\mathcal{G}$ be a *sub-σ-algebra* of $\mathcal{F}$ so that:

(1) $\mathcal{G}$ is a σ-algebra;
(2) $\mathcal{G} \subseteq \mathcal{F}$.

Let X be an $\mathbb{R}$-valued random variable with $\mathbb{E}(|X|) < \infty$, and for now assume that $X \geq 0$. We define a finite measure Q_X on $(\Omega, \mathcal{G})$ by the prescription $Q_X(A) = \mathbb{E}(X\chi_A)$ for $A \in \mathcal{G}$; then $Q_X \ll P$, and we write

$$\mathbb{E}(X|\mathcal{G}) = \frac{dQ_X}{dP}.$$

We call $\mathbb{E}(X|\mathcal{G})$ the *conditional expectation* of X with respect to $\mathcal{G}$. It is a random variable on $(\Omega, \mathcal{G}, P)$ and is uniquely defined up to sets of P-measure zero. For arbitrary real-valued X with $\mathbb{E}(|X|) < \infty$, we define

$$\mathbb{E}(X|\mathcal{G}) = \mathbb{E}(X^+|\mathcal{G}) - \mathbb{E}(X^-|\mathcal{G}).$$

When $X = (X_1, X_2, \ldots, X_d)$ takes values in $\mathbb{R}^d$ with $\mathbb{E}(|X|) < \infty$, we define

$$\mathbb{E}(X|\mathcal{G}) = (\mathbb{E}(X_1|\mathcal{G}), E(X_2|\mathcal{G}), \ldots, \mathbb{E}(X_d|\mathcal{G})).$$

We sometimes write $\mathbb{E}_\mathcal{G}(\cdot) = \mathbb{E}(\cdot|\mathcal{G})$.

We now list a number of key properties of the conditional expectation.

- $\mathbb{E}(\mathbb{E}(X|\mathcal{G})) = \mathbb{E}(X)$.
- $|\mathbb{E}(X|\mathcal{G})| \leq \mathbb{E}(|X||\mathcal{G})$ a.s.
- If Y is a $\mathcal{G}$-measurable random variable and $\mathbb{E}(|(X, Y)|) < \infty$ then

$$\mathbb{E}((X, Y)|\mathcal{G}) = (\mathbb{E}(X|\mathcal{G}), Y) \quad \text{a.s.}$$

- If $\mathcal{H}$ is a sub-σ-algebra of $\mathcal{G}$ then

$$\mathbb{E}(\mathbb{E}(X|\mathcal{G})|\mathcal{H}) = \mathbb{E}(X|\mathcal{H}) \quad \text{a.s.}$$

- The mapping $\mathbb{E}_\mathcal{G} : L^2(\Omega, \mathcal{F}, P) \to L^2(\Omega, \mathcal{G}, P)$ is an orthogonal projection.

In particular, given any $\mathcal{G}$-measurable random variable Y such that $\mathbb{E}(|Y|) < \infty$ and for which

$$\mathbb{E}(Y\chi_A) = \mathbb{E}(X\chi_A)$$

for all $A \in \mathcal{G}$, then $Y = \mathbb{E}(X|\mathcal{G})$ (a.s.).

The monotone and dominated convergence theorems and also Jensen's inequality all have natural conditional forms (see e.g. Dudley [84], pp. 266 and 274).

The following result is, in fact, a special case of the convergence theorem for reversed martingales. This is proved, in full generality, in Dudley [84], p. 290.

Proposition 1.1.5 *If Y is a random variable with $\mathbb{E}(|Y|) < \infty$ and $(\mathcal{G}_n, n \in \mathbb{N})$ is a decreasing sequence of sub-σ-algebras of $\mathcal{F}$, then*

$$\lim_{n \to \infty} \mathbb{E}(Y|\mathcal{G}_n) = \mathbb{E}(Y|\mathcal{G}) \quad \text{a.s.,}$$

where $\mathcal{G} = \bigcap_{n \in \mathbb{N}} \mathcal{G}_n$.

If Y is a random variable defined on the same probability space as X we write $\mathbb{E}(X|Y) = \mathbb{E}(X|\sigma(Y))$, and if $A \in \mathcal{F}$ we write $\mathbb{E}(X|A) = \mathbb{E}(X|\sigma(A))$ where $\sigma(A) = \{A, A^c, \Omega, \emptyset\}$.

If $A \in \mathcal{F}$ we define $P(A|\mathcal{G}) = \mathbb{E}(\chi_A|\mathcal{G})$. We call $P(A|\mathcal{G})$ the *conditional probability* of A given $\mathcal{G}$. Note that it is not, in general, a probability measure on $\mathcal{F}$ (not even a.s.) although it does satisfy each of the requisite axioms with probability 1. Let Y be an $\mathbb{R}^d$-valued random variable on Ω and define the *conditional distribution* of Y, given $\mathcal{G}$ to be the mapping $P_{Y|\mathcal{G}} : \mathcal{B}(\mathbb{R}^d) \times \Omega \to [0, 1]$ for which

$$P_{Y|\mathcal{G}}(B, \omega) = P(Y^{-1}(B)|\mathcal{G})(\omega)$$

for each $B \in \mathcal{B}(\mathbb{R}^d)$, $\omega \in \Omega$. Then $P_{Y|\mathcal{G}}$ is a probability measure on $\mathcal{B}(\mathbb{R}^d)$ for almost all $\omega \in \Omega$. Moreover, for each $g : \mathbb{R}^d \to \mathbb{R}^d$ with $|\mathbb{E}(g(Y))| < \infty$ we have

$$\mathbb{E}(g \circ Y|\mathcal{G}) = \int_{\mathbb{R}^d} g(y) P_{Y|\mathcal{G}}(dy, \cdot) \quad \text{a.s.} \quad (1.1)$$

1.1.4 Independence and product measures

Let $(\Omega, \mathcal{F}, P)$ be a probability space. A sequence $(\mathcal{F}_n, n \in \mathbb{N})$ of sub-σ-algebras of $\mathcal{F}$ is said to be *independent* if, for any n-tuple $i_1, i_2, \ldots, i_n$ and any $A_{i_j} \in \mathcal{F}_j, 1 \leq j \leq n$,

$$P(A_{i_1} \cap A_{i_2} \cap \cdots \cap A_{i_n}) = \prod_{j=1}^{n} P(A_{i_j}).$$

In particular, a sequence of random variables $(X_n, n \in \mathbb{N})$ is said to be *independent* if $(\sigma(X_n), n \in \mathbb{N})$ is independent in the above sense. Such a sequence is said to be i.i.d. if the random variables are independent and also identically distributed, i.e. the laws $(p_{X_n}, n \in \mathbb{N})$ are identical probability measures. We say that a random variable X and a sub-σ-algebra $\mathcal{G}$ of $\mathcal{F}$ are independent if $\sigma(X)$ and $\mathcal{G}$ are independent. In this case we have

$$\mathbb{E}(X|\mathcal{G}) = \mathbb{E}(X) \quad \text{a.s.}$$

Now let $\{(S_1, \mathcal{F}_1, \mu_1), \ldots, (S_n, \mathcal{F}_n, \mu_n)\}$ be a family of measure spaces. We define their *product* to be the space $(S, \mathcal{F}, \mu)$, where S is the Cartesian product $S_1 \times S_2 \times \cdots \times S_n$, $\mathcal{F} = \mathcal{F}_1 \otimes \mathcal{F}_2 \otimes \cdots \otimes \mathcal{F}_n$ is the smallest σ-algebra containing all sets of the form $A_1 \times A_2 \times \cdots \times A_n$ for which each $A_i \in \mathcal{F}_i$ and $\mu = \mu_1 \times \mu_2 \times \cdots \times \mu_n$ is the *product measure* for which

$$\mu(A_1 \times A_2 \times \cdots \times A_n) = \prod_{i=1}^{n} \mu(A_i).$$

To ease the notation, we state the following key result only in the case $n = 2$.

Theorem 1.1.6 (Fubini) *If $(S_i, \mathcal{F}_i, \mu_i)$ are measure spaces for $i = 1, 2$ and if $f : S_1 \times S_2 \to \mathbb{R}$ is $\mathcal{F}_1 \otimes \mathcal{F}_2$-measurable with*

$$\iint |f(x,y)| \mu_1(dx) \mu_2(dy) < \infty,$$

then

$$\int_{S_1 \times S_2} f(x,y)(\mu_1 \times \mu_2)(dx, dy) = \int_{S_2} \left[\int_{S_1} f(x,y) \mu_1(dx) \right] \mu_2(dy)$$

$$= \int_{S_1} \left[\int_{S_2} f(x,y) \mu_2(dy) \right] \mu_1(dx).$$

The functions $y \to \int f(x,y) \mu_1(dx)$ and $x \to \int f(x,y) \mu_2(dy)$ are defined μ_2 (a.e.) and μ_1 (a.e.), respectively.

For $1 \leq j \leq n$, let X_j be a random variable defined on a probability space $(\Omega, \mathcal{F}, P)$ and form the random vector $X = (X_1, X_2, \ldots, X_n)$; then the X_n are independent if and only if $p_X = p_{X_1} \times p_{X_2} \times \cdots \times p_{X_n}$.

At various times in this book, we will require a conditional version of Fubini's theorem. Since this is not included in many standard texts, we give a statement and proof of the precise result we require.

Theorem 1.1.7 (Conditional Fubini) *Let $(\Omega, \mathcal{F}, P)$ be a probability space and $\mathcal{G}$ be a sub-σ-algebra of $\mathcal{F}$. If (S, Σ, μ) is a measure space and $F \in L^1(S \times \Omega, \Sigma \otimes \mathcal{F}, \mu \times P)$, then*

$$\mathbb{E}\left(\left|\int_S \mathbb{E}_{\mathcal{G}}(F(s, \cdot))\mu(ds)\right|\right) < \infty,$$

and

$$\mathbb{E}_{\mathcal{G}}\left(\int_S F(s, \cdot)\mu(ds)\right) = \int_S \mathbb{E}_{\mathcal{G}}(F(s, \cdot))\mu(ds) \qquad \text{a.s.}$$

Proof Using the usual Fubini theorem, we find that

$$\mathbb{E}\left(\left|\int_S \mathbb{E}_{\mathcal{G}}(F(s, \cdot))\mu(ds)\right|\right) \leq \int_S \mathbb{E}(|\mathbb{E}_{\mathcal{G}}(F(s, \cdot))|)\mu(ds)$$

$$\leq \int_S \mathbb{E}(\mathbb{E}_{\mathcal{G}}(|F(s, \cdot)|))\mu(ds)$$

$$= \int_S \mathbb{E}(|F(s, \cdot)|)\mu(ds) < \infty$$

and, for each $A \in \mathcal{G}$,

$$\mathbb{E}\left(\chi_A \int_S F(s, \cdot)\mu(ds)\right) = \int_S \mathbb{E}(\chi_A F(s, \cdot))\mu(ds)$$

$$= \int_S \mathbb{E}(\chi_A \mathbb{E}_{\mathcal{G}}(F(s, \cdot)))\mu(ds)$$

$$= \mathbb{E}\left(\chi_A \int_S \mathbb{E}_{\mathcal{G}}(F(s, \cdot))\mu(ds)\right),$$

from which the required result follows. $\square$

The following result gives a nice interplay between conditioning and independence and is extremely useful for proving the Markov property, as we will see later. For a proof, see Sato [274], p. 7.

Lemma 1.1.8 *Let $\mathcal{G}$ be a sub-σ-algebra of $\mathcal{F}$. If X and Y are $\mathbb{R}^d$-valued random variables such that X is $\mathcal{G}$-measurable and Y is independent of $\mathcal{G}$ then*

$$\mathbb{E}(f(X,Y)|\mathcal{G}) = G_f(X) \quad \text{a.s.}$$

for all $f \in B_b(\mathbb{R}^{2d})$, where $G_f(x) = \mathbb{E}(f(x,Y))$ for each $x \in \mathbb{R}^d$.

1.1.5 Convergence of random variables

Let $(X(n), n \in \mathbb{N})$ be a sequence of $\mathbb{R}^d$-valued random variables and X be an $\mathbb{R}^d$-valued random variable. We say that:

- $X(n)$ converges to X *almost surely* if $\lim_{n\to\infty} X(n)(\omega) = X(\omega)$ for all $\omega \in \Omega - \mathcal{N}$, where $\mathcal{N} \in \mathcal{F}$ satisfies $P(\mathcal{N}) = 0$;
- $X(n)$ converges to X in L^p ($1 \le p < \infty$) if $\lim_{n\to\infty} \mathbb{E}(|X(n) - X|^p) = 0$. The case $p = 2$ is often called *convergence in mean square* and in this case we sometimes write $L^2 - \lim X(n)_{n\to\infty} = X$;
- $X(n)$ converges to X *in probability* if, for all $a > 0$, $\lim_{n\to\infty} P(|X(n) - X| > a) = 0$;
- $X(n)$ converges to X *in distribution* if

$$\lim_{n\to\infty} \int_{\mathbb{R}^d} f(x) p_{X(n)}(dx) = \int_{\mathbb{R}^d} f(x) p_X(dx) \quad \text{for all} \quad f \in C_b(\mathbb{R}^d).$$

In the case $d = 1$, convergence in distribution is equivalent to the requirement on distribution functions that $\lim_{n\to\infty} F_{X(n)}(x) = F_X(x)$ at all continuity points of F_X.

The following relations between modes of convergence are important:

almost-sure convergence $\Rightarrow$ convergence in probability $\Rightarrow$
convergence in distribution;
L^p-convergence $\Rightarrow$ convergence in probability $\Rightarrow$
convergence in distribution.

Conversely, if $X(n)$ converges in probability to X then we can always find a subsequence that converges almost surely to X.

Let $L^0 = L^0(\Omega, \mathcal{F}, P)$ denote the linear space of all equivalence classes of $\mathbb{R}^d$-valued random variables that agree almost surely; then L^0 becomes a complete metric space with respect to the *Ky Fan metric*

$$d(X, Y) = \inf\{\epsilon > 0, P(|X - Y| > \epsilon) \le \epsilon\}$$

for $X, Y \in L^0$. The function d metrises convergence in probability in that a sequence $(X(n), n \in \mathbb{N})$ in L^0 converges in probability to $X \in L^0$ if and only if $\lim_{n \to \infty} d(X(n), X) = 0$.

We will find the following result of use later on.

Proposition 1.1.9 *If $(X(n), n \in \mathbb{N})$ and $(Y(n), n \in \mathbb{N})$ are sequences of random variables for which $X(n) \to X$ in probability and $Y(n) \to 0$ almost surely, then $X(n)Y(n) \to 0$ in probability.*

Proof We will make use of the following elementary inequality for random variables W and Z, where $a > 0$:

$$P(|W + Z| > a) \leq P\left(|W| > \frac{a}{2}\right) + P\left(|Z| > \frac{a}{2}\right).$$

We then find that, for all $n \in \mathbb{N}$,

$$P(|X(n)Y(n)| > a) = P(|X(n)Y(n) - XY(n) + XY(n)| > a)$$
$$\leq P\left(|X(n)Y(n) - XY(n)| > \frac{a}{2}\right) + P\left(|XY(n)| > \frac{a}{2}\right).$$

Now $Y(n) \to 0$ (a.s.) $\Rightarrow XY(n) \to 0$ (a.s.) $\Rightarrow XY(n) \to 0$ in probability.

For each $k > 0$ let $\mathcal{N}_k = \{\omega \in \Omega; |Y(n)(\omega)| \leq k\}$ and assume, without loss of generality, that $P(\{\omega \in \Omega; Y(n)(\omega) = 0\}) = 0$ for all sufficiently large n; then

$$P\left(|X(n)Y(n) - XY(n)| > \frac{a}{2}\right) \leq P\left(|Y(n)||X(n) - X| > \frac{a}{2}\right)$$
$$= P\left(|Y(n)||X(n) - X| > \frac{a}{2}, \mathcal{N}_k\right)$$
$$+ P\left(|Y(n)||X(n) - X| > \frac{a}{2}, \mathcal{N}_k^c\right)$$
$$\leq P\left(|X(n) - X| > \frac{a}{2k}\right) + P(|Y(n)| > k)$$
$$\to 0 \quad \text{as} \quad n \to \infty,$$

and the result follows. $\square$

As well as random variables, we will also want to consider the convergence of probability measures. A sequence $(\mu(n), n \in \mathbb{N})$ of such measures on $\mathbb{R}^d$ is said to converge *weakly* to a probability measure μ if

$$\lim_{n \to \infty} \int f(x)\mu(n)(dx) = \int f(x)\mu(dx)$$

1.1 Review of measure and probability

for all $f \in C_b(\mathbb{R}^d)$. A sufficient (but not necessary) condition for this to hold is that $\mu(n)(E) \to \mu(E)$ as $n \to \infty$, for every $E \in \mathcal{B}(\mathbb{R}^d)$.

1.1.6 Characteristic functions

Let X be a random variable defined on $(\Omega, \mathcal{F}, P)$ and taking values in $\mathbb{R}^d$ with probability law p_X. Its *characteristic function* $\phi_X : \mathbb{R}^d \to \mathbb{C}$ is defined by

$$\phi_X(u) = \mathbb{E}\big(e^{i(u,X)}\big) = \int_\Omega e^{i(u,X(\omega))} P(d\omega)$$

$$= \int_{\mathbb{R}^d} e^{i(u,y)} p_X(dy),$$

for each $u \in \mathbb{R}^d$. More generally, if p is a probability measure on $\mathbb{R}^d$ then its characteristic function is the map $u \to \int_{\mathbb{R}^d} e^{i(u,y)} p(dy)$, and it can be shown that this mapping uniquely determines the measure p.

The following properties of ϕ_X are elementary:

- $|\phi_X(u)| \leq 1$;
- $\phi_X(-u) = \overline{\phi_X(u)}$;
- X is symmetric if and only if ϕ_X is real-valued;
- if $X = (X_1, \ldots, X_d)$ and $\mathbb{E}(|X_j^n|) < \infty$ for some $1 \leq j \leq d$ and $n \in \mathbb{N}$ then

$$\mathbb{E}(X_j^n) = i^{-n} \left. \frac{\partial^n}{\partial u_j^n} \phi_X(u) \right|_{u=0}.$$

If $M_X(u) = \phi_X(-iu)$ exists, at least in a neighbourhood of $u = 0$, then M_X is called the *moment generating function* of X. In this case all the moments of X exist and can be obtained by partial differentiation of M_X as above.

For fixed $u_1, \ldots, u_d \in \mathbb{R}^d$, we denote as Φ_X the $d \times d$ matrix whose (i, j)th entry is $\phi_X(u_i - u_j)$. Further properties of ϕ_X are collected in the following lemma.

Lemma 1.1.10

(1) Φ_X is positive definite for all $u_1, \ldots, u_d \in \mathbb{R}^d$.
(2) $\phi_X(0) = 1$.
(3) The map $u \to \phi_X(u)$ is continuous at the origin.

Proof Parts (2) and (3) are straightforward.
For (1) we need to show that $\sum_{j,k=1}^d c_j \overline{c_k} \phi_X(u_j - u_k) \geq 0$ for all $u_1, \ldots, u_d \in \mathbb{R}^d$ and all $c_1, \ldots, c_d \in \mathbb{C}$.

Define $f : \mathbb{R}^d \to \mathbb{C}$ by $f(x) = \sum_{j=1}^d c_j e^{i(u_j, x)}$ for each $x \in \mathbb{R}^d$; then $f \in L^2(\mathbb{R}^d, p_X)$ and we find that

$$\sum_{i=1}^d \sum_{j=1}^d c_i \overline{c_j} \phi_X(u_i - u_j) = \int_{\mathbb{R}^d} \sum_{i,j=1}^d c_i \overline{c_j} e^{i(u_i - u_j, x)} p_X(dx)$$

$$= \int_{\mathbb{R}^d} |f(x)|^2 p_X(dx) = \|f\|^2 \geq 0.$$

□

A straightforward application of dominated convergence verifies that ϕ_X is, in fact, uniformly continuous on the whole of $\mathbb{R}^d$. Nonetheless the weaker statement (3) is sufficient for the following powerful theorem.

Theorem 1.1.11 (Bochner's theorem) *If $\phi : \mathbb{R}^d \to \mathbb{C}$ satisfies parts (1), (2) and (3) of Lemma 1.1.10, then ϕ is the characteristic function of a probability distribution.*

We will sometimes want to apply Bochner's theorem to functions of the form $\phi(u) = e^{t\psi(u)}$ where $t > 0$ and, in this context, it is useful to have a condition on ψ that is equivalent to the positive definiteness of ϕ.

We say that $\psi : \mathbb{R}^d \to \mathbb{C}$ is *conditionally positive definite* if for all $n \in \mathbb{N}$ and $c_1, \ldots, c_n \in \mathbb{C}$ for which $\sum_{j=1}^n c_j = 0$ we have

$$\sum_{j,k=1}^n c_j \bar{c}_k \psi(u_j - u_k) \geq 0$$

for all $u_1, \ldots, u_n \in \mathbb{R}^d$. The mapping $\psi : \mathbb{R}^d \to \mathbb{C}$ is said to be *hermitian* if $\overline{\psi(u)} = \psi(-u)$ for all $u \in \mathbb{R}^d$.

Theorem 1.1.12 (Schoenberg correspondence) *The mapping $\psi : \mathbb{R}^d \to \mathbb{C}$ is hermitian and conditionally positive definite if and only if $e^{t\psi}$ is positive definite for each $t > 0$.*

Proof We give only the easy part here. For the full story see Berg and Forst [35], p. 41, or Parthasarathy and Schmidt [246], pp. 1–4.

Suppose that $e^{t\psi}$ is positive definite for all $t > 0$. Fix $n \in \mathbb{N}$ and choose $c_1, \ldots, c_n$ and $u_1, \ldots, u_n$ as above. We then find that, for each $t > 0$,

$$\frac{1}{t} \sum_{j,k=1}^n c_j \bar{c}_k \left[e^{t\psi(u_j - u_k)} - 1 \right] \geq 0,$$

and so

$$\sum_{j,k=1}^{n} c_j \bar{c}_k \psi(u_j - u_k) = \lim_{t \to 0} \frac{1}{t} \sum_{j,k=1}^{n} c_j \bar{c}_k \left[e^{t\psi(u_j - u_k)} - 1 \right] \geq 0.$$

$\square$

To see the need for ψ to be hermitian, define $\tilde{\psi}(\cdot) = \psi(\cdot) + ix$, where ψ is hermitian and conditionally positive definite and $x \in \mathbb{R}, x \neq 0$. $\tilde{\psi}$ is clearly conditionally positive definite but not hermitian, and it is then easily verified that $e^{t\tilde{\psi}}$ cannot be positive definite for any $t > 0$.

Note that Berg and Forst [35] adopt the analyst's convention of using $-\psi$, which they call 'negative definite', rather than the hermitian, conditionally positive definite ψ.

Two important convergence results are the following:

Theorem 1.1.13 (Glivenko) *If ϕ_n and ϕ are the characteristic functions of probability distributions p_n and p (respectively), for each $n \in \mathbb{N}$, then $\phi_n(u) \to \phi(u)$ for all $u \in \mathbb{R}^d \Rightarrow p_n \to p$ weakly as $n \to \infty$.*

Theorem 1.1.14 (Lévy continuity theorem) *If $(\phi_n, n \in \mathbb{N})$ is a sequence of characteristic functions and there exists a function $\psi : \mathbb{R}^d \to \mathbb{C}$ such that, for all $u \in \mathbb{R}^d$, $\phi_n(u) \to \psi(u)$ as $n \to \infty$ and ψ is continuous at 0 then ψ is the characteristic function of a probability distribution.*

Now let $X_1, \ldots, X_n$ be a family of random variables all defined on the same probability space. Our final result in this subsection is

Theorem 1.1.15 (Kac's theorem) *The random variables $X_1, \ldots, X_n$ are independent if and only if*

$$\mathbb{E}\left(\exp\left[i \sum_{j=1}^{n} (u_j, X_j) \right] \right) = \phi_{X_1}(u_1) \cdots \phi_{X_n}(u_n)$$

for all $u_1, \ldots, u_n \in \mathbb{R}^d$.

1.1.7 Stochastic processes

To model the evolution of chance in time we need the notion of a *stochastic process*. This is a family of random variables $X = (X(t), t \geq 0)$ that are all defined on the same probability space.

Two stochastic processes $X = (X(t), t \geq 0)$ and $Y = (Y(t), t \geq 0)$ are *independent* if, for all $m, n \in \mathbb{N}$, all $0 \leq t_1 < t_2 < \cdots < t_n < \infty$ and all

$0 \leq s_1 < s_2 < \cdots < s_m < \infty$, the σ-algebras $\sigma(X(t_1), X(t_2), \ldots, X(t_n))$ and $\sigma(Y(s_1), Y(s_2), \ldots, Y(s_m))$ are independent.

Similarly, a stochastic process $X = (X(t), t \geq 0)$ and a sub-σ-algebra $\mathcal{G}$ are *independent* if $\mathcal{G}$ and $\sigma(X(t_1), X(t_2), \ldots, X(t_n))$ are independent for all $n \in \mathbb{N}, 0 \leq t_1 < t_2 < \cdots < t_n < \infty$.

The *finite-dimensional distributions* of a stochastic process X are the collection of probability measures $(p_{t_1, t_2, \ldots, t_n}, 0 \leq t_1 < t_2 < \cdots < t_n < \infty, n \in \mathbb{N})$ defined on $\mathbb{R}^{dn}$ for each $n \in \mathbb{N}$ by

$$p_{t_1, t_2, \ldots, t_n}(H) = P((X(t_1), X(t_2), \ldots, X(t_n)) \in H)$$

for each $H \in \mathcal{B}(\mathbb{R}^{dn})$.

Let π be a permutation of $\{1, 2, \ldots, n\}$; then it is clear that, for each $H_1, H_2, \ldots, H_n \in \mathcal{B}(\mathbb{R}^d)$,

$$p_{t_1, t_2, \ldots, t_n}(H_1 \times H_2 \times \cdots \times H_n)$$
$$= p_{t_{\pi(1)}, t_{\pi(2)}, \ldots, t_{\pi(n)}}(H_{\pi(1)} \times H_{\pi(2)} \times \cdots \times H_{\pi(n)}); \quad (1.2)$$
$$p_{t_1, t_2, \ldots, t_n, t_{n+1}}(H_1 \times H_2 \times \cdots \times H_n \times \mathbb{R}^d)$$
$$= p_{t_1, t_2, \ldots, t_n}(H_1 \times H_2 \times \cdots \times H_n). \quad (1.3)$$

Equations (1.2) and (1.3) are called *Kolmogorov's consistency criteria*.

Now suppose that we are given a family of probability measures

$$(p_{t_1, t_2, \ldots, t_n}, 0 \leq t_1 < t_2 < \cdots < t_n < \infty, n \in \mathbb{N})$$

satisfying these criteria. Kolmogorov's construction, which we will now describe, allows us to build a stochastic process for which these are the finite-dimensional distributions. The procedure is as follows.

Let Ω be the set of all mappings from $\mathbb{R}^+$ into $\mathbb{R}^d$ and $\mathcal{F}$ be the smallest σ-algebra containing all *cylinder sets* of the form

$$I^H_{t_1, t_2, \ldots, t_n} = \{\omega \in \Omega; (\omega(t_1), \omega(t_2), \ldots, \omega(t_n)) \in H\},$$

where $H \in \mathcal{B}(\mathbb{R}^{dn})$.

Define the *co-ordinate process* $X = (X(t), t \geq 0)$ by

$$X(t)(\omega) = \omega(t)$$

for each $t \geq 0, \omega \in \Omega$.

The main result is

1.1 Review of measure and probability

Theorem 1.1.16 (Kolmogorov's existence theorem) *Given a family of probability measures* $(p_{t_1, t_2, \ldots, t_n}, 0 \leq t_1 < t_2 < \cdots < t_n < \infty, n \in \mathbb{N})$ *satisfying the Kolmogorov consistency criteria, there exists a probability measure P on* $(\Omega, \mathcal{F})$ *such that the co-ordinate process X is a stochastic process on* $(\Omega, \mathcal{F}, P)$ *having the* $p_{t_1, t_2, \ldots, t_n}$ *as its finite-dimensional distributions.*

A stochastic process $X = (X(t), t \geq 0)$ is said to be *separable* if there exists a countable subset $D \subset \mathbb{R}^+$ such that, for each $t \geq 0$, there exists a sequence $(t(n), n \in \mathbb{N})$ in D with each $t(n) \neq t$ such that $\lim_{n \to \infty} t(n) = t$ and $\lim_{n \to \infty} X(t(n)) = X(t)$.

Kolmogorov's theorem can be extended to show that, given a family $(p_{t_1, t_2, \ldots, t_n}, 0 \leq t_1 < t_2 < \cdots < t_n < \infty, n \in \mathbb{N})$ of probability measures satisfying the Kolmogorov consistency criteria, we can always construct a separable process $X = (X(t), t \geq 0)$ on some $(\Omega, \mathcal{F}, P)$ having the $p_{t_1, t_2, \ldots, t_n}$ as its finite-dimensional distributions. Bearing this in mind, we will suffer no loss in generality if we assume all stochastic processes considered in this book to be separable.

The maps from $\mathbb{R}^+$ to $\mathbb{R}^d$ given by $t \to X(t)(\omega)$, where $\omega \in \Omega$ are called the *sample paths* of the stochastic process X. We say that a process is continuous, bounded, increasing etc. if almost all its sample paths have this property.

Let G be a group of matrices acting on $\mathbb{R}^d$. We say that a stochastic process $X = (X(t), t \geq 0)$ is *G-invariant* if the law $p_{X(t)}$ is G-invariant for all $t \geq 0$. Clearly X is G-invariant if and only if

$$\phi_{X(t)}(g^T u) = \phi_{X(t)}(u)$$

for all $t \geq 0, u \in \mathbb{R}^d, g \in G$.

In the case where $G = O(d)$, the group of all $d \times d$ orthogonal matrices acting in $\mathbb{R}^d$, we say that the process X is *rotationally invariant* and when G is the normal subgroup of $O(d)$ comprising the two points $\{-I, I\}$ we say that X is *symmetric*.

1.1.8 Random fields

A random field is a natural generalisation of a stochastic process in which the time interval is replaced by a different set E. Here we will assume that $E \in \mathcal{B}(\mathbb{R}^d)$ and define a *random field* on E to be a family of random variables $X = (X(y), y \in E)$. We will only use random fields on one occasion, in Chapter 6, and there it will be important for us to be able to show that they are (almost surely) continuous. Fortunately, we have the celebrated Kolmogorov criterion to facilitate this.

Theorem 1.1.17 (Kolmogorov's continuity criterion) *Let X be a random field on E and suppose that there exist strictly positive constants γ, C and ϵ such that*

$$\mathbb{E}(|X(y_2) - X(y_1)|^\gamma) \leq C|y_2 - y_1|^{d+\epsilon}$$

for all $y_1, y_2 \in E$. Then there exists another random field $\tilde{X}$ on E such that $\tilde{X}(y) = X(y)$ (a.s.), for all $y \in E$, and $\tilde{X}$ is almost surely continuous.

For a proof of this result, see Revuz and Yor [260], Section 1.2, or Kunita [182], Section 1.4.

1.2 Infinite divisibility

1.2.1 Convolution of measures

Let $\mathcal{M}_1(\mathbb{R}^d)$ denote the set of all Borel probability measures on $\mathbb{R}^d$. We define the *convolution* of two probability measures as follows:

$$(\mu_1 * \mu_2)(A) = \int_{\mathbb{R}^d} \mu_1(A - x)\mu_2(dx) \tag{1.4}$$

for each $\mu_i \in \mathcal{M}_1(\mathbb{R}^d)$, $i = 1, 2$, and each $A \in \mathcal{B}(\mathbb{R}^d)$, where we note that $A - x = \{y - x, y \in A\}$.

Proposition 1.2.1 *The convolution $\mu_1 * \mu_2$ is a probability measure on $\mathbb{R}^d$.*

Proof First we show that convolution is a measure. Let $(A_n, n \in \mathbb{N})$ be a sequence of disjoint sets in $\mathcal{B}(\mathbb{R}^d)$; then, for each $x \in \mathbb{R}^d$, the members of the sequence $(A_n - x, n \in \mathbb{N})$ are also disjoint and

$$(\mu_1 * \mu_2)\left(\bigcup_{n \in \mathbb{N}} A_n\right) = \int_{\mathbb{R}^d} \mu_1\left[\left(\bigcup_{n \in \mathbb{N}} A_n\right) - x\right] \mu_2(dx)$$

$$= \int_{\mathbb{R}^d} \mu_1\left[\bigcup_{n \in \mathbb{N}}(A_n - x)\right] \mu_2(dx)$$

$$= \int_{\mathbb{R}^d} \sum_{n \in \mathbb{N}} \mu_1(A_n - x)\mu_2(dx)$$

$$= \sum_{n \in \mathbb{N}} \int_{\mathbb{R}^d} \mu_1(A_n - x)\mu_2(dx)$$

$$= \sum_{n \in \mathbb{N}} (\mu_1 * \mu_2)(A_n),$$

where the interchange of sum and integral is justified by dominated convergence.

The fact that $\mu_1 * \mu_2$ is a probability measure now follows easily from the observation that the map from $\mathbb{R}^d$ to itself given by the translation $y \to y - x$ is a bijection, and so $\mathbb{R}^d = \mathbb{R}^d - x$. □

From the above proposition, we see that convolution is a binary operation on $\mathcal{M}_1(\mathbb{R}^d)$.

Proposition 1.2.2 *If $f \in B_b(\mathbb{R}^d)$, then for all $\mu_i \in \mathcal{M}_1(\mathbb{R}^d)$, $i = 1, 2, 3$,*

(1)
$$\int_{\mathbb{R}^d} f(y)(\mu_1 * \mu_2)(dy) = \int_{\mathbb{R}^d} \int_{\mathbb{R}^d} f(x+y)\mu_1(dy)\mu_2(dx),$$

(2)
$$\mu_1 * \mu_2 = \mu_2 * \mu_1,$$

(3)
$$(\mu_1 * \mu_2) * \mu_3 = \mu_1 * (\mu_2 * \mu_3).$$

Proof (1) We can easily verify the result for indicator functions by the fact that, for any $A \in \mathcal{B}(\mathbb{R}^d)$ and $x, y \in \mathbb{R}^d$,

$$\chi_A(x+y) = \chi_{A-x}(y).$$

The result is then extended by linearity to simple functions. The general result is settled by approximation as follows.

Let $M = \sup_{x \in \mathbb{R}^d} |f(x)|$, fix $\epsilon > 0$ and, for each $n \in \mathbb{N}$, let $a_0^{(n)} < a_1^{(n)} < \cdots < a_{m_n}^{(n)}$ be such that the collection of intervals $\{(a_{i-1}^{(n)}, a_i^{(n)}]; 1 \leq i \leq m_n\}$ covers $[-M, M]$ with $\max_{1 \leq i \leq m_n} |a_i^{(n)} - a_{i-1}^{(n)}| < \epsilon$, for sufficiently large n.

Define a sequence of simple functions by $f_n = \sum_{i=1}^{m_n} a_{i-1}^{(n)} \chi_{A_i^{(n)}}$, where each $A_i^{(n)} = f^{-1}((a_{i-1}^{(n)}, a_i^{(n)}])$. Then for sufficiently large n we have

$$\int_{\mathbb{R}^d} |f_n(x) - f(x)|(\mu_1 * \mu_2)(dx) \leq \sup_{x \in \mathbb{R}^d} |f_n(x) - f(x)|$$
$$= \max_{1 \leq i \leq m_n} \sup_{x \in A_i^{(n)}} |f_n(x) - f(x)| < \epsilon.$$

If we define $g_n(x, y) = f_n(x+y)$ and $g(x, y) = f(x+y)$ for each $n \in \mathbb{N}$, $x, y \in \mathbb{R}^d$, then an argument similar to the above shows that $\lim_{n \to \infty} g_n = g$ in $L^1(\mathbb{R}^d \times \mathbb{R}^d, \mu_1 \times \mu_2)$. The required result now follows from use of the dominated convergence theorem.

(2) By Fubini's theorem in (1),
$$\int_{\mathbb{R}^d} f(z)(\mu_1 * \mu_2)(dz) = \int_{\mathbb{R}^d} f(z)(\mu_2 * \mu_1)(dz).$$
Then take $f = \chi_A$ where A is a Borel set and the result follows.

(3) Use Fubini's theorem again to show that both expressions are equal to
$$\int_{\mathbb{R}^d} \int_{\mathbb{R}^d} \int_{\mathbb{R}^d} f(x+y+z)\mu_1(dx)\mu_2(dy)\mu_3(dz).$$
□

Now let X_1 and X_2 be independent random variables defined on a probability space $(\Omega, \mathcal{F}, P)$ with joint distribution p and marginals μ_1 and μ_2 respectively.

Corollary 1.2.3 *For each $f \in B_b(\mathbb{R}^n)$,*
$$\mathbb{E}(f(X_1 + X_2)) = \int_{\mathbb{R}^d} f(z)(\mu_1 * \mu_2)(dz).$$

Proof Using part (1) of Proposition 1.2.2,
$$\mathbb{E}(f(X_1 + X_2)) = \int_{\mathbb{R}^d} \int_{\mathbb{R}^d} f(x+y) p(dx, dy)$$
$$= \int_{\mathbb{R}^d} \int_{\mathbb{R}^d} f(x+y) \mu_1(dx) \mu_2(dy)$$
$$= \int_{\mathbb{R}^d} f(z)(\mu_1 * \mu_2)(dz).$$
□

By Corollary 1.2.3, we see that convolution gives the probability law for the sum of two independent random variables X_1 and X_2, i.e.
$$P(X_1 + X_2 \in A) = \mathbb{E}(\chi_A(X_1 + X_2)) = (\mu_1 * \mu_2)(A).$$
Proposition 1.2.2 also tells us that $\mathcal{M}_1(\mathbb{R}^d)$ is an abelian semigroup under $*$ in which the identity element is given by the Dirac measure δ_0, where we recall that in general, for $x \in \mathbb{R}^d$,
$$\delta_x(A) = \begin{cases} 1 & \text{if } x \in A, \\ 0 & \text{otherwise,} \end{cases}$$
for any Borel set A, so we have $\delta_0 * \mu = \mu * \delta_0 = \mu$ for all $\mu \in \mathcal{M}_1(\mathbb{R}^d)$.

We define $\mu^{*^n} = \mu * \cdots * \mu$ (n times) and say that μ has a *convolution nth root*, if there exists a measure $\mu^{1/n} \in \mathcal{M}_1(\mathbb{R}^d)$ for which $(\mu^{1/n})^{*^n} = \mu$.

Exercise 1.2.4 If X and Y are independent random variables having probability density functions (pdfs) f_X and f_Y respectively, show that $X + Y$ has density

$$f_{X+Y}(x) = \int_{\mathbb{R}^d} f_X(x - y) f_Y(y) dy,$$

where $x \in \mathbb{R}^d$.

Exercise 1.2.5 Let X have a gamma distribution with parameters $n \in \mathbb{N}$ and $\lambda > 0$, so that X has pdf

$$f(x) = \frac{\lambda^n x^{n-1} e^{-\lambda x}}{(n-1)!} \quad \text{for } x > 0.$$

Show that X has a convolution nth root given by the exponental distribution with parameter λ and pdf $f_X^{1/n}(x) = \lambda e^{-\lambda x}$.

Note In general, the convolution nth root of a probability measure may not be unique. However, it is always unique when the measure is infinitely divisible (see e.g. Sato [274], p. 34).

1.2.2 Definition of infinite divisibility

Let X be a random variable taking values in $\mathbb{R}^d$ with law μ_X. We say that X is *infinitely divisible* if, for all $n \in \mathbb{N}$, there exist i.i.d. random variables $Y_1^{(n)}, \ldots, Y_n^{(n)}$ such that

$$X \stackrel{d}{=} Y_1^{(n)} + \cdots + Y_n^{(n)}. \tag{1.5}$$

Let $\phi_X(u) = \mathbb{E}(e^{i(u,X)})$ denote the characteristic function of X, where $u \in \mathbb{R}^d$. More generally, if $\mu \in \mathcal{M}_1(\mathbb{R}^d)$ then $\phi_\mu(u) = \int_{\mathbb{R}^d} e^{i(u,y)} \mu(dy)$.

Proposition 1.2.6 *The following are equivalent:*

(1) X *is infinitely divisible;*
(2) μ_X *has a convolution nth root that is itself the law of a random variable, for each $n \in \mathbb{N}$;*
(3) ϕ_X *has an nth root that is itself the characteristic function of a random variable, for each $n \in \mathbb{N}$.*

Proof (1) $\Rightarrow$ (2). The common law of the $Y_j^{(n)}$ is the required convolution nth root.

(2) ⇒ (3). Let Y be a random variable with law $(\mu_X)^{1/n}$. We have by Proposition 1.2.2(1), for each $u \in \mathbb{R}^d$,

$$\phi_X(u) = \int \cdots \int e^{i(u,\, y_1+\cdots+y_n)}(\mu_X)^{1/n}(dy_1)\cdots(\mu_X)^{1/n}(dy_n)$$
$$= \psi_Y(u)^n$$

where $\psi_Y(u) = \int_{\mathbb{R}^d} e^{i(u,y)}(\mu_X)^{1/n}(dy)$, and the required result follows.

(3) ⇒ (1). Choose $Y_1^{(n)}, \ldots, Y_n^{(n)}$ to be independent copies of the given random variable; then we have

$$\mathbb{E}(e^{i(u,X)}) = \mathbb{E}\bigl(e^{i(u,Y_1^{(n)})}\bigr) \cdots \mathbb{E}\bigl(e^{i(u,Y_n^{(n)})}\bigr) = \mathbb{E}\bigl(e^{i(u,(Y_1^{(n)}+\cdots+Y_n^{(n)}))}\bigr),$$

from which we deduce (1.5) as required. □

Proposition 1.2.6(2) suggests that we generalise the definition of infinite divisibility as follows: $\mu \in \mathcal{M}_1(\mathbb{R}^d)$ is *infinitely divisible* if it has a convolution nth root in $\mathcal{M}_1(\mathbb{R}^d)$ for each $n \in \mathbb{N}$.

Exercise 1.2.7 Show that $\mu \in \mathcal{M}_1(\mathbb{R}^d)$ is infinitely divisible if and only if for each $n \in \mathbb{N}$ there exists $\mu^{1/n} \in \mathcal{M}_1(\mathbb{R}^d)$ for which

$$\phi_\mu(x) = \bigl[\phi_{\mu^{1/n}}(x)\bigr]^n$$

for each $x \in \mathbb{R}^d$.

Note As remarked above, the convolution nth root $\mu^{1/n}$ in Exercise 1.2.7 is unique when μ is infinitely divisible. Moreover, in this case the complex-valued function ϕ_μ always has a 'distinguished' nth root, which we denote by $\phi_\mu^{1/n}$; this is the characteristic function of $\mu^{1/n}$ (see Sato [274], pp. 32–4, for further details).

1.2.3 Examples of infinite divisibility

Example 1.2.8 (Gaussian random variables) Let $X = (X_1, \ldots, X_d)$ be a random vector.

We say that it is *(non-degenerate) Gaussian*, or *normal*, if there exists a vector $m \in \mathbb{R}^d$ and a strictly positive definite symmetric $d \times d$ matrix A such that X has a pdf of the form

$$f(x) = \frac{1}{(2\pi)^{d/2}\sqrt{\det(A)}} \exp\bigl[-\tfrac{1}{2}(x-m, A^{-1}(x-m))\bigr], \quad (1.6)$$

for all $x \in \mathbb{R}^d$.

In this case we will write $X \sim N(m, A)$. The vector m is the mean of X, so that $m = \mathbb{E}(X)$, and A is the covariance matrix, so that $A = \mathbb{E}((X - m)(X - m)^T)$. A standard calculation yields

$$\phi_X(u) = \exp\left[i(m, u) - \tfrac{1}{2}(u, Au)\right], \tag{1.7}$$

and hence

$$\left[\phi_X(u)\right]^{1/n} = \exp\left[i(\tfrac{m}{n}, u) - \tfrac{1}{2}(u, \tfrac{1}{n}Au)\right],$$

so we see that X is infinitely divisible with $Y_j^{(n)} \sim N(m/n, (1/n)A)$ for each $1 \leq j \leq n$.

We say that X is a *standard normal* whenever $X \sim N(0, \sigma^2 I)$ for some $\sigma > 0$.

Remark: Degenerate Gaussians Suppose that the matrix A is only required to be positive definite; then we may have $\det(A) = 0$, in which case the density (1.6) does not exist. Let $\phi(u)$ denote the quantity appearing on the right-hand side of (1.7); if we now replace A therein by $A + (1/n)I$ and take the limit as $n \to \infty$, it follows from Lévy's convergence theorem that ϕ is again the characteristic function of a probability measure μ. Any random variable X with such a law μ is called a *degenerate Gaussian*, and we again write $X \sim N(m, A)$.

Let S denote the linear subspace of $\mathbb{R}^n$ that is the linear span of those eigenvectors corresponding to non-zero eigenvalues of A; then the restriction of A to S is strictly positive definite and so is associated with a non-degenerate Gaussian density of the form (1.6). On $S^\perp$ we have $\phi(u) = e^{imu}$, which corresponds to a random variable taking the constant value m, almost surely. Thus we can understand degenerate Gaussians as the embeddings of non-degenerate Gaussians into higher-dimensional spaces.

Example 1.2.9 (Poisson random variables) In this case, we take $d = 1$ and consider a random variable X taking values in the set $n \in \mathbb{N} \cup \{0\}$. We say that is *Poisson* if there exists $c > 0$ for which

$$P(X = n) = \frac{c^n}{n!}e^{-c}.$$

In this case we will write $X \sim \pi(c)$. We have $\mathbb{E}(X) = \text{Var}(X) = c$. It is easy to verify that

$$\phi_X(u) = \exp[c(e^{iu} - 1)],$$

from which we deduce that X is infinitely divisible with each $Y_j^{(n)} \sim \pi(c/n)$, for $1 \leq j \leq n, n \in \mathbb{N}$.

Example 1.2.10 (Compound Poisson random variables) Suppose that $(Z(n), n \in \mathbb{N})$ is a sequence of i.i.d. random variables taking values in $\mathbb{R}^d$ with common law μ_Z and let $N \sim \pi(c)$ be a Poisson random variable that is independent of all the $Z(n)$. The *compound Poisson random variable* X is defined as follows: $X = Z(1) + \cdots + Z(N)$, so we can think of X as a random walk with a random number of steps, which are controlled by the Poisson random variable N.

Proposition 1.2.11 *For each $u \in \mathbb{R}^d$,*
$$\phi_X(u) = \exp\left[\int_{\mathbb{R}^d} (e^{i(u,y)} - 1) c \mu_Z(dy)\right].$$

Proof Let ϕ_Z be the common characteristic function of the Z_n. By conditioning and using independence we find that
$$\phi_X(u) = \sum_{n=0}^{\infty} \mathbb{E}\big(\exp[i(u, Z(1) + \cdots + Z(N))]\big|N = n\big) P(N = n)$$
$$= \sum_{n=0}^{\infty} \mathbb{E}\big(\exp[i(u, Z(1) + \cdots + Z(n))]\big) e^{-c}\frac{c^n}{n!}$$
$$= e^{-c} \sum_{n=0}^{\infty} \frac{[c\phi_Z(u)]^n}{n!}$$
$$= \exp[c(\phi_Z(u) - 1)],$$
and the result follows on writing $\phi_Z(u) = \int_{\mathbb{R}^d} e^{i(u,y)} \mu_Z(dy)$. □

Note We have employed the convention that $Z(0) = 0$ (a.s.).

If X is compound Poisson as above, we write $X \sim \pi(c, \mu_Z)$. It is clearly infinitely divisible with each $Y_j^{(n)} \sim \pi(c/n, \mu_Z)$, for $1 \leq j \leq n$.

The quantity X will have a finite mean if and only if each Z_n does. Indeed, in this case, straightforward differentiation of ϕ_X yields $\mathbb{E}(X) = cm_Z$, where m_Z is the common value of the $\mathbb{E}(Z_n)$. Similar remarks apply to higher-order moments of X.

Exercise 1.2.12

(1) Verify that the sum of two independent infinitely divisible random variables is itself infinitely divisible.
(2) Show that the weak limit of a sequence of infinitely divisible probability measures is itself infinitely divisible. (Hint: use Lévy's continuity theorem.)

We will frequently meet examples of the following type. Let $X = X_1 + X_2$, where X_1 and X_2 are independent with $X_1 \sim N(m, A)$ and $X_2 \sim \pi(c, \mu_Z)$; then, for each $u \in \mathbb{R}^d$,

$$\phi_X(u) = \exp\left[i(m, u) - \tfrac{1}{2}(u, Au) + \int_{\mathbb{R}^d} (e^{i(u,y)} - 1)c\mu_Z(dy)\right]. \quad (1.8)$$

1.2.4 The Lévy–Khintchine formula

In this section, we will present a beautiful formula, first established by Paul Lévy and A.Ya. Khintchine in the 1930s, which gives a characterisation of infinitely divisible random variables through their characteristic functions. First we need a definition.

Let ν be a Borel measure defined on $\mathbb{R}^d - \{0\} = \{x \in \mathbb{R}^d, x \neq 0\}$. We say that it is a *Lévy measure* if

$$\int_{\mathbb{R}^d - \{0\}} (|y|^2 \wedge 1)\nu(dy) < \infty. \quad (1.9)$$

Since $|y|^2 \wedge \epsilon \leq |y|^2 \wedge 1$ whenever $0 < \epsilon \leq 1$, it follows from (1.9) that

$$\nu((-\epsilon, \epsilon)^c) < \infty \quad \text{for all } \epsilon > 0.$$

Exercise 1.2.13 Show that every Lévy measure on $\mathbb{R}^d - \{0\}$ is σ-finite.

Alternative characterisations of Lévy measures can be found in the literature. One of the most popular replaces (1.9) by

$$\int_{\mathbb{R}^d - \{0\}} \frac{|y|^2}{1 + |y|^2} \nu(dy) < \infty. \quad (1.10)$$

To see that (1.9) and (1.10) are equivalent, it is sufficient to verify the inequalities

$$\frac{|y|^2}{1 + |y|^2} \leq |y|^2 \wedge 1 \leq 2\frac{|y|^2}{1 + |y|^2}$$

for each $y \in \mathbb{R}^d$.

Note that any finite measure on $\mathbb{R}^d - \{0\}$ is a Lévy measure. Also, if the reader so wishes, the alternative convention may be adopted of defining Lévy measures on the whole of $\mathbb{R}^d$ via the assignment $\nu(\{0\}) = 0$; see e.g. Sato [274].

The result given below is usually called the *Lévy–Khintchine formula* and it is the cornerstone for much of what follows.

Theorem 1.2.14 (Lévy–Khintchine) $\mu \in \mathcal{M}_1(\mathbb{R}^d)$ *is infinitely divisible if there exists a vector* $b \in \mathbb{R}^d$, *a positive definite symmetric* $d \times d$ *matrix A and a Lévy measure* ν *on* $\mathbb{R}^d - \{0\}$ *such that, for all* $u \in \mathbb{R}^d$,

$$\phi_\mu(u) = \exp\left\{i(b,u) - \tfrac{1}{2}(u, Au) + \int_{\mathbb{R}^d - \{0\}} [e^{i(u,y)} - 1 - i(u,y)\chi_{\hat{B}}(y)]\nu(dy)\right\}, \quad (1.11)$$

where $\hat{B} = B_1(0)$.

Conversely, any mapping of the form (1.11) *is the characteristic function of an infinitely divisible probability measure on* $\mathbb{R}^d$.

Proof We are only going to prove the second part of the theorem here; the more difficult first part will be proved as a by-product of the Lévy–Itô decomposition in Chapter 2. First we need to show that the right-hand side of (1.11) is a characteristic function. To this end, let $(\alpha(n), n \in \mathbb{N})$ be a sequence in $\mathbb{R}^d$ that is monotonic decreasing to zero and define for all $u \in \mathbb{R}^d$, $n \in \mathbb{N}$,

$$\phi_n(u) = \exp\left[i\left(b - \int_{[-\alpha(n),\alpha(n)]^c \cap \hat{B}} y\nu(dy), u\right) - \tfrac{1}{2}(u, Au) + \int_{[-\alpha(n),\alpha(n)]^c} (e^{i(u,y)} - 1)\nu(dy)\right].$$

Then each ϕ_n represents the convolution of a normal distribution with an independent compound Poisson distribution, as in (1.8), and thus is the characteristic function of a probability measure μ_n. We clearly have

$$\phi_\mu(u) = \lim_{n \to \infty} \phi_n(u).$$

The fact that ϕ_μ is a characteristic function will follow by Lévy's continuity theorem if we can show that it is continuous at zero. This boils down to proving the continuity at 0 of ψ_μ, where, for each $u \in \mathbb{R}^d$,

$$\psi_\mu(u) = \int_{\mathbb{R}^d - \{0\}} [e^{i(u,y)} - 1 - i(u,y)\chi_{\hat{B}}(y)]\nu(dy)$$

$$= \int_{\hat{B}} [e^{i(u,y)} - 1 - i(u,y)]\nu(dy) + \int_{\hat{B}^c} (e^{i(u,y)} - 1)\nu(dy).$$

Now using Taylor's theorem, the Cauchy–Schwarz inequality, (1.9) and dominated convergence, we obtain

$$|\psi_\mu(u)| \leq \frac{1}{2}\int_{\hat{B}}|(u,y)|^2 v(dy) + \int_{\hat{B}^c}|e^{i(u,y)} - 1|v(dy)$$

$$\leq \frac{|u|^2}{2}\int_{\hat{B}}|y|^2 v(dy) + \int_{\hat{B}^c}|e^{i(u,y)} - 1|v(dy)$$

$$\to 0 \quad \text{as} \quad u \to 0.$$

It is now easy to verify directly that μ is infinitely divisible. □

Notes

(1) The technique used in the proof above of taking the limits of sequences composed of sums of Gaussians with independent compound Poissons will recur frequently.
(2) The proof of the 'only if' part involves much more work. See e.g. Sato ([274]), pp. 41–5, for one way of doing this. An alternative approach will be given in Chapter 2, as a by-product of the Lévy–Itô decomposition.
(3) There is nothing special about the 'cut-off' function $c(y) = y\chi_B$ that occurs within the integral in (1.11). An alternative that is often used is $c(y) = y/(1 + |y|^2)$. The only constraint in choosing c is that the function $g_c(y) = e^{i(u,y)} - 1 - i(c(y), u)$ should be ν-integrable for each $u \in \mathbb{R}^d$. Note that if you adopt a different c then you must change the vector b accordingly in (1.11).
(4) Relative to the choice of c that we have taken, the members of the triple (b, A, ν) are called the *characteristics* of the infinitely divisible random variable X. Examples of these are as follows.
  - Gaussian case: b is the mean, m, A is the covariance matrix, $\nu = 0$.
  - Poisson case: $b = 0$, $A = 0$, $\nu = c\delta_1$.
  - Compound Poisson case: $b = 0$, $A = 0$, $\nu = c\mu$, where $c > 0$ and μ is a probability measure on $\mathbb{R}^d$.
(5) It is important to be aware that the interpretation of b and A as mean and covariance, respectively, is particular to the Gaussian case; e.g. in (1.8),

$$\mathbb{E}(X) = m + c\int_{\mathbb{R}^d} y\mu_Z(dy),$$

when the integral is finite.

In the proof of Theorem 1.2.14, we wrote down the characteristic function $\phi_\mu(u) = e^{\eta(u)}$. We will call the map $\eta : \mathbb{R}^d \to \mathbb{C}$ a *Lévy symbol*, as it is the symbol for a pseudo-differential operator (see Chapter 3). Many other authors call η the *characteristic exponent* or *Lévy exponent*.

Since, for all $u \in \mathbb{R}^d$, $|\phi_\mu(u)| \leq 1$ for any probability measure μ and $\phi_\mu(u) = e^{\eta(u)}$, when μ is infinitely divisible we deduce that $\Re \eta(u) \leq 0$.

Exercise 1.2.15 Show that η is continuous at every $u \in \mathbb{R}^d$ (and uniformly so in a neighbourhood of the origin).

Exercise 1.2.16 Establish the useful inequality

$$|\eta(u)| \leq C(1 + |u|^2)$$

for each $u \in \mathbb{R}^d$, where $C > 0$.

The following theorem gives an interesting analytic insight into the Lévy–Khintchine formula.

Theorem 1.2.17 *The map η is a Lévy symbol if and only if it is a continuous, hermitian, conditionally positive definite function for which $\eta(0) = 0$.*

Proof Suppose that η is a Lévy symbol; then so is $t\eta$, for each $t > 0$. Then there exists a probability measure $\mu(t)$ for each $t \geq 0$, such that $\phi_{\mu(t)}(u) = e^{t\eta(u)}$ for each $u \in \mathbb{R}^d$. But η is continuous by Exercise 1.2.15 and $\eta(0) = 0$. Since ϕ_μ is positive definite then η is hermitian and conditionally positive definite by the Schoenberg correspondence.

Conversely, suppose that η is continuous, hermitian and conditionally positive definite with $\eta(0) = 0$. By the Schoenberg correspondence (Theorem 1.1.2) and Bochner's theorem, there exists a probability measure μ for which $\phi_\mu(u) = e^{\eta(u)}$ for each $u \in \mathbb{R}^d$. Since η/n is, for each $n \in \mathbb{N}$, another continuous, hermitian, conditionally positive definite function that vanishes at the origin, we see that μ is infinitely divisible and the result follows. $\square$

We will gain more insight into the meaning of the Lévy–Khintchine formula when we consider Lévy processes. For now it is important to be aware that all infinitely divisible distributions can be constructed as weak limits of convolutions of Gaussians with independent compound Poissons, as the proof of Theorem 1.2.14 indicated. In the next section we will see that some very interesting examples occur as such limits. The final result of this section shows that in fact the compound Poisson distribution is enough for a weak approximation.

1.2 Infinite divisibility

Theorem 1.2.18 *Any infinitely divisible probability measure can be obtained as the weak limit of a sequence of compound Poisson distributions.*

Proof Let ϕ be the characteristic function of an arbitrary infinitely divisible probability measure μ, so that $\phi^{1/n}$ is the characteristic function of $\mu^{1/n}$; then for each $n \in \mathbb{N}$, $u \in \mathbb{R}^d$, we may define

$$\phi_n(u) = \exp\left\{n[\phi^{1/n}(u) - 1]\right\} = \exp\left[\int_{\mathbb{R}^d}(e^{i(u,y)} - 1)n\mu^{1/n}(dy)\right],$$

so that ϕ_n is the characteristic function of a compound Poisson distribution. We then have

$$\phi_n(u) = \exp\left[n(e^{(1/n)\log[\phi(u)]} - 1)\right]$$
$$= \exp\left\{\log[\phi(u)] + n\,o\left(\frac{1}{n}\right)\right\} \to \phi(u) \quad \text{as} \quad n \to \infty,$$

where 'log' is the principal value of the logarithm; the result follows by Glivenko's theorem. $\square$

Corollary 1.2.19 *The set of all infinitely divisible probability measures on $\mathbb{R}^d$ coincides with the weak closure of the set of all compound Poisson distributions on $\mathbb{R}^d$.*

Proof This follows directly from Theorem 1.2.18 and Exercise 1.2.12(2). $\square$

Although the Lévy–Khintchine formula represents all infinitely divisible random variables as arising through the interplay between Gaussian and Poisson distributions, a vast array of different behaviour appears between these two extreme cases. A large number of examples are given in Chapter 1 of Sato ([274]). We will be content to focus on a subclass of great importance and then look at two rather diverse and interesting cases that originate from outside probability theory.[1]

1.2.5 Stable random variables

We consider the general central limit problem in dimension $d = 1$, so let $(Y_n, n \in \mathbb{N})$ be a sequence of real-valued i.i.d. random variables and construct the sequence $(S_n, n \in \mathbb{N})$ of rescaled partial sums

$$S_n = \frac{Y_1 + Y_2 + \cdots + Y_n - b_n}{\sigma_n},$$

---

[1] Readers with an interest in statistics will be pleased to know that the gamma distribution (of which the chi-squared distribution is a special case) is infinitely divisible. We will say more about this is Subsection 1.3.2. The t-distribution is also infinitely divisible; see Grosswald [124].

where $(b_n, n \in \mathbb{N})$ is an arbitrary sequence of real numbers and $(\sigma_n, n \in \mathbb{N})$ an arbitrary sequence of positive numbers. We are interested in the case where there exists a random variable X for which

$$\lim_{n \to \infty} P(S_n \leq x) = P(X \leq x) \qquad (1.12)$$

for all $x \in \mathbb{R}$, i.e. $(S_n, n \in \mathbb{N})$ converges in distribution to X. If each $b_n = nm$ and $\sigma_n = \sqrt{n}\sigma$ for fixed $m \in \mathbb{R}, \sigma > 0$, then $X \sim N(m, \sigma^2)$ by the usual Laplace–de-Moivre central limit theorem.

More generally a random variable is said to be *stable* if it arises as a limit, as in (1.12). It is not difficult (see e.g. Breiman [58], Gnedenko and Kolmogorov [121]) to show that (1.12) is equivalent to the following. There exist real-valued sequences $(c_n, n \in \mathbb{N})$ and $(d_n, n \in \mathbb{N})$ with each $c_n > 0$ such that

$$X_1 + X_2 + \cdots + X_n \stackrel{d}{=} c_n X + d_n, \qquad (1.13)$$

where $X_1, X_2, \ldots, X_n$ are independent copies of X. In particular, X is said to be *strictly stable* if each $d_n = 0$.

To see that (1.13) $\Rightarrow$ (1.12) take each $Y_j = X_j, b_n = d_n$ and $\sigma_n = c_n$. In fact it can be shown (see Feller [102], p. 166) that the only possible choice of c_n in (1.13) is of the form $\sigma n^{1/\alpha}$, where $0 < \alpha \leq 2$. The parameter α plays a key role in the investigation of stable random variables and is called the *index of stability*.

Note that (1.13) can be expressed in the equivalent form

$$\phi_X(u)^n = e^{iud_n}\phi_X(c_n u),$$

for each $u \in \mathbb{R}$.

It follows immediately from (1.13) that all stable random variables are infinitely divisible. The characteristics in the Lévy–Khintchine formula are given by the following result.

Theorem 1.2.20 *If X is a stable real-valued random variable, then its characteristics must take one of the two following forms:*

(1) *when $\alpha = 2, \nu = 0$, so $X \sim N(b, A)$;*
(2) *when $\alpha \neq 2, A = 0$ and*

$$\nu(dx) = \frac{c_1}{x^{1+\alpha}}\chi_{(0,\infty)}(x)dx + \frac{c_2}{|x|^{1+\alpha}}\chi_{(-\infty,0)}(x)dx,$$

where $c_1 \geq 0, c_2 \geq 0$ and $c_1 + c_2 > 0$.

A proof can be found in Sato [274], p. 80.

1.2 Infinite divisibility

A careful transformation of the integrals in the Lévy–Khintchine formula gives a different form for the characteristic function, which is often more convenient (see Sato [274], p. 86).

Theorem 1.2.21 *A real-valued random variable X is stable if and only if there exist $\sigma > 0, -1 \leq \beta \leq 1$ and $\mu \in \mathbb{R}$ such that for all $u \in \mathbb{R}$:*

(1) *when $\alpha = 2$,*

$$\phi_X(u) = \exp\left(i\mu u - \frac{1}{2}\sigma^2 u^2\right);$$

(2) *when $\alpha \neq 1, 2$,*

$$\phi_X(u) = \exp\left\{i\mu u - \sigma^\alpha |u|^\alpha \left[1 - i\beta \operatorname{sgn}(u) \tan\left(\frac{\pi\alpha}{2}\right)\right]\right\};$$

(3) *when $\alpha = 1$,*

$$\phi_X(u) = \exp\left\{i\mu u - \sigma|u|\left[1 + i\beta\frac{2}{\pi} \operatorname{sgn}(u) \log(|u|)\right]\right\}.$$

It can be shown that $\mathbb{E}(X^2) < \infty$ if and only if $\alpha = 2$ (i.e. X is Gaussian) and that $\mathbb{E}(|X|) < \infty$ if and only if $1 < \alpha \leq 2$.

All stable random variables have densities f_X, which can in general be expressed in series form (see Feller [102], Chapter 17, Section 6). In three important cases, there are closed forms.

The normal distribution

$$\alpha = 2, \quad X \sim N(\mu, \sigma^2).$$

The Cauchy distribution

$$\alpha = 1, \ \beta = 0, \quad f_X(x) = \frac{\sigma}{\pi[(x-\mu)^2 + \sigma^2]}.$$

The Lévy distribution

$$\alpha = \frac{1}{2}, \ \beta = 1,$$

$$f_X(x) = \left(\frac{\sigma}{2\pi}\right)^{1/2} \frac{1}{(x-\mu)^{3/2}} \exp\left[-\frac{\sigma}{2(x-\mu)}\right] \quad \text{for } x > \mu.$$

Exercise 1.2.22 (The Cauchy distribution) Prove directly that

$$\int_{-\infty}^{\infty} e^{iux} \frac{\sigma}{\pi[(x-\mu)^2 + \sigma^2]} dx = e^{i\mu u - \sigma|u|}.$$

(Hint: One approach is to use the calculus of residues. Alternatively, by integrating from $-\infty$ to 0 and then 0 to ∞, separately, deduce that

$$\int_{-\infty}^{\infty} e^{-itx} e^{-|x|} dx = \frac{2}{1+t^2}.$$

Now use Fourier inversion; see Subsection 3.8.4.)

Exercise 1.2.23 Let X and Y be independent standard normal random variables. Show that Z has a Cauchy distribution, where $Z = Y/X$ when $X \neq 0$ and $Z = 0$ otherwise. Show also that W has a Lévy distribution, where $W = 1/X^2$ when $X \neq 0$ and $W = 0$ otherwise.

Note that if a stable random variable is symmetric then Theorem 1.2.21 yields

$$\phi_X(u) = \exp(-\rho^\alpha |u|^\alpha) \qquad \text{for all } 0 < \alpha \leq 2, \tag{1.14}$$

where $\rho = \sigma$ for $0 < \alpha < 2$ and $\rho = \sigma/\sqrt{2}$ when $\alpha = 2$; we will write $X \sim S\alpha S$ in this case.

Although it does not have a closed-form density, the symmetric stable distribution with $\alpha = 3/2$ is of considerable practical importance. It is called the *Holtsmark distribution* and its three-dimensional generalisation has been used to model the gravitational field of stars: see Feller [102], p. 173 and Zolotarev [312].

One of the reasons why stable laws are so important in applications is the nice decay properties of the tails. The case $\alpha = 2$ is special in that we have exponential decay; indeed, for a standard normal X there is the elementary estimate

$$P(X > y) \sim \frac{e^{-y^2/2}}{\sqrt{2\pi} y} \qquad \text{as } y \to \infty;$$

see Feller [101], Chapter 7, Section 1.

When $\alpha \neq 2$ we have a slower, polynomial, decay as expressed in the following:

$$\lim_{y \to \infty} y^\alpha P(X > y) = C_\alpha \frac{1+\beta}{2} \sigma^\alpha,$$

$$\lim_{y \to \infty} y^\alpha P(X < -y) = C_\alpha \frac{1-\beta}{2} \sigma^\alpha,$$

1.2 Infinite divisibility

where $C_\alpha > 1$; see Samorodnitsky and Taqqu [271], pp. 16–18, for a proof and an explicit expression for the constant C_α. The relatively slow decay of the tails of non-Gaussian stable laws makes them ideally suited for modelling a wide range of interesting phenomena, some of which exhibit 'long-range dependence'; see Taqqu [294] for a nice survey of such applications. The mathematical description of 'heavy tails' is intimately related to the concept of regular variation. For a detailed account of this, see Bingham *et al.* [46], particularly Chapter 8, or Resnick [259].

The generalisation of stability to random vectors is straightforward: just replace $X_1, \ldots, X_n, X$ and each d_n in (1.13) by vectors, and the formula in Theorem 1.2.20 extends directly. Note however that when $\alpha \neq 2$ in the random vector version of Theorem 1.2.20, the Lévy measure takes the form

$$\nu(dx) = \frac{c}{|x|^{d+\alpha}} dx$$

where $c > 0$.

The corresponding extension of Theorem 1.2.21 is as follows (see Sato [274], p. 83 for a proof).

Theorem 1.2.24 *A random variable X taking values in $\mathbb{R}^d$ is stable if and only if for all $u \in \mathbb{R}^d$ there exists a vector $m \in \mathbb{R}^d$ and*

(1) *there exists a positive definite symmetric $d \times d$ matrix A such that, when $\alpha = 2$,*

$$\phi_X(u) = \exp\left[i(m, u) - \frac{1}{2}(u, Au)\right];$$

(2) *there exists a finite measure ρ on S^{d-1} such that, when $\alpha \neq 1, 2$,*

$$\phi_X(u) = \exp\left\{i(m, u) - \int_{S^{d-1}} |(u, s)|^\alpha \left[1 - i \tan\left(\frac{\pi\alpha}{2}\right)\right] \operatorname{sgn}(u, s) \rho(ds)\right\};$$

(3) *there exists a finite measure ρ on S^{d-1} such that, when $\alpha = 1$,*

$$\phi_X(u) = \exp\left\{i(m, u) - \int_{S^{d-1}} |(u, s)| \left[1 + i\frac{2}{\pi} \operatorname{sgn}(u, s) \log |(u, s)|\right] \rho(ds)\right\}.$$

Note that, for $0 < \alpha < 2$, X is symmetric if and only if

$$\phi_X(u) = \exp\left[-\int_{S^{d-1}} |(u,s)|^\alpha \rho(ds)\right]$$

for each $u \in \mathbb{R}^d$ and X is rotationally invariant for $0 < \alpha \leq 2$ if and only if the $\mathbb{R}^d$-version of equation (1.14) holds.

We can generalise the definition of stable random variables if we weaken the conditions on the random variables $(Y(n), n \in \mathbb{N})$ in the general central limit problem by requiring these to be independent but no longer necessarily identically distributed. In this case the limiting random variables are called *self-decomposable* (or of class L) and they are also infinitely divisible. Alternatively, a random variable X is self-decomposable if and only if for each $0 < a < 1$ there exists a random variable Y_a that is independent of X and such that

$$X \stackrel{d}{=} aX + Y_a \quad \Leftrightarrow \quad \phi_X(u) = \phi_X(au)\phi_{Y_a}(u),$$

for all $u \in \mathbb{R}^d$. Self-decomposable distributions are discussed in Sato [274], p. 90–9, where it is shown that an infinitely divisible law on $\mathbb{R}$ is self-decomposable if and only if the Lévy measure is of the form

$$\nu(dx) = \frac{k(x)}{|x|} dx,$$

where k is decreasing on $(0, \infty)$ and increasing on $(-\infty, 0)$. There has recently been increasing interest in these distributions from both the theoretical and applied perspectives; see for example Bingham and Keisel [49] or the article by Z. Jurek in [19] and references therein.

1.2.6 Diversion: Number theory and relativity

We will look at two interesting examples of infinitely divisible distributions.

The Riemann zeta distribution

The *Riemann zeta function* ζ is defined, initially for complex numbers $z = u + iv$ where $u > 1$, by the (absolutely) convergent series expansion

$$\zeta(z) = \sum_{n=1}^{\infty} \frac{1}{n^z},$$

1.2 Infinite divisibility

which is equivalent to the Euler product formula

$$\zeta(z) = \prod_{p \in \mathcal{P}} \frac{1}{1 - p^{-z}}, \qquad (1.15)$$

$\mathcal{P}$ being the set of all prime numbers.

Riemann showed that ζ can be extended by analytic continuation to a meromorphic function on the whole of $\mathbb{C}$, having a single (simple) pole at $z = 1$. He also investigated the zeros of ζ and showed that that these are at $\{-2n, n \in \mathbb{N}\}$ and in the strip $|u| \leq 1$. The celebrated *Riemann hypothesis* is that all the latter class are in fact on the line $u = 1/2$, and this question remains unresolved although Hardy has shown that an infinite number of zeros are of this form. For more about this and related issues see e.g. Chapter 9 of Patterson [250] and references therein.

We will now look at a remarkable connection between the Riemann zeta function and infinite divisibility that is originally due to A. Khintchine (see [121], pp. 75–6), although it has its antecedents in work by Jessen and Wintner [159].

Fix $u \in \mathbb{R}$ with $u > 1$ and define $\phi_u : \mathbb{R} \to \mathbb{C}$ by

$$\phi_u(v) = \frac{\zeta(u + iv)}{\zeta(u + i0)},$$

for all $v \in \mathbb{R}$.

Proposition 1.2.25 (Khintchine) *For each $u > 1$, ϕ_u is the characteristic function of an infinitely divisible probability measure.*

Proof Using (1.15) and the Taylor series expansion of the complex function $\log(1 + w)$, where $|w| < 1$, we find for all $v \in \mathbb{R}$ that (taking the principal value of the logarithm),

$$\log[\phi_u(v)] = \log[\zeta(u+iv)] - \log[\zeta(u+i0)]$$

$$= \sum_{p \in \mathcal{P}} \log(1 - p^{-u}) - \sum_{p \in \mathcal{P}} \log(1 - p^{-(u+iv)})$$

$$= \sum_{p \in \mathcal{P}} \sum_{m=1}^{\infty} \left(\frac{p^{-m(u+iv)}}{m} - \frac{p^{-mu}}{m} \right)$$

$$= \sum_{p \in \mathcal{P}} \sum_{m=1}^{\infty} \frac{p^{-mu}}{m} (e^{-im \log(p) v} - 1)$$

$$= \sum_{p \in \mathcal{P}} \sum_{m=1}^{\infty} \int_{\mathbb{R}} (e^{i\alpha v} - 1) \frac{e^{u\alpha}}{m} \delta_{-m \log(p)}(d\alpha).$$

Hence we see that ϕ_u is the limit of a sequence of characteristic functions of Poisson laws. It follows by the Lévy continuity theorem that ϕ_u is the characteristic function of a probability measure that is infinitely divisible, by Glivenko's theorem and Exercise 1.2.12(2). □

After many years of neglect, some investigations into this distribution have recently appeared in Lin and Hu [197]. Other developments involving number-theoretic aspects of infinite divisibility can be found in Jurek [163], where the relationship between Dirichlet series and self-decomposable distributions is explored, and in the survey article by Biane, Pitman and Yor [41].

A relativistic distribution

We will consider an example that originates in Einstein's theory of relativity. A particle of rest mass $m > 0$ has momentum $p = (p_1, p_2, p_3) \in \mathbb{R}^3$. According to relativity theory, its total energy is

$$E(p) = \sqrt{m^2 c^4 + c^2 |p|^2},$$

where $c > 0$ is the velocity of light (see e.g. Born [54], p. 291) and, if we subtract the energy mc^2 that is tied up in the rest mass, we obtain the kinetic energy, i.e. the energy due to motion,

$$E_{m,c}(p) = \sqrt{m^2 c^4 + c^2 |p|^2} - mc^2.$$

Although m and c are 'fixed' by physics we have indicated an explicit dependence of the energy on these 'parameters' for reasons that will become clearer below. Define

$$\phi_{m,c}(p) = e^{-E_{m,c}(p)},$$

where we now take $p \in \mathbb{R}^d$ for greater generality.

Theorem 1.2.26 $\phi_{m,c}$ *is the characteristic function of an infinitely divisible probability distribution.*

Proof The fact that $\phi_{m,c}$ is a characteristic function follows by Bochner's theorem once we have shown that it is positive definite. Since $E_{m,c}$ is clearly hermitian, demonstrating this latter fact is equivalent, by the Schoenberg correspondence, to demonstrating that

$$\sum_{i,j=1}^{n} \alpha_i \bar{\alpha}_j E_{m,c}(p_i - p_j) \leq 0$$

for all $n \in \mathbb{N}, \alpha_i \in \mathbb{C}, p_i \in \mathbb{R}^d, 1 \leq i \leq n$, with $\sum_{i=1}^n \alpha_i = 0$. Now

$$\sum_{i,j=1}^n \alpha_i \bar{\alpha}_j E_{m,c}(p_i - p_j) = mc^2 \sum_{i,j=1}^n \alpha_i \bar{\alpha}_j \left[\left(1 + \frac{|p_i - p_j|^2}{m^2 c^2}\right)^{1/2} - 1 \right]$$

$$= mc^2 \sum_{i,j=1}^n \alpha_i \bar{\alpha}_j \left(1 + \frac{|p_i - p_j|^2}{m^2 c^2}\right)^{1/2}$$

$$\leq mc^2 \sum_{i,j=1}^n \alpha_i \bar{\alpha}_j \left(1 + \frac{|p_i - p_j|^2}{m^2 c^2}\right)$$

$$= \frac{1}{m} \sum_{i,j=1}^n \alpha_i \bar{\alpha}_j |p_i - p_j|^2 \leq 0,$$

where the last assertion follows from the fact that the mapping $p \to -|p|^2$ is the Lévy symbol of a normal distribution and so is itself conditionally positive definite.

To verify that the associated probability measure is infinitely divisible, it is sufficient to observe that

$$\left[\phi_{m,c}(p)\right]^{1/n} = \phi_{nm,c/n}(p)$$

for all $p \in \mathbb{R}^d, n \in \mathbb{N}$. □

We will meet this example again in Chapter 3 in 'quantised' form.

1.3 Lévy processes

Let $X = (X(t), t \geq 0)$ be a stochastic process defined on a probability space $(\Omega, \mathcal{F}, P)$. We say that it has *independent increments* if for each $n \in \mathbb{N}$ and each $0 \leq t_1 < t_2 \leq \cdots < t_{n+1} < \infty$ the random variables $(X(t_{j+1}) - X(t_j), 1 \leq j \leq n)$ are independent and that it has *stationary increments* if each $X(t_{j+1}) - X(t_j) \stackrel{d}{=} X(t_{j+1} - t_j) - X(0)$.

We say that X is a *Lévy process* if:

(L1) $X(0) = 0$ (a.s);
(L2) X has independent and stationary increments;
(L3) X is *stochastically continuous*, i.e. for all $a > 0$ and for all $s \geq 0$

$$\lim_{t \to s} P(|X(t) - X(s)| > a) = 0.$$

Note that in the presence of (L1) and (L2), (L3) is equivalent to the condition

$$\lim_{t \downarrow 0} P(|X(t)| > a) = 0$$

for all $a > 0$.

We are now going to explore the relationship between Lévy processes and infinite divisibility.

Proposition 1.3.1 *If X is a Lévy process, then $X(t)$ is infinitely divisible for each $t \geq 0$.*

Proof For each $n \in \mathbb{N}$, we can write

$$X(t) = Y_1^{(n)}(t) + \cdots + Y_n^{(n)}(t)$$

where each

$$Y_k^{(n)}(t) = X\left(\frac{kt}{n}\right) - X\left(\frac{(k-1)t}{n}\right).$$

The $Y_k^{(n)}(t)$ are i.i.d. by (L2). □

By Proposition 1.3.1, we can write $\phi_{X(t)}(u) = e^{\eta(t,u)}$ for each $t \geq 0$, $u \in \mathbb{R}^d$, where each $\eta(t, \cdot)$ is a Lévy symbol. We will see below that $\eta(t, u) = t\eta(1, u)$ for each $t \geq 0$, $u \in \mathbb{R}^d$, but first we need the following lemma.

Lemma 1.3.2 *If $X = (X(t), t \geq 0)$ is stochastically continuous, then the map $t \to \phi_{X(t)}(u)$ is continuous for each $u \in \mathbb{R}^d$.*

Proof For each $s, t \geq 0$ with $t \neq s$, write $X(s, t) = X(t) - X(s)$. Fix $u \in \mathbb{R}^d$. Since the map $y \to e^{i(u,y)}$ is continuous at the origin, given any $\epsilon > 0$ we can find $\delta_1 > 0$ such that

$$\sup_{0 \leq |y| < \delta_1} |e^{i(u,y)} - 1| < \frac{\epsilon}{2}$$

and, by stochastic continuity, we can find $\delta_2 > 0$ such that whenever $0 < |t - s| < \delta_2$, $P(|X(s, t)| > \delta_1) < \epsilon/4$.

Hence for all $0 < |t-s| < \delta_2$ we have

$$|\phi_{X(t)}(u) - \phi_{X(s)}(u)| = \left|\int_\Omega e^{i(u,X(s)(\omega))}\left[e^{i(u,X(s,t)(\omega))} - 1\right] P(d\omega)\right|$$

$$\leq \int_{\mathbb{R}^d} |e^{i(u,y)} - 1| p_{X(s,t)}(dy)$$

$$= \int_{B_{\delta_1}(0)} |e^{i(u,y)} - 1| p_{X(s,t)}(dy)$$

$$+ \int_{B_{\delta_1}(0)^c} |e^{i(u,y)} - 1| p_{X(s,t)}(dy)$$

$$\leq \sup_{0 \leq |y| < \delta_1} |e^{i(u,y)} - 1| + 2P(|X(s,t)| > \delta_1)$$

$$< \epsilon,$$

and the required result follows. □

Theorem 1.3.3 *If X is a Lévy process, then*

$$\phi_{X(t)}(u) = e^{t\eta(u)}$$

for each $u \in \mathbb{R}^d$, $t \geq 0$, where η is the Lévy symbol of $X(1)$.

Proof Suppose that X is a Lévy process and that, for each $u \in \mathbb{R}^d$, $t \geq 0$. Define $\phi_u(t) = \phi_{X(t)}(u)$; then by (L2) we have for all $s \geq 0$

$$\phi_u(t+s) = \mathbb{E}\left(e^{i(u,X(t+s))}\right)$$
$$= \mathbb{E}\left(e^{i(u,X(t+s)-X(s))} e^{i(u,X(s))}\right)$$
$$= \mathbb{E}\left(e^{i(u,X(t+s)-X(s))}\right) \mathbb{E}\left(e^{i(u,X(s))}\right)$$
$$= \phi_u(t)\phi_u(s) \qquad (1.16)$$

Now

$$\phi_u(0) = 1 \qquad (1.17)$$

by (L1), and from (L3) and Lemma 1.3.2 we have that the map $t \to \phi_u(t)$ is continuous. However, the unique continuous solution to (1.16) and (1.17) is given by $\phi_u(t) = e^{t\alpha(u)}$, where $\alpha : \mathbb{R}^d \to \mathbb{C}$ (see e.g. Bingham et al. [46], pp. 4–6). Now by Proposition 1.3.1 $X(1)$ is infinitely divisible, hence α is a Lévy symbol and the result follows. □

We now have the Lévy–Khinchine formula for a Lévy process $X = (X(t), t \geq 0)$,

$$\mathbb{E}(e^{i(u,X(t))}) = \exp\left(t\left\{i(b,u) - \tfrac{1}{2}(u, Au)\right.\right.$$
$$\left.\left. + \int_{\mathbb{R}^d - \{0\}} [e^{i(u,y)} - 1 - i(u,y)\chi_{\hat{B}}(y)]\nu(dy)\right\}\right) \quad (1.18)$$

for each $t \geq 0$, $u \in \mathbb{R}^d$, where (b, A, ν) are the characteristics of $X(1)$.

We will define the Lévy symbol and the characteristics of a Lévy process X to be those of the random variable $X(1)$. We will sometimes write the former as η_X when we want to emphasise that it belongs to the process X.

Exercise 1.3.4 If X is a Lévy process with characteristics (b, A, ν), show that $-X = (-X(t), t \geq 0)$ is also a Lévy process and has characteristics $(-b, A, \tilde{\nu})$, where $\tilde{\nu}(A) = \nu(-A)$ for each $A \in \mathcal{B}(\mathbb{R}^d)$. Show also that for each $c \in \mathbb{R}$ the process, $(X(t) + tc, t \geq 0)$ is a Lévy process, and find its characteristics.

Exercise 1.3.5 Show that if X and Y are stochastically continuous processes then so is their sum $X + Y = (X(t) + Y(t), t \geq 0)$. (Hint: Use the elementary inequality

$$P(|A + B| > c) \leq P\left(|A| > \frac{c}{2}\right) + P\left(|B| > \frac{c}{2}\right),$$

where A and B are random variables and $c > 0$.)

Exercise 1.3.6 Show that the sum of two independent Lévy processes is again a Lévy process. (Hint: Use Kac's theorem to establish independent increments.)

Theorem 1.3.7 *If $X = (X(t), t \geq 0)$ is a stochastic process and there exists a sequence of Lévy processes $(X_n, n \in \mathbb{N})$ with each $X_n = (X_n(t), t \geq 0)$ such that $X_n(t)$ converges in probability to $X(t)$ for each $t \geq 0$ and $\lim_{n \to \infty} \limsup_{t \to 0} P(|X_n(t) - X(t)| > a) = 0$ for all $a > 0$, then X is a Lévy process.*

Proof (L1) follows immediately from the fact that $(X_n(0), n \in \mathbb{N})$ has a subsequence converging to 0 almost surely. For (L2) we obtain stationary increments by observing that for each $u \in \mathbb{R}^d, 0 \leq s < t < \infty$,

$$\begin{aligned}\mathbb{E}(e^{i(u,X(t)-X(s))}) &= \lim_{n\to\infty} \mathbb{E}(e^{i(u,X_n(t)-X_n(s))}) \\ &= \lim_{n\to\infty} \mathbb{E}(e^{i(u,X_n(t-s))}) \\ &= \mathbb{E}(e^{i(u,X(t-s))}),\end{aligned}$$

where the convergence of the characteristic functions follows by the argument used in the proof of Lemma 1.3.2. The independence of the increments is proved similarly.

Finally, to establish (L3), for each $a > 0, t \geq 0, n \in \mathbb{N}$ we have

$$\begin{aligned}P(|X(t)| > a) &\leq P(|X(t) - X_n(t)| + |X_n(t)| > a) \\ &\leq P\left(|X(t) - X_n(t)| > \frac{a}{2}\right) + P\left(|X_n(t)| > \frac{a}{2}\right)\end{aligned}$$

and hence

$$\limsup_{t\to 0} P(|X(t)| > a)$$
$$\leq \limsup_{t\to 0} P\left(|X(t) - X_n(t)| > \frac{a}{2}\right) + \limsup_{t\to 0} P\left(|X_n(t)| > \frac{a}{2}\right). \tag{1.19}$$

But each X_n is a Lévy process and so

$$\limsup_{t\to 0} P\left(|X_n(t)| > \frac{a}{2}\right) = \lim_{t\to 0} P\left(|X_n(t)| > \frac{a}{2}\right) = 0,$$

and the result follows on taking $\lim_{n\to\infty}$ in (1.19). □

1.3.1 Examples of Lévy processes

Example 1.3.8 (Brownian motion and Gaussian processes) A *(standard) Brownian motion* in $\mathbb{R}^d$ is a Lévy process $B = (B(t), t \geq 0)$ for which

(B1) $B(t) \sim N(0, tI)$ for each $t \geq 0$,
(B2) B has continuous sample paths.

It follows immediately from (B1) that if B is a standard Brownian motion then its characteristic function is given by

$$\phi_{B(t)}(u) = \exp\left(-\tfrac{1}{2}t|u|^2\right)$$

for each $u \in \mathbb{R}^d, t \geq 0$.

We introduce the marginal processes $B_i = (B_i(t), t \geq 0)$, where each $B_i(t)$ is the ith component of $B(t)$; then it is not difficult to verify that the B_i are mutually independent Brownian motions in $\mathbb{R}$. We will henceforth refer to these as *one-dimensional Brownian motions*.

Brownian motion has been the most intensively studied Lévy process. In the early years of the twentieth century, it was introduced as a model for the physical phenomenon of Brownian motion by Einstein and Smoluchowski and as a description of the dynamical evolution of stock prices by Bachelier. Einstein's papers on the subject are collected in [92] while Bachelier's thesis can be found in [16]. The theory was placed on a rigorous mathematical basis by Norbert Wiener [302] in the 1920s; see also [301]. The first part of Nelson [236] contains a historical account of these developments from the physical point of view.

We could try to use the Kolmogorov existence theorem (Theorem 1.1.16) to construct one-dimensional Brownian motion from the following prescription on cylinder sets of the form $I_{t_1,t_2,\ldots,t_n}^H$:

$$P(I_{t_1,t_2,\ldots,t_n}^H)$$
$$= \int_H \frac{1}{(2\pi)^{n/2}\sqrt{t_1(t_2-t_1)\cdots(t_n-t_{n-1})}}$$
$$\times \exp\left\{-\frac{1}{2}\left[\frac{x_1^2}{t_1} + \frac{(x_2-x_1)^2}{t_2-t_1} + \cdots + \frac{(x_n-x_{n-1})^2}{t_n-t_{n-1}}\right]\right\} dx_1 \cdots dx_n.$$

However, the resulting canonical process lives on the space of all mappings from $\mathbb{R}^+$ to $\mathbb{R}$ and there is then no guarantee that the paths are continuous. A nice account of Wiener's solution to this problem can be found in [301].

The literature contains a number of ingenious methods for constructing Brownian motion. One of the most delightful of these, originally due to Paley and Wiener [244], obtains Brownian motion in the case $d = 1$ as a random Fourier series

$$B(t) = \frac{\sqrt{2}}{\pi} \sum_{n=0}^{\infty} \frac{\sin\left[\pi t(n+\frac{1}{2})\right]}{n+\frac{1}{2}} \xi(n)$$

for each $t \geq 0$, where $(\xi(n), n \in \mathbb{N} \cup \{0\})$ is a sequence of i.i.d. $N(0,1)$ random variables; see Chapter 1 of Knight [171]) for a modern account. A construction of Brownian motion from a wavelet point of view can be found in Steele [288], pp. 35–9.

1.3 Lévy processes

We list a number of useful properties of Brownian motion in the case $d = 1$; this is far from exhaustive and, for further examples as well as details of the proofs, the reader is advised to consult works such as Sato [274], pp. 22–8, Revuz and Yor [260], Rogers and Williams [261], Karatzas and Shreve [167], Knight [171] and Itô and McKean [141].

- Brownian motion is locally Hölder continuous with exponent α for every $0 < \alpha < 1/2$, i.e. for every $T > 0$, $\omega \in \Omega$, there exists $K = K(T, \omega)$ such that
$$|B(t)(\omega) - B(s)(\omega)| \leq K |t - s|^\alpha$$
for all $0 \leq s < t \leq T$.
- The sample paths $t \to B(t)(\omega)$ are almost surely nowhere differentiable.
- For any sequence $(t_n, n \in \mathbb{N})$ in $\mathbb{R}^+$ with $t_n \uparrow \infty$,
$$\liminf_{n \to \infty} B(t_n) = -\infty \text{ a.s.}, \quad \limsup_{n \to \infty} B(t_n) = \infty \quad \text{a.s.}$$
- The law of the iterated logarithm,
$$P\left(\limsup_{t \downarrow 0} \frac{B(t)}{\{2t \log[\log(1/t)]\}^{1/2}} = 1\right) = 1$$
holds.

For deeper properties of Brownian motion, the reader should consult two volumes by Marc Yor, [307], [308].

Let A be a positive definite symmetric $d \times d$ matrix and let σ be a square root of A, so that σ is a $d \times m$ matrix for which $\sigma \sigma^T = A$. Now let $b \in \mathbb{R}^d$ and let B be a Brownian motion in $\mathbb{R}^m$. We construct a process $C = (C(t), t \geq 0)$ in $\mathbb{R}^d$ by

$$C(t) = bt + \sigma B(t); \tag{1.20}$$

then C is a Lévy process with each $C(t) \sim N(tb, tA)$. It is not difficult to verify that C is also a Gaussian process, i.e. that all its finite-dimensional distributions are Gaussian. It is sometimes called *Brownian motion with drift*. The Lévy symbol of C is

$$\eta_C(u) = i(b, u) - \tfrac{1}{2}(u, Au).$$

In the case $b = 0$, we sometimes write $B_A(t) = C(t)$, for each $t \geq 0$, and call the process *Brownian motion with covariance A*.

We will show in the next chapter that a Lévy process has continuous sample paths if and only if it is of the form (1.20).

Example 1.3.9 (The Poisson process) The Poisson process of intensity $\lambda > 0$ is a Lévy process N taking values in $\mathbb{N} \cup \{0\}$ wherein each $N(t) \sim \pi(\lambda t)$, so that we have

$$P(N(t) = n) = \frac{(\lambda t)^n}{n!} e^{-\lambda t}$$

for each $n = 0, 1, 2, \ldots$. The Poisson process is widely used in applications and there is a wealth of literature concerning it and its generalisations; see e.g. Kingman [169] and references therein. We define non-negative random variables $(T_n, \mathbb{N} \cup \{0\})$, usually called *waiting times*, by $T_0 = 0$ and for $n \in \mathbb{N}$

$$T_n = \inf\{t \geq 0; N(t) = n\};$$

it is well known that the T_n are gamma distributed. Moreover, the inter-arrival times $T_n - T_{n-1}$ for $n \in \mathbb{N}$ are i.i.d. and each has exponential distribution with mean $1/\lambda$; see e.g. Grimmett and Stirzaker [123], Section 6.8. The sample paths of N are clearly piecewise constant on finite intervals with 'jump' discontinuities of size 1 at each of the random times $(T_n, n \in \mathbb{N})$.

For later work it is useful to introduce the *compensated Poisson process* $\tilde{N} = (\tilde{N}(t), t \geq 0)$ where each $\tilde{N}(t) = N(t) - \lambda t$. Note that $\mathbb{E}(\tilde{N}(t)) = 0$ and $\mathbb{E}(\tilde{N}(t)^2) = \lambda t$ for each $t \geq 0$.

Example 1.3.10 (The compound Poisson process) Let $(Z(n), n \in \mathbb{N})$ be a sequence of i.i.d. random variables taking values in $\mathbb{R}^d$ with common law μ_Z and let N be a Poisson process of intensity λ that is independent of all the $Z(n)$. The *compound Poisson process* Y is defined as follows:

$$Y(t) = Z(1) + \cdots + Z(N(t)) \qquad (1.21)$$

for each $t \geq 0$, so each $Y(t) \sim \pi(\lambda t, \mu_Z)$.

Proposition 1.3.11 *The compound Poisson process Y is a Lévy process.*

1.3 Lévy processes

Proof To verify (L1) and (L2) is straightforward. To establish (L3), let $a > 0$; then by conditioning and independence we have

$$P(|Y(t)| > a) = \sum_{n=0}^{\infty} P(|Z(1) + \cdots + Z(n)| > a) P(N(t) = n),$$

and the required result follows by dominated convergence. □

By Proposition 1.2.11 we see that Y has Lévy symbol

$$\eta_Y(u) = \left[\int_{\mathbb{R}^d} (e^{i(u,y)} - 1) \lambda \mu_Z(dy) \right].$$

Again the sample paths of Y are piecewise constant on finite intervals with 'jump discontinuities' at the random times $(T(n), n \in \mathbb{N})$; however, this time the size of the jumps is itself random, and the jump at $T(n)$ can take any value in the range of the random variable $Z(n)$.

The compound Poisson process has important applications to models of insurance risk; see e.g. Chapter 1 of Embrechts *et al.* [94].

Clearly a compound Poisson process is Poisson if and only if $d = 1$ and each $Z(n) = 1$ (a.s.), so $\mu_Z = \delta_1$. The following proposition tells us that two independent Poisson processes must jump at distinct times (a.s.).

Proposition 1.3.12 *If $(N_1(t), t \geq 0)$ and $(N_2(t), t \geq 0)$ are two independent Poisson processes defined on the same probability space, with arrival times $(T_n^{(j)}, n \in \mathbb{N})$ for $j = 1, 2$, respectively, then*

$$P\left(T_m^{(1)} = T_n^{(2)} \text{ for some } m, n \in \mathbb{N}\right) = 0.$$

Proof Let $N(t) = N_1(t) + N_2(t)$ for each $t \geq 0$; then it follows from Exercise 1.3.6 and a straightforward computation of the characteristic function that N is another Poisson process. Hence, for each $t \geq 0$, we can write $N(t) = Z(1) + \cdots + Z(N(t))$ where $(Z(n), n \in \mathbb{N})$ is i.i.d. with each $Z(n) = 1$ (a.s.). Now let $m, n \in \mathbb{N}$ be such that $T_m^{(1)} = T_n^{(2)}$ (a.s.); if these are the first times at which such an event occurs, it follows that $Z(m + n - 1) = 2$ (a.s.), and we have our required contradiction. □

Example 1.3.13 (Interlacing processes) Let C be a Gaussian Lévy process as in Example 1.3.8 and Y be a compound Poisson process, as in Example 1.3.10, that is independent of C. Define a new process X by

$$X(t) = C(t) + Y(t),$$

for all $t \geq 0$; then it is not difficult to verify that X is a Lévy process with Lévy symbol of the form (1.8). The paths of X have jumps of random size occurring at random times. In fact, using the notation of Examples 1.3.9 and 1.3.10, we have

$$X(t) = \begin{cases} C(t) & \text{for } 0 \leq t < T_1, \\ C(T_1) + Z_1 & \text{for } t = T_1, \\ X(T_1) + C(t) - C(T_1) & \text{for } T_1 < t < T_2, \\ X(T_2) + Z_2 & \text{for } t = T_2, \end{cases}$$

and so on recursively. We call this procedure an *interlacing*, since a continuous-path process is 'interlaced' with random jumps. This type of construction will recur throughout the book. In particular, if we examine the proof of Theorem 1.2.14, it seems reasonable that the most general Lévy process might arise as the limit of a sequence of such interlacings, and we will investigate this further in the next chapter.

Example 1.3.14 (Stable Lévy processes) A *stable Lévy process* is a Lévy process X in which each $X(t)$ is a stable random variable. So the Lévy symbol is given by Theorem 1.2.24. Of particular interest is the rotationally invariant case, where the Lévy symbol is given by

$$\eta(u) = -\sigma^\alpha |u|^\alpha;$$

here $0 < \alpha \leq 2$ is the index of stability and $\sigma > 0$.

One reason why stable Lévy processes are important in applications is that they display self-similarity. In general, a stochastic process $Y = (Y(t), t \geq 0)$ is *self-similar with Hurst index* $H > 0$ if the two processes $(Y(at), t \geq 0)$ and $(a^H Y(t), t \geq 0)$ have the same finite-dimensional distributions for all $a \geq 0$. By examining characteristic functions, it is easily verified that a rotationally invariant stable Lévy process is self-similar with Hurst index $H = 1/\alpha$, so that e.g. Brownian motion is self-similar with $H = 1/2$. A nice general account of self-similar processes can be found in Embrechts and Maejima [95]. In particular, it is shown therein that a Lévy process X is self-similar if and only if each $X(t)$ is strictly stable.

Just as with Gaussian processes, we can extend the notion of stability beyond the class of stable Lévy processes. In general, then, we say that a stochastic process $X = (X(t), t \geq 0)$ is *stable* if all its finite-dimensional distributions are stable. For a comprehensive introduction to such processes, see Samorodnitsky and Taqqu [271], Chapter 3.

1.3.2 Subordinators

A *subordinator* is a one-dimensional Lévy process that is non-decreasing (a.s.). Such processes can be thought of as a random model of time evolution, since if $T = (T(t), t \geq 0)$ is a subordinator we have

$$T(t) \geq 0 \quad \text{a.s.} \quad \text{for each } t > 0,$$

and

$$T(t_1) \leq T(t_2) \quad \text{a.s.} \quad \text{whenever } t_1 \leq t_2.$$

Now since for $X(t) \sim N(0, At)$ we have $P(X(t) \geq 0) = P(X(t) \leq 0) = 1/2$, it is clear that such a process cannot be a subordinator. More generally we have

Theorem 1.3.15 *If T is a subordinator, then its Lévy symbol takes the form*

$$\eta(u) = ibu + \int_0^\infty (e^{iuy} - 1)\lambda(dy), \tag{1.22}$$

where $b \geq 0$ and the Lévy measure λ satisfies the additional requirements

$$\lambda(-\infty, 0) = 0 \quad \text{and} \quad \int_0^\infty (y \wedge 1)\lambda(dy) < \infty.$$

Conversely, any mapping from $\mathbb{R}^d \to \mathbb{C}$ of the form (1.22) is the Lévy symbol of a subordinator.

A proof of this can be found in Bertoin [37], Theorem 1.2 (see also Rogers and Williams [261], pp. 78–9).

We call the pair (b, λ) the *characteristics* of the subordinator T.

Exercise 1.3.16 Show that the additional constraint on Lévy measures of subordinators is equivalent to the requirement

$$\int_0^\infty \frac{y}{1+y} \lambda(dy) < \infty.$$

Now for each $t \geq 0$ the map $u \to \mathbb{E}(e^{iuT(t)})$ can clearly be analytically continued to the region $\{iu, u > 0\}$, and we then obtain the following expression for the Laplace transform of the distribution:

$$\mathbb{E}(e^{-uT(t)}) = e^{-t\psi(u)},$$

where

$$\psi(u) = -\eta(iu) = bu + \int_0^\infty (1 - e^{-uy})\lambda(dy) \quad (1.23)$$

for each $u > 0$. We observe that this is much more useful for both theoretical and practical applications than the characteristic function.

The function ψ is usually called the *Laplace exponent* of the subordinator.

Examples

Example 1.3.17 (The Poisson case) Poisson processes are clearly subordinators. More generally, a compound Poisson process will be a subordinator if and only if the $Z(n)$ in equation (1.21) are all $\mathbb{R}^+$-valued.

Example 1.3.18 (α-stable subordinators) Using straightforward calculus (see the appendix at the end of this chapter if you need a hint), we find that for $0 < \alpha < 1, u \geq 0$,

$$u^\alpha = \frac{\alpha}{\Gamma(1-\alpha)} \int_0^\infty (1 - e^{-ux}) \frac{dx}{x^{1+\alpha}}.$$

Hence by (1.23), Theorem 1.3.15 and Theorem 1.2.20 we see that for each $0 < \alpha < 1$ there exists an α-stable subordinator T with Laplace exponent

$$\psi(u) = u^\alpha,$$

and the characteristics of T are $(0, \lambda)$ where

$$\lambda(dx) = \frac{\alpha}{\Gamma(1-\alpha)} \frac{dx}{x^{1+\alpha}}.$$

Note that when we analytically continue this to obtain the Lévy symbol we obtain the form given in Theorem 1.2.21(2), with $\mu = 0$, $\beta = 1$ and $\sigma^\alpha = \cos(\alpha\pi/2)$.

Example 1.3.19 (The Lévy subordinator) The $\frac{1}{2}$-stable subordinator has a density given by the Lévy distribution (with $\mu = 0$ and $\sigma = t^2/2$)

$$f_{T(t)}(s) = \left(\frac{t}{2\sqrt{\pi}}\right) s^{-3/2} e^{-t^2/(4s)},$$

for $s \geq 0$. The Lévy subordinator has a nice probabilistic interpretation as a first hitting time for one-dimensional standard Brownian motion $(B(t), t \geq 0)$.

More precisely:

$$T(t) = \inf\left\{s > 0; B(s) = \frac{t}{\sqrt{2}}\right\}. \quad (1.24)$$

For details of this see Revuz and Yor [260], p. 109, and Rogers and Williams [261], p. 133. We will prove this result by using martingale methods in the next chapter (Theorem 2.2.9).

Exercise 1.3.20 Show directly that, for each $t \geq 0$,

$$\mathbb{E}(e^{-uT(t)}) = \int_0^\infty e^{-us} f_{T(t)}(s) ds = e^{-tu^{1/2}},$$

where $(T(t), t \geq 0)$ is the Lévy subordinator. (Hint: Write $g_t(u) = \mathbb{E}(e^{-uT(t)})$. Differentiate with respect to u and make the substitution $x = t^2/(4us)$ to obtain the differential equation $g'_t(u) = -(t/2\sqrt{u})g_t(u)$. Via the substitution $y = t/(2\sqrt{s})$ we see that $g_t(0) = 1$, and the result follows; see also Sato [274] p. 12.)

Example 1.3.21 (Inverse Gaussian subordinators) We generalise the Lévy subordinator by replacing the Brownian motion by the Gaussian process $C = (C(t), t \geq 0)$ where each $C(t) = B(t) + \gamma t$ and $\gamma \in \mathbb{R}$. The *inverse Gaussian subordinator* is defined by

$$T(t) = \inf\{s > 0; C(s) = \delta t\},$$

where $\delta > 0$, and is so-called because $t \to T(t)$ is the generalised inverse of a Gaussian process.

Again by using martingale methods, as in Theorem 2.2.9, we can show that for each $t, u > 0$,

$$\mathbb{E}(e^{-uT(t)}) = \exp\left[-t\delta(\sqrt{2u + \gamma^2} - \gamma)\right] \quad (1.25)$$

(see Exercise 2.2.10). In fact each $T(t)$ has a density, and we can easily compute these from (1.25) and the result of Exercise 1.3.20, obtaining

$$f_{T(t)}(s) = \frac{\delta t}{\sqrt{2\pi}} e^{\delta t \gamma} s^{-3/2} \exp\left[-\tfrac{1}{2}(t^2\delta^2 s^{-1} + \gamma^2 s)\right] \quad (1.26)$$

for each $s, t \geq 0$.

In general any random variable with density $f_{T(1)}$ is called an *inverse Gaussian* and denoted as $IG(\delta, \gamma)$.

Example 1.3.22 (Gamma subordinators) Let $(T(t), t \geq 0)$ be a *gamma process* with parameters $a, b > 0$, so that each $T(t)$ has density

$$f_{T(t)}(x) = \frac{b^{at}}{\Gamma(at)} x^{at-1} e^{-bx},$$

for $x \geq 0$; then it is easy to verify that, for each $u \geq 0$,

$$\int_0^\infty e^{-ux} f_{T(t)}(x) dx = \left(1 + \frac{u}{b}\right)^{-at} = \exp\left[-ta \log\left(1 + \frac{u}{b}\right)\right].$$

From here it is a straightforward exercise in calculus to show that

$$\int_0^\infty e^{-ux} f_{T(t)}(x) dx = \int_0^\infty (1 - e^{-ux}) a x^{-1} e^{-bx} dx;$$

see Sato [274] p. 45 if you need a hint.

From this we see that $(T(t), t \geq 0)$ is a subordinator with $b = 0$ and $\lambda(dx) = ax^{-1} e^{-bx} dx$. Moreover, $\psi(u) = a \log(1 + u/b)$ is the associated Bernstein function (see below).

Before we go further into the probabilistic properties of subordinators we will make a quick diversion into analysis.

Let $f \in C^\infty((0, \infty))$ with $f \geq 0$. We say f is *completely monotone* if $(-1)^n f^{(n)} \geq 0$ for all $n \in \mathbb{N}$ and a *Bernstein function* if $(-1)^n f^{(n)} \leq 0$ for all $n \in \mathbb{N}$. We then have the following:

Theorem 1.3.23

(1) f is a Bernstein function if and only if the mapping $x \to e^{-tf(x)}$ is completely monotone for all $t \geq 0$.
(2) f is a Bernstein function if and only if it has the representation

$$f(x) = a + bx + \int_0^\infty (1 - e^{-yx}) \lambda(dy)$$

for all $x > 0$, where $a, b \geq 0$ and $\int_0^\infty (y \wedge 1) \lambda(dy) < \infty$.
(3) g is completely monotone if and only if there exists a measure μ on $[0, \infty)$ for which

$$g(x) = \int_0^\infty e^{-xy} \mu(dy).$$

A proof of this theorem can be found in Berg and Forst [35], pp. 61–72.

To interpret this theorem, first consider the case $a = 0$. In this case, if we compare the statement in theorem 1.3.23(2) with equation (1.23), we see that

1.3 Lévy processes

there is a one-to-one correspondence between Bernstein functions for which $\lim_{x \to 0} f(x) = 0$ and Laplace exponents of subordinators. The Laplace transforms of the laws of subordinators are always completely monotone functions, and a subclass of all possible measures μ appearing in Theorem 1.3.23(3) is given by all possible laws $p_{T(t)}$ associated with subordinators. Now let f be a general Bernstein function with $a > 0$. We can give it a probabilistic interpretation as follows. Let T be the subordinator with Laplace exponent $\psi(u) = f(u) - a$ for each $u \geq 0$ and let S be an exponentially distributed random variable with parameter a independent of T, so that S has the pdf $g_S(x) = ae^{-ax}$ for each $x \geq 0$.

Now define a process $T_S = (T_S(t), t \geq 0)$, which takes values in $\mathbb{R}^+ \cup \{\infty\}$ and which we will call a *killed subordinator*, by the prescription

$$T_S(t) = \begin{cases} T(t) & \text{for } 0 \leq t < S, \\ \infty & \text{for } t \geq S. \end{cases}$$

Proposition 1.3.24 *There is a one-to-one correspondence between killed subordinators $T(S)$ and Bernstein functions f, given by*

$$\mathbb{E}(e^{-uT_S(t)}) = e^{-tf(u)}$$

for each $t, u \geq 0$.

Proof By independence, we have

$$\begin{aligned}\mathbb{E}(e^{-uT_S(t)}) &= \mathbb{E}(e^{-uT_S(t)} \chi_{[0,S)}(t)) + \mathbb{E}(e^{-uT_S(t)} \chi_{[S,\infty)}(t)) \\ &= \mathbb{E}(e^{-uT(t)}) P(S \geq t) \\ &= e^{-t[\psi(u)+a]},\end{aligned}$$

where we have adopted the convention $e^{-\infty} = 0$. □

One of the most important probabilistic applications of subordinators is to 'time changing'. Let X be an arbitrary Lévy process and let T be a subordinator defined on the same probability space as X such that X and T are independent. We define a new stochastic process $Z = (Z(t), t \geq 0)$ by the prescription

$$Z(t) = X(T(t)),$$

for each $t \geq 0$, so that for each $\omega \in \Omega$, $Z(t)(\omega) = X(T(t)(\omega))(\omega)$. The key result is then the following.

Theorem 1.3.25 *Z is a Lévy process.*

Proof (L1) is trivial. To establish (L2) we first prove stationary increments. Let $0 \leq t_1 < t_2 < \infty$ and $A \in \mathcal{B}(\mathbb{R}^d)$. We denote as p_{t_1,t_2} the joint probability law of $T(t_1)$ and $T(t_2)$; then by the independence of X and T and the fact that X has stationary increments we find that

$$P(Z(t_2) - Z(t_1) \in A) = P(X(T(t_2)) - X(T(t_1)) \in A)$$
$$= \int_0^\infty \int_0^\infty P(X(s_2) - X(s_1) \in A) p_{t_1,t_2}(ds_1, ds_2)$$
$$= \int_0^\infty \int_0^\infty P(X(s_2 - s_1) \in A) p_{t_1,t_2}(ds_1, ds_2)$$
$$= P(Z(t_2 - t_1) \in A).$$

For independent increments, let $0 \leq t_1 < t_2 < t_3 < \infty$. We write p_{t_1,t_2,t_3} for the joint probability law of $T(t_1), T(t_2)$ and $T(t_3)$. For arbitrary $y \in \mathbb{R}^d$, define $h_y : \mathbb{R}^+ \to \mathbb{C}$ by $h_y(s) = \mathbb{E}(e^{i(y, X(s))})$ and, for arbitrary $y_1, y_2 \in \mathbb{R}^d$, define $f_{y_1,y_2} : \mathbb{R}^+ \times \mathbb{R}^+ \times \mathbb{R}^+ \to \mathbb{C}$ by

$$f_{y_1,y_2}(u_1, u_2, u_3) = \mathbb{E}\big(\exp[i(y_1, X(u_2) - X(u_1))]\big)$$
$$\times \mathbb{E}(\exp[i(y_2, X(u_3) - X(u_2))]),$$

where $0 \leq u_1 < u_2 < u_3 < \infty$. By conditioning, using the independence of X and T and the fact that X has independent increments we obtain

$$\mathbb{E}\big(\exp\{i[(y_1, Z(t_2) - Z(t_1)) + (y_2, Z(t_3) - Z(t_2))]\}\big)$$
$$= \mathbb{E}\big(f_{y_1,y_2}(T(t_1), T(t_2), T(t_3))\big).$$

However, since X has stationary increments, we have that

$$f_{y_1,y_2}(u_1, u_2, u_3) = h_{y_1}(u_2 - u_1) h_{y_2}(u_3 - u_2)$$

for each $0 \leq u_1 < u_2 < u_3 < \infty$.

Hence, by the independent increments property of T, we obtain

$$\mathbb{E}\big(\exp\{i[(y_1, Z(t_2) - Z(t_1)) + (y_2, Z(t_3) - Z(t_2))]\}\big)$$
$$= \mathbb{E}(h_{y_1}(T_2 - T_1) h_{y_2}(T_3 - T_2))$$
$$= \mathbb{E}(h_{y_1}(T_2 - T_1)) \mathbb{E}(h_{y_2}(T_3 - T_2))$$
$$= \mathbb{E}\big(\exp[i(y_1, Z(t_2 - t_1))]\big) \mathbb{E}\big(\exp[i(y_2, Z(t_3 - t_2))]\big),$$

by the independence of X and T.

1.3 Lévy processes

The fact that $Z(t_2) - Z(t_1)$ and $Z(t_3) - Z(t_2)$ are independent now follows by Kac's theorem from the fact that Z has stationary increments, which was proved above. The extension to n time intervals is by a similar argument; see also Sato [274], pp. 199–200.

We now establish (L3). Since X and T are stochastically continuous, we know that, for any $a \in \mathbb{R}^d$, if we are given any $\epsilon > 0$ then we can find $\delta > 0$ such that $0 < h < \delta \Rightarrow P(|X(h)| > a) < \epsilon/2$, and we can find $\delta' > 0$ such that $0 < h < \delta' \Rightarrow P(T(h) > \delta) < \epsilon/2$.

Now, for all $t \geq 0$ and all $0 \leq h < \min\{\delta, \delta'\}$, we have

$$P(|Z(h)| > a)$$
$$= P(|X(T(h))| > a) = \int_0^\infty P(|X(u)| > a) p_{T(h)}(du)$$
$$= \int_{[0,\delta)} P(|X(u)| > a) p_{T(h)}(du) + \int_{[\delta,\infty)} P(|X(u)| > a) p_{T(h)}(du)$$
$$\leq \sup_{0 \leq u < \delta} P(|X(u)| > a) + P(T(h) \geq \delta)$$
$$< \frac{\epsilon}{2} + \frac{\epsilon}{2} = \epsilon.$$

$\square$

Exercise 1.3.26 Show that for each $A \in \mathcal{B}(\mathbb{R}^d)$, $t \geq 0$,

$$p_{Z(t)}(A) = \int_0^\infty p_{X(u)}(A) p_{T(t)}(du).$$

We now compute the Lévy symbol η_Z of the subordinated process Z.

Proposition 1.3.27

$$\eta_Z = -\psi_T \circ (-\eta_X).$$

Proof For each $u \in \mathbb{R}^d$, $t \geq 0$,

$$\mathbb{E}(e^{i(u,Z(t))}) = \mathbb{E}(e^{i(u,X(T(t)))}) = \int_0^\infty \mathbb{E}(e^{i(u,X(s))}) p_{T(t)}(ds)$$
$$= \int_0^\infty e^{-s(-\eta_X(u))} p_{T(t)}(ds) = \mathbb{E}(e^{-\eta_X(u)T(t)})$$
$$= e^{-t\psi_T(-\eta_X(u))}.$$

$\square$

Note The penultimate step in the above proof necessitates analytic continuation of the map $u \to \mathbb{E}(e^{iuT(t)})$ to the region Ran (η_X).

Example 1.3.28 (From Brownian motion to 2α-stable processes) Let T be an α-stable subordinator, with $0 < \alpha < 1$, and X be a d-dimensional Brownian motion with covariance $A = 2I$ that is independent of T. Then for each $s \geq 0$, $u \in \mathbb{R}^d$, $\psi_T(s) = s^\alpha$ and $\eta_X(u) = -|u|^2$, and hence $\eta_Z(u) = -|u|^{2\alpha}$, i.e. Z is a rotationally invariant 2α-stable process.

In particular, if $d = 1$ and T is the Lévy subordinator then Z is the *Cauchy process*, so each $Z(t)$ has a symmetric Cauchy distribution with parameters $\mu = 0$ and $\sigma = 1$. It is interesting to observe from (1.24) that Z is constructed from two independent standard Brownian motions.

Example 1.3.29 (From Brownian motion to relativity) Let T be the Lévy subordinator, and for each $t \geq 0$ define

$$f_{c,m}(s;t) = \exp(-m^2c^4s + mc^2t)f_{T(t)}(s)$$

for each $s \geq 0$, where $m, c > 0$.

It is then an easy exercise in calculus to deduce that

$$\int_0^\infty e^{-us} f_{c,m}(s;t)ds = \exp\{-t[(u + m^2c^4)^{1/2} - mc^2]\}.$$

Since the map $u \to (u + m^2c^4)^{1/2} - mc^2$ is a Bernstein function that vanishes at the origin, we deduce that there is a subordinator $T_{c,m} = (T_{c,m}(t), t \geq 0)$ for which each $T_{c,m}(t)$ has density $f_{c,m}(\cdot\,;t)$. Now let B be a Brownian motion, with covariance $A = 2c^2I$, that is independent of $T_{c,m}$; then for the subordinated process we find, for all $p \in \mathbb{R}^d$,

$$\eta_Z(p) = -[(c^2|p|^2 + m^2c^4)^{1/2} - mc^2]$$

so that Z is a relativistic process, i.e. each $Z(t)$ has a relativistic distribution as in Subsection 1.2.6.

Exercise 1.3.30 Generalise this last example to the case where T is an α-stable subordinator with $0 < \alpha < 1$; see Ryznar [268] for more about such subordinated processes.

Examples of subordinated processes have recently found useful applications in mathematical finance and we will discuss this again in Chapter 5. We briefly mention two interesting cases.

Example 1.3.31 (The variance gamma process) In this case $Z(t) = B(T(t))$ for each $t \geq 0$, where B is a standard Brownian motion and T is an independent gamma subordinator. The name derives from the fact that, in a formal sense, each $Z(t)$ arises by replacing the variance of a normal random variable by a gamma random variable. Using Proposition 1.3.27, a simple calculation yields

$$\Phi_{Z(t)}(u) = \left(1 + \frac{u^2}{2b}\right)^{-at}$$

for each $t \geq 0$, $u \in \mathbb{R}$, where a and b are the usual parameters determining the gamma process. For further details, see Madan and Seneta [206].

Example 1.3.32 (The normal inverse Gaussian process) In this case $Z(t) = C(T(t)) + \mu t$ for each $t \geq 0$, where each $C(t) = B(t) + \beta t$ with $\beta \in \mathbb{R}$. Here T is an inverse Gaussian subordinator which is independent of B and in which we write the parameter $\gamma = \sqrt{\alpha^2 - \beta^2}$, where $\alpha \in \mathbb{R}$ with $\alpha^2 \geq \beta^2$. Z depends on four parameters and has characteristic function

$$\Phi_{Z(t)}(\alpha, \beta, \delta, \mu)(u)$$
$$= \exp\left\{\delta t\left[\sqrt{\alpha^2 - \beta^2} - \sqrt{\alpha^2 - (\beta + iu)^2}\right] + i\mu tu\right\}$$

for each $u \in \mathbb{R}$, $t \geq 0$. Here $\delta > 0$ is as in (1.25). Note that the relativistic process considered in Example 1.3.29 above is a special case of this wherein $\mu = \beta = 0$, $\delta = c$ and $\alpha = mc$.

Each $Z(t)$ has a density given by

$$f_{Z(t)}(x) = C(\alpha, \beta, \delta, \mu; t) q\left(\frac{x - \mu t}{\delta t}\right)^{-1} K_1\left(\delta t \alpha q\left(\frac{x - \mu t}{\delta t}\right)\right) e^{\beta x},$$

for each $x \in \mathbb{R}$, where $q(x) = \sqrt{1 + x^2}$;

$$C(\alpha, \beta, \delta, \mu; t) = \pi^{-1} \alpha e^{\delta t \sqrt{\alpha^2 - \beta^2} - \beta \mu t}$$

and K_1 is a Bessel function of the third kind (see Subsection 5.6).

For further details see Barndorff-Nielsen [27, 28] and Carr et al. [67].

We now return to general considerations and probe a little more deeply into the structure of η_Z. To this end we define a Borel measure $m_{X,T}$ on $\mathbb{R}^d - \{0\}$ by

$$m_{X,T}(A) = \int_0^\infty p_{X(t)}(A) \lambda(dt)$$

for each $A \in \mathcal{B}(\mathbb{R}^d - \{0\})$; λ is the Lévy measure of the subordinator T. In fact, $m_{X,T}$ is a Lévy measure satisfying the stronger condition $\int_0^\infty (|y| \wedge 1) m_{X,T}(dy) < \infty$. You can derive this from the fact that for any Lévy process X there exists a constant $C \geq 0$ such that for each $t \geq 0$

$$\left| \mathbb{E}(X(t); |X(t)| \leq 1) \right| \leq C(t \wedge 1);$$

see Sato [274] p. 198, Lemma 30.3, for a proof of this.

Theorem 1.3.33 *For each $u \in \mathbb{R}^d$,*

$$\eta_Z(u) = b\eta_X + \int_{\mathbb{R}^d} (e^{i(u,y)} - 1) m_{X,T}(dy).$$

Proof By Proposition 1.3.27, (1.23) and Fubini's theorem we find that

$$\eta_Z(u) = b\eta_X(u) + \int_0^\infty (e^{s\eta_X(u)} - 1) \lambda(ds)$$

$$= b\eta_X(u) + \int_0^\infty [\mathbb{E}(e^{i(u,X(s))}) - 1] \lambda(ds)$$

$$= b\eta_X(u) + \int_0^\infty \int_{\mathbb{R}^d} (e^{i(u,y)} - 1) p_{X(s)}(dy) \lambda(ds)$$

$$= b\eta_X(u) + \int_{\mathbb{R}^d} (e^{i(u,y)} - 1) m_{X,T}(dy).$$

$\square$

More results about subordinators can be found in Bertoin's Saint-Flour lectures on the subject [37], Chapter 6 of Sato [274] and Chapter 3 of Bertoin [36].

1.4 Convolution semigroups of probability measures

In this section, we look at an important characterisation of Lévy processes. We begin with a definition. Let $(p_t, t \geq 0)$ be a family of probability measures on $\mathbb{R}^d$. We say that it is *weakly convergent to δ_0* if

$$\lim_{t \downarrow 0} \int_{\mathbb{R}^d} f(y) p_t(dy) = f(0)$$

for all $f \in C_b(\mathbb{R}^d)$.

1.4 Convolution semigroups of probability measures

Proposition 1.4.1 *If X is a stochastic process wherein $X(t)$ has law p_t for each $t \geq 0$ and $X(0) = 0$ (a.s.) then $(p_t, t \geq 0)$ is weakly convergent to δ_0 if and only if X is stochastically continuous at $t = 0$.*

Proof First assume that X is stochastically continuous at $t = 0$ and suppose that $f \in C_b(\mathbb{R}^d)$ with $f \neq 0$; then given any $\epsilon > 0$ there exists $\delta > 0$ such that $\sup_{x \in B_\delta(0)} |f(x) - f(0)| \leq \epsilon/2$ and there exists $\delta' > 0$ such that $0 < t < \delta' \Rightarrow P(|X(t)| > \delta) < \epsilon/(4M)$, where $M = \sup_{x \in \mathbb{R}^d} |f(x)|$. For such t we then find that

$$\left| \int_{\mathbb{R}^d} [f(x) - f(0)] p_t(dx) \right|$$
$$\leq \int_{B_\delta(0)} |f(x) - f(0)| p_t(dx) + \int_{B_\delta(0)^c} |f(x) - f(0)| p_t(dx)$$
$$\leq \sup_{x \in B_\delta(0)} |f(x) - f(0)| + 2M P(X(t) \in B_\delta(0)^c)$$
$$< \epsilon.$$

Conversely, suppose that $(p_t, t \geq 0)$ is weakly convergent to δ_0. We use the argument of Malliavin *et al.* [207], pp. 98–9. Fix $r > 0$ and $\epsilon > 0$. Let $f \in C_b(\mathbb{R}^d)$ with support in $B_r(0)$ be such that $0 \leq f \leq 1$ and $f(0) > 1 - (\epsilon/2)$. By weak convergence we can find $t_0 > 0$ such that

$$0 \leq t < t_0 \Rightarrow \left| \int_{\mathbb{R}^d} [f(y) - f(0)] p_t(dy) \right| < \frac{\epsilon}{2}.$$

We then find that

$$P(|X(t)| > r) = 1 - P(|X(t)| \leq r)$$
$$\leq 1 - \int_{B_r(0)} f(y) p_t(dy) = 1 - \int_{\mathbb{R}^d} f(y) p_t(dy)$$
$$= 1 - f(0) + \int_{\mathbb{R}^d} [f(0) - f(y)] p_t(dy)$$
$$< \frac{\epsilon}{2} + \frac{\epsilon}{2} = \epsilon.$$

$\square$

A family of probability measures $(p_t, t \geq 0)$ with $p_0 = \delta_0$ is said to be a *convolution semigroup* if

$$p_{s+t} = p_s * p_t \qquad \text{for all } s, t \geq 0,$$

and such a semigroup is said to be *weakly continuous* if it is weakly convergent to δ_0.

Exercise 1.4.2 Show that a convolution semigroup is weakly continuous if and only if

$$\lim_{s \downarrow t} \int_{\mathbb{R}^d} f(y) p_s(dy) = \int_{\mathbb{R}^d} f(y) p_t(dy)$$

for all $f \in C_b(\mathbb{R}^d), t \geq 0$.

Exercise 1.4.3 Show directly that the Gauss semigroup defined by

$$p_t(dx) = \frac{1}{\sqrt{2\pi t}} e^{-x^2/(2t)} dx$$

for each $x \in \mathbb{R}, t \geq 0$, is a weakly continuous convolution semigroup.

Of course, we can recognise the Gauss semigroup in the last example as giving the law of a standard one-dimensional Brownian motion. More generally we have the following:

Proposition 1.4.4 *If $X = (X(t), t \geq 0)$ is a Lévy process wherein $X(t)$ has law p_t for each $t \geq 0$ then $(p_t, t \geq 0)$ is a weakly continuous convolution semigroup.*

Proof This is straightforward once you have Proposition 1.4.1. □

We will now aim to establish a partial converse to Proposition 1.4.4.

1.4.1 Canonical Lévy processes

Let $(p_t, t \geq 0)$ be a weakly continuous convolution semigroup of probability measures on $\mathbb{R}^d$. Define $\Omega = \{\omega : \mathbb{R}^+ \to \mathbb{R}^d; \omega(0) = 0\}$. We construct a σ-algebra of subsets of Ω as follows: for each $n \in \mathbb{N}$, choose $0 \leq t_1 < t_2 < \cdots < t_n < \infty$ and choose $A_1, A_2, \ldots, A_n \in \mathcal{B}(\mathbb{R}^d)$. As in Subsection 1.1.7, we define cylinder sets $I_{t_1, t_2, \ldots, t_n}^{A_1, A_2, \ldots, A_n}$ by

$$I_{t_1, t_2, \ldots, t_n}^{A_1, A_2, \ldots, A_n} = \{\omega \in \Omega; \omega(t_1) \in A_1, \omega(t_2) \in A_2, \ldots, \omega(t_n) \in A_n\}.$$

1.4 Convolution semigroups of probability measures

Let $\mathcal{F}$ denote the smallest σ-algebra containing all cylinder sets of Ω. We define a set-function P on the collection of all cylinder sets by the prescription

$$P(I_{t_1,t_2,\ldots,t_n}^{A_1,A_2,\ldots,A_n})$$
$$= \int_{A_1} p_{t_1}(dy_1) \int_{A_2} p_{t_2-t_1}(dy_2 - y_1) \cdots \int_{A_n} p_{t_n-t_{n-1}}(dy_n - y_{n-1})$$
$$= \int_{\mathbb{R}^d} \int_{\mathbb{R}^d} \cdots \int_{\mathbb{R}^d} \chi_{A_1}(y_1) \chi_{A_2}(y_1 + y_2) \cdots \chi_{A_n}(y_1 + y_2 + \cdots + y_n)$$
$$\times p_{t_1}(dy_1) p_{t_2-t_1}(dy_2) \cdots p_{t_n-t_{n-1}}(dy_n). \tag{1.27}$$

By a slight variation on Kolmogorov's existence theorem (Theorem 1.1.16) P extends uniquely to a probability measure on $(\Omega, \mathcal{F})$. Furthermore if we define $X = (X(t), t \geq 0)$ by

$$X(t)(\omega) = \omega(t)$$

for all $\omega \in \Omega$, $t \geq 0$, then X is a stochastic process on Ω whose finite-dimensional distributions are given by

$$P(X(t_1) \in A_1, X(t_2) \in A_2, \ldots, X(t_n) \in A_n) = P(I_{t_1,t_2,\ldots,t_n}^{A_1,A_2,\ldots,A_n}),$$

so that in particular each $X(t)$ has law p_t. We will show that X is a Lévy process. First note that (L1) and (L3) are immediate (via Proposition 1.4.1). To obtain (L2) we remark that, for any $f \in B_b(\mathbb{R}^{dn})$,

$$\mathbb{E}(f(X(t_1), X(t_2), \ldots, X(t_n)))$$
$$= \int_{\mathbb{R}^d} \int_{\mathbb{R}^d} \cdots \int_{\mathbb{R}^d} f(y_1, y_1 + y_2, \ldots, y_1 + y_2 + \cdots + y_n)$$
$$\times p_{t_1}(dy_1) p_{t_2-t_1}(dy_2) \cdots p_{t_n-t_{n-1}}(dy_n).$$

In fact this gives precisely equation (1.27) when f is an indicator function, and the more general result then follows by linearity, approximation and dominated convergence. (L2) can now be deduced by fixing $u \in \mathbb{R}^n$ and setting

$$f(x_1, x_2, \ldots, x_n) = \exp\left[i \sum_{j=1}^n (u_j, x_j - x_{j-1})\right]$$

for each $x \in \mathbb{R}^n$; see Sato [274], p. 36 for more details. So we have proved the following theorem.

Theorem 1.4.5 *If $(p(t), t \geq 0)$ is a weakly continuous convolution semigroup of probability measures, then there exists a Lévy process X such that, for each $t \geq 0$, $X(t)$ has law $p(t)$.*

We call X as constructed above the *canonical Lévy process*. Note that Kolmogorov's construction ensures that

$$\mathcal{F} = \sigma\{X(t), t \geq 0\}.$$

We thus obtain the following result, which makes the correspondence between infinitely divisible distributions and Lévy processes quite explicit:

Corollary 1.4.6 *If μ is an infinitely divisible probability measure on $\mathbb{R}^d$ with Lévy symbol η, then there exists a Lévy process X such that μ is the law of $X(1)$.*

Proof Suppose that μ has characteristics (b, A, ν); then for each $t \geq 0$ the mapping from $\mathbb{R}^d$ to $\mathbb{C}$ given by $u \to e^{t\eta(u)}$ has the form (1.11) and hence, by Theorem 1.2.14, for each $t \geq 0$ it is the characteristic function of an infinitely divisible probability measure $p(t)$. Clearly $p(0) = \delta_0$ and, by the unique correspondence between measures and characteristic functions, we obtain $p(s+t) = p(s) * p(t)$ for each $s, t \geq 0$. By Glivenko's theorem we have the weak convergence $p(t) \to \delta_0$ as $t \downarrow 0$, and the required result now follows from Theorem 1.4.5. □

Note Let $(p_t, t \geq 0)$ be a family of probability measures on $\mathbb{R}^d$. We say that it is *vaguely convergent to* δ_0 if

$$\lim_{t \downarrow 0} \int_{\mathbb{R}^d} f(y) p_t(dy) = f(0),$$

for all $f \in C_c(\mathbb{R}^d)$. We leave it to the reader to verify that Propositions 1.4.1 and 1.4.4 and Theorem 1.4.5 continue to hold if weak convergence is replaced by vague convergence. We will need this in Chapter 3.

1.4.2 Modification of Lévy processes

Let $X = (X(t), t \geq 0)$ and $Y = (Y(t), t \geq 0)$ be stochastic processes defined on the same probability space; then Y is said to be a *modification* of X if, for each $t \geq 0$, $P(X(t) \neq Y(t)) = 0$. It then follows that X and Y have the same finite-dimensional distributions.

Lemma 1.4.7 *If X is a Lévy process and Y is a modification of X, then Y is a Lévy process with the same characteristics as X.*

Proof (L1) is immediate. For (L2), fix $0 \leq s < t < \infty$ and let

$$\mathcal{N}(s, t) = \{\omega \in \Omega; X(s)(\omega) = Y(s)(\omega) \text{ and } X(t)(\omega) = Y(t)(\omega)\}.$$

It follows that $P(\mathcal{N}(s, t)) = 1$ since

$$P(\mathcal{N}(s, t)^c) = \{\omega \in \Omega; X(s)(\omega) \neq Y(s)(\omega) \text{ or } X(t)(\omega) \neq Y(t)(\omega)\}$$
$$\leq P(X(s) \neq Y(s)) + P(X(t) \neq Y(t)) = 0.$$

To see that Y has stationary increments, let $A \in \mathcal{B}(\mathbb{R}^d)$; then

$$P(Y(t) - Y(s) \in A)$$
$$= P(Y(t) - Y(s) \in A, \mathcal{N}(s, t)) + P(Y(t) - Y(s) \in A, \mathcal{N}(s, t)^c)$$
$$= P(X(t) - X(s) \in A, \mathcal{N}(s, t))$$
$$\leq P(X(t) - X(s) \in A)$$
$$= P(X(t - s) \in A) = P(Y(t - s) \in A).$$

The reverse inequality is obtained in similar fashion. Similar arguments can be used to show that Y has independent increments and to establish (L3).

We then see easily that X and Y have the same characteristic functions and hence the same characteristics. □

Note In view of Lemma 1.4.7, we lose nothing in replacing a Lévy process by a modification if the latter has nicer properties. For example, in Chapter 2 we will show that we can always find a modification that is right-continuous with left limits.

1.5 Some further directions in Lévy processes

This is not primarily a book about Lévy processes themselves. Our main aim is to study stochastic integration with respect to Lévy processes and to investigate new processes that can be built from them. Nonetheless, it is worth taking a short diversion from our main task in order to survey briefly some of the more advanced properties of Lévy processes, if only to stimulate the reader to learn more from the basic texts by Bertoin [36] and Sato [274]. We emphasise that the remarks in this section are of necessity somewhat brief and incomplete. The first two topics we will consider rely heavily on the perspective of Lévy processes as continuous-time analogues of random walks.

1.5.1 Recurrence and transience

Informally, an $\mathbb{R}^d$-valued stochastic process $X = (X(t), t \geq 0)$ is recurrent at $x \in \mathbb{R}^d$ if it visits every neighbourhood of x an infinite number of times (almost surely) and transient if it makes only a finite number of visits there (almost surely). If X is a Lévy process then if X is recurrent (transient) at some $x \in \mathbb{R}^d$, it is recurrent (transient) at every $x \in \mathbb{R}^d$; thus it is sufficient to concentrate on behaviour at the origin. We also have the useful dichotomy that 0 must be either recurrent or transient.

More precisely, we can make the following definitions. A Lévy process X is *recurrent* (at the origin) if

$$\liminf_{t \to \infty} |X(t)| = 0 \quad \text{a.s.}$$

and *transient* (at the origin) if

$$\lim_{t \to \infty} |X(t)| = \infty \quad \text{a.s.}$$

A remarkable fact about Lévy processes is that we can test for recurrence or transience using the Lévy symbol η alone. More precisely, we have the following two key results.

Theorem 1.5.1 (Chung–Fuchs criterion) *Fix $a > 0$. Then the following are equivalent:*

(1) X *is recurrent;*

(2)
$$\lim_{q \downarrow 0} \int_{B_a(0)} \Re\left(\frac{1}{q - \eta(u)}\right) du = \infty;$$

(3)
$$\limsup_{q \downarrow 0} \int_{B_a(0)} \Re\left(\frac{1}{q - \eta(u)}\right) du = \infty.$$

Theorem 1.5.2 (Spitzer criterion) X *is recurrent if and only if*

$$\int_{B_a(0)} \Re\left(\frac{1}{-\eta(u)}\right) du = \infty,$$

for any $a > 0$.

The Chung–Fuchs criterion is proved in Sato [274], pp. 252–3, as is the 'only if' part of the Spitzer criterion. For the full story, see the original papers by Port and Stone [252], but readers should be warned that these are demanding.

By application of the Spitzer criterion, we see immediately that Brownian motion is recurrent for $d = 1, 2$ and that for $d = 1$ every α-stable process is recurrent if $1 \leq \alpha < 2$ and transient if $0 < \alpha < 1$. For $d = 2$, all strictly α-stable processes are transient when $0 < \alpha < 2$. For $d \geq 3$, every Lévy process is transient. Further results with detailed proofs can be found in Chapter 7 of Sato [274].

A spectacular application of the recurrence and transience of Lévy processes to quantum physics can be found in Carmona *et al.* [64]. Here the existence of bound states for relativistic Schrödinger operators is shown to be intimately connected with the recurrence of a certain associated Lévy process, whose Lévy symbol is precisely that of the relativistic distribution discussed in Subsection 1.2.6.

1.5.2 Wiener–Hopf factorisation

Let X be a one-dimensional Lévy process with càdlàg paths (see Chapter 2 for more about these) and define the extremal processes $M = (M(t), t \geq 0)$ and $N = (N(t), t \geq 0)$ by

$$M(t) = \sup_{0 \leq s \leq t} X(s) \quad \text{and} \quad N(t) = \inf_{0 \leq s \leq t} X(s).$$

Fluctuation theory for Lévy processes studies the behaviour of a Lévy process in the neighbourhood of its suprema (or equivalently its infima) and a nice introduction to this subject is given in Chapter 6 of Bertoin [36]. One of the most fundamental and beautiful results in the area is the Wiener–Hopf factorisation, which we now describe. First we fix $q > 0$; then there exist two infinitely divisible characteristic functions ϕ_q^+ and ϕ_q^-, defined as follows:

$$\phi_q^+(u) = \exp\left[\int_0^\infty t^{-1} e^{-qt} \int_0^\infty (e^{iux} - 1) p_{X(t)}(dx) dt\right],$$

$$\phi_q^-(u) = \exp\left[\int_0^\infty t^{-1} e^{-qt} \int_{-\infty}^0 (e^{iux} - 1) p_{X(t)}(dx) dt\right],$$

for each $u \in \mathbb{R}$. The Wiener–Hopf factorisation identities yield a remarkable factorisation of the Laplace transform of the joint characteristic function of X and $M - X$ in terms of ϕ_q^+ and ϕ_q^-. More precisely we have:

Theorem 1.5.3 (Wiener–Hopf factorisation) *For each $q, t > 0, x, y \in \mathbb{R}$,*

$$q \int_0^\infty e^{-qt} \mathbb{E}\big(\exp(i\{xM(t) + y[X(t) - M(t)]\})\big)dt$$
$$= q \int_0^\infty e^{-qt} \mathbb{E}\big(\exp(i\{yN(t) + x[X(t) - N(t)]\})\big)dt$$
$$= \phi_q^+(x)\phi_q^-(y).$$

For a proof and related results, see Chapter 9 of Sato [274].

In Prabhu [253], Wiener–Hopf factorisation and other aspects of fluctuation theory for Lévy processes are applied to a class of 'storage problems' that includes models for the demand of water from dams, insurance risk, queues and inventories.

1.5.3 Local times

The *local time* of a Lévy process is a random field that, for each $x \in \mathbb{R}^d$, describes the amount of time spent by the process at x in the interval $[0, t]$. More precisely we define a measurable mapping $L : \mathbb{R}^d \times \mathbb{R}^+ \times \Omega \to \mathbb{R}^+$ by

$$L(x, t) = \limsup_{\epsilon \downarrow 0} \frac{1}{2\epsilon} \int_0^t \chi_{\{|X(s)-x|<\epsilon\}} ds,$$

and we have the 'occupation density formula'

$$\int_0^t f(X(s))ds = \int_{-\infty}^\infty f(x)L(x, t)dx \qquad \text{a.s.}$$

for all non-negative $f \in B_b(\mathbb{R}^d)$. From this we gain a pleasing intuitive understanding of local time as a random distribution, i.e.

$$L(x, t) = \int_0^t \delta(|x - X(s)|) \, ds,$$

where δ is the Dirac delta function.

It is not difficult to show that the map $t \to L(x, t)$ is continuous almost surely; see e.g. Bertoin [36], pp. 128–9. A more difficult problem, which was the subject of a great deal of work in the 1980s and 1990s, concerns the joint continuity of L as a function of x and t. A necessary and sufficient condition for this, which we do not state here as it is quite complicated, was established by Barlow [18] and Barlow and Hawkes [17] and is described in Chapter 5 of Bertoin [36], pp. 143–50. The condition is much simpler in the case where X

is symmetric. In this case, we define the 1-potential density of X by

$$u(y) = \int_0^\infty e^{-t} p_{X(t)}(y)\, dt$$

for each $u \in \mathbb{R}^d$ and consider the centred Gaussian field $(G(x), x \in \mathbb{R}^d)$ with covariance structure determined by u, so that for each $x, y \in \mathbb{R}^d$,

$$\mathbb{E}(G(x)G(y)) = u(x - y).$$

The main result in this case is that the almost-sure continuity of G is a necessary and sufficient condition for the almost-sure joint continuity of L. This result is due to Barlow and Hawkes [17] and was further developed by Marcus and Rosen [212]. The brief account by Marcus in [19] indicates many other interesting consequences of this approach.

Another useful property of local times concerns the generalised inverse process at the origin, i.e. the process $L_0^{-1} = (L_0^{-1}(t), t \geq 0)$, where each $L_0^{-1}(t) = \inf\{s \geq 0; L(0, s) \geq t\}$. When the origin is 'regular', so that X returns to 0 with probability 1 at arbitrary small times, then L_0^{-1} is a killed subordinator and this fact plays an important role in fluctuation theory; see e.g. Chapters 5 and 6 of Bertoin [36].

1.6 Notes and further reading

The Lévy–Khintchine formula was established independently by Paul Lévy and Alexander Khintchine in the 1930s. Earlier, both B. de Finetti and A. Kolmogorov had established special cases of the formula. The book by Gnedenko and Kolmogorov [121] was one of the first texts to appear on this subject and it is still highly relevant today. Proofs of the Lévy–Khintchine formula often appear in standard graduate texts in probability; see e.g. Fristedt and Gray [106] or Stroock [291]. An alternative approach, based on distributions, which has been quite influential in applications of infinite divisibility to mathematical physics, may be found in Gelfand and Vilenkin [113]. Another recent proof, which completely avoids probability theory, is given by Jacob and Schilling [147]. Aficionados of convexity can find the Lévy–Khintchine formula deduced from the Krein–Milman theorem in Johansen [160] or in Appendix 1 of Linnik and Ostrovskii [199].

As the Fourier transform generalises in a straightforward way to general locally compact abelian groups, the Lévy–Khintchine formula can be generalised to that setting; see Parthasarathy [247]. Further generalisations to the non-abelian case require the notion of a semigroup of linear operators (see Chapter 3 below); a classic reference for this is Heyer [128]. A more recent survey by the author of Lévy processes in Lie groups and Riemannian manifolds can be found in the volume [23], pp. 111–39. Lévy processes are also studied in more exotic structures based on extensions of the group concept. For processes in quantum groups (or Hopf algebras) see Schürmann [281], while the case of hypergroups can be found in Bloom and Heyer [52].

Another interesting generalisation of the Lévy–Khintchine formula is to the infinite-dimensional linear setting, and in the context of a Banach space the relevant references are Araujo and Giné [11] and Linde [198]. The Hilbert-space case can again be found in Parthasarathy [247].

The notion of a stable law is also due to Paul Lévy, and there is a nice early account of the theory in Gnedenko and Kolmogorov [121]. A number of books and papers appearing in the 1960s and 1970s contained errors in either the statement or proof of the key formulae in Theorem 1.2.21 and these are analysed by Hall in [126]. Accounts given in modern texts such as Sato [274] and Samorodnitsky and Taqqu [271] are fully trustworthy.

In Subsection 1.2.5, we discussed how stable and self-decomposable laws arise naturally as limiting distributions for certain generalisations of the central limit problem. More generally, the class of all infinitely divisible distributions coincides with those distributions that arise as limits of row sums of uniformly asymptotically negligible triangular arrays of random variables. The one-dimensional case is one of the main themes of Gnedenko and Kolmogorov [121]. The multi-dimensional case is given a modern treatment in Meerschaert and Scheffler [216]; see also Chapter VII of Jacod and Shiryaev [151].

The concept of a Lévy process was of course due to Paul Lévy and readers may consult his books [192, 193] for his own account of these. The key modern references are Bertoin [36] and Sato [274]. For a swift overview of the scope of the theory, see Bertoin [38]. Fristedt [107] is a classic source for sample-path properties of Lévy processes. Note that many books and articles, particularly those written before the 1990s, call Lévy processes 'stochastic processes with stationary and independent increments'. In French, this is sometimes shortened to 'PSI'.

A straightforward generalisation of a Lévy process just drops the requirement of stationary increments from the axioms. You then get an *additive process*. The theory of these is quite similar to that of Lévy processes, e.g. the Lévy–Khintchine formula has the same structure but the characteristics are no longer constant in time. For more details see Sato [274] and also the monograph by Skorohod [287]. Another interesting generalisation is that of an *infinitely divisible process*, i.e. a process all of whose finite-dimensional distributions are infinitely divisible. Important special cases are the Gaussian and stable processes, whose finite-dimensional distributions are always Gaussian and stable, respectively. Again there is a Lévy–Khinchine formula in the general case, but now the characteristics are indexed by finite subsets of $[0, \infty)$. For further details see Lee [190] and Maruyama [215].

Subordination was introduced by S. Bochner and is sometimes called 'subordination in the sense of Bochner' in the literature. His approach is outlined in his highly influential book [53]. The application of these to subordinate Lévy processes was first studied systematically by Huff [136]. If you want to learn more about the inverse Gaussian distribution, there is a very interesting book devoted to it by V. Seshradi [282].

Lévy processes are sometimes called 'Lévy flights' in physics: [285] is a volume based on applications of these, and the related concept of the 'Lévy walk' (i.e. a random walk in which the steps are stable random variables), to a range of topics including turbulence, dynamical systems, statistical mechanics and biology.

1.7 Appendix: An exercise in calculus

Here we establish the identity

$$u^\alpha = \frac{\alpha}{\Gamma(1-\alpha)} \int_0^\infty (1 - e^{-ux}) \frac{dx}{x^{1+\alpha}},$$

where $u \geq 0, 0 < \alpha < 1$.

This was applied to study α-stable subordinators in Subsection 1.3.2. We follow the method of Sato [274], p. 46, and employ the well-known trick of writing a repeated integral as a double integral and then changing the order of integration. We thus obtain

$$\int_0^\infty (1 - e^{-ux}) x^{-1-\alpha} dx = -\int_0^\infty \left(\int_0^x u e^{-uy} dy \right) x^{-1-\alpha} dx$$

$$= -\int_0^\infty \left(\int_y^\infty x^{-1-\alpha} dx \right) u e^{-uy} dy$$

$$= \frac{u}{\alpha} \int_0^\infty e^{-uy} y^{-\alpha} dy = \frac{u^\alpha}{\alpha} \int_0^\infty e^{-x} x^{-\alpha} dx$$

$$= \frac{u^\alpha}{\alpha} \Gamma(1-\alpha),$$

and the result follows immediately.

2

Martingales, stopping times and random measures

Summary We begin by introducing the important concepts of filtration, martingale and stopping time. These are then applied to establish the strong Markov property for Lévy processes and to prove that every Lévy process has a càdlàg modification. We then meet random measures, particularly those of Poisson type, and the associated Poisson integrals, which track the jumps of a Lévy process. The most important result of this chapter is the Lévy–Itô decomposition of a Lévy process into a Brownian motion with drift (the continuous part), a Poisson integral (the large jumps) and a compensated Poisson integral (the small jumps). As a corollary, we complete the proof of the Lévy–Khintchine formula. Finally, we establish the interlacing construction, whereby a Lévy process is obtained as the almost-sure limit of a sequence of Brownian motions with drift wherein random jump discontinuities are inserted at random times.

In this chapter we will frequently encounter stochastic processes with càdlàg paths (i.e. paths that are continuous on the right and always have limits on the left). Readers requiring background knowledge in this area should consult the appendix at the end of the chapter.

Before you start reading this chapter, be aware that parts of it are quite technical. If you are mainly interested in applications, feel free to skim it, taking note of the results of the main theorems without worrying too much about the proofs. However, make sure you get a 'feel' for the important ideas: Poisson integration, the Lévy–Itô decomposition and interlacing.

2.1 Martingales

2.1.1 Filtrations and adapted processes

Let $\mathcal{F}$ be a σ-algebra of subsets of a given set Ω. A family $(\mathcal{F}_t, t \geq 0)$ of sub σ-algebras of $\mathcal{F}$ is called a *filtration* if

$$\mathcal{F}_s \subseteq \mathcal{F}_t \quad \text{whenever} \quad s \leq t.$$

2.1 Martingales

A probability space $(\Omega, \mathcal{F}, P)$ that comes equipped with such a family $(\mathcal{F}_t, t \geq 0)$ is said to be *filtered*. We write $\mathcal{F}_\infty = \bigvee_{t \geq 0} \mathcal{F}_t$. A family of σ-algebras $(\mathcal{G}_t, t \geq 0)$ is called a *subfiltration* of $(\mathcal{F}_t, t \geq 0)$ if $\mathcal{G}_t \subseteq \mathcal{F}_t$ for each $t \geq 0$.

Now let $X = (X(t), t \geq 0)$ be a stochastic process defined on a filtered probability space $(\Omega, \mathcal{F}, P)$. We say that it is *adapted* to the filtration (or $\mathcal{F}_t$-*adapted*) if

$$X(t) \text{ is } \mathcal{F}_t\text{-measurable for each } t \geq 0.$$

Any process X is adapted to its own filtration $\mathcal{F}_t^X = \sigma\{X(s); 0 \leq s \leq t\}$ and this is usually called the *natural filtration*.

Clearly, if X is adapted we have

$$\mathbb{E}(X(s)|\mathcal{F}_s) = X(s) \quad \text{a.s.}$$

The intuitive idea behind an adapted process is that $\mathcal{F}_t$ should contain all the information needed to predict the behaviour of X up to and including time t.

Exercise 2.1.1 If X is $\mathcal{F}_t$-adapted show that, for all $t \geq 0$, $\mathcal{F}_t^X \subseteq \mathcal{F}_t$.

Exercise 2.1.2 Let X be a Lévy process, which we will take to be adapted to its natural filtration. Show that for any $f \in B_b(\mathbb{R}^d)$, $0 \leq s < t < \infty$,

$$\mathbb{E}(f(X(t))|\mathcal{F}_s) = \int_{\mathbb{R}^d} f(X(s) + y) p_{t-s}(dy),$$

where p_t is the law of $X(t)$ for each $t \geq 0$. (Hint: Use Lemma 1.1.8.)

Hence deduce that any Lévy process is a *Markov process*, i.e.

$$\mathbb{E}\big(f(X(t))\big|\mathcal{F}_s^X\big) = \mathbb{E}(f(X(t))|X(s)) \quad \text{a.s.}$$

The theme of this example will be developed considerably in Chapter 3.

Let X and Y be $\mathcal{F}_t$-adapted processes and let $\alpha, \beta \in \mathbb{R}$; then it is a simple consequence of measurability that the following are also adapted processes:

- $\alpha X + \beta Y = (\alpha X(t) + \beta Y(t), t \geq 0)$;
- $XY = (X(t)Y(t), t \geq 0)$;
- $f(X) = (f(X(t)), t \geq 0)$ where f is a Borel measurable function on $\mathbb{R}^d$;
- $\lim_{n \to \infty} X_n = (\lim_{n \to \infty} X_n(t), t \geq 0)$, where $(X_n, n \in \mathbb{N})$ is a sequence of adapted processes wherein $X_n(t)$ converges pointwise almost surely for each $t \geq 0$.

When we deal with a filtration $(\mathcal{F}_t, t \geq 0)$ we will frequently want to compute conditional expectations $\mathbb{E}(\cdot|\mathcal{F}_s)$ for some $s \geq 0$, and we will often find it convenient to use the more compact notation $\mathbb{E}_s(\cdot)$ for this.

It is convenient to require some further conditions on a filtration, and we refer to the following pair of conditions as the *usual hypotheses*. These are precisely:

(1) (completeness) $\mathcal{F}_0$ contains all sets of P-measure zero (see Section 1.1);
(2) (right continuity) $\mathcal{F}_t = \mathcal{F}_{t+}$, where $\mathcal{F}_{t+} = \bigcap_{\epsilon>0} \mathcal{F}_{t+\epsilon}$.

Given a filtration $(\mathcal{F}_t, t \geq 0)$ we can always enlarge it to satisfy the completeness property (1) by the following trick. Let $\mathcal{N}$ denote the collection of all sets of P-measure zero in $\mathcal{F}$ and define $\mathcal{G}_t = \mathcal{F}_t \vee \mathcal{N}$ for each $t \geq 0$; then $(\mathcal{G}_t, t \geq 0)$ is another filtration of $\mathcal{F}$, which we call the *augmented filtration*. The following then hold:

- any $\mathcal{F}_t$-adapted stochastic process is $\mathcal{G}_t$-adapted;
- for any integrable random variable Y defined on $(\Omega, \mathcal{F}, P)$, we have $\mathbb{E}(Y|\mathcal{G}_t) = \mathbb{E}(Y|\mathcal{F}_t)$ (a.s.) for each $t \geq 0$.

If X is a stochastic process with natural filtration $\mathcal{F}^X$ then we denote the augmented filtration as $\mathcal{G}^X$ and call it the *augmented natural filtration*.

The right-continuity property (2) is more problematic than (1) and needs to be established on a case by case basis. In the next section we will show that it always holds for the augmented natural filtration of a Lévy process, but we will need to employ martingale techniques.

2.1.2 Martingales and Lévy processes

Let X be an adapted process defined on a filtered probability space that also satisfies the integrability requirement $\mathbb{E}(|X(t)|) < \infty$ for all $t \geq 0$. We say that it is a *martingale* if, for all $0 \leq s < t < \infty$,

$$\mathbb{E}(X(t)|\mathcal{F}_s) = X(s) \quad \text{a.s.}$$

Note that if X is a martingale then the map $t \to \mathbb{E}(X(t))$ is constant.

We will find the martingale described in the following proposition to be of great value later.

Proposition 2.1.3 *If X is a Lévy process with Lévy symbol η, then, for each $u \in \mathbb{R}^d$, $M_u = (M_u(t), t \geq 0)$ is a complex martingale with respect to $\mathcal{F}^X$, where each*

$$M_u(t) = \exp\big[i(u, X(t)) - t\eta(u)\big].$$

2.1 Martingales

Proof $\mathbb{E}(|M_u(t)|) = \exp[-t\eta(u)] < \infty$ for each $t \geq 0$.

For each $0 \leq s \leq t$, write $M_u(t) = M_u(s)\exp[i(u, X(t) - X(s)) - (t-s)\eta(u)]$; then by (L2) and Theorem 1.3.3

$$\mathbb{E}(M_u(t)|\mathcal{F}_s^X) = M_u(s)\,\mathbb{E}\big(\exp[i(u, X(t-s))]\big) \cdot \exp\big[-(t-s)\eta(u)\big]$$
$$= M_u(s)$$

as required. □

Exercise 2.1.4 Show that the following processes, whose values at each $t \geq 0$ are given below, are all martingales.

(1) $C(t) = \sigma B(t)$, where $B(t)$ is a standard Brownian motion and σ is an $r \times d$ matrix.
(2) $|C(t)|^2 - \text{tr}(A)\,t$, where $A = \sigma^T\sigma$.
(3) $\exp[(u, C(t)) - \frac{1}{2}(u, Au)]$ where $u \in \mathbb{R}^d$.
(4) $\tilde{N}(t)$ where $\tilde{N}$ is a compensated Poisson process with intensity λ (see Subsection 1.3.1).
(5) $\tilde{N}(t)^2 - \lambda t$.
(6) $(\mathbb{E}(Y|\mathcal{F}_t), t \geq 0)$ where Y is an arbitrary random variable in a filtered probability space for which $\mathbb{E}(|Y|) < \infty$.

Martingales that are of the form (6) above are called *closed*. Note that in (1) to (5) the martingales have mean zero. In general, martingales with this latter property are said to be *centred*. A martingale $M = (M(t), t \geq 0)$ is said to be L^2 (or *square-integrable*) if $\mathbb{E}(|M(t)|^2) < \infty$ for each $t \geq 0$ and is *continuous* if it has almost surely continuous sample paths.

One useful generalisation of the martingale concept is the following.

An adapted process X for which $\mathbb{E}(|X(t)|) < \infty$ for all $t \geq 0$ is a *submartingale* if, for all $0 \leq s < t < \infty$, $1 \leq i \leq d$,

$$\mathbb{E}(X_i(t)|\mathcal{F}_s) \geq X_i(s) \quad \text{a.s.}$$

$E[f(x_0)|\mathcal{F}_s] \geq f(E[X_t|\mathcal{F}_s]) = f(X_s)$

We call X a *supermartingale* if $-X$ is a submartingale.

By a straightforward application of the conditional form of Jensen's inequality we see that if X is a real-valued martingale and if $f : \mathbb{R} \to \mathbb{R}$ is convex with $\mathbb{E}(|f(X(t))|) < \infty$ for all $t \geq 0$ then $f(X)$ is a submartingale. In particular, if each $X(t) \geq 0$ (a.s.) then $(X(t)^p, t \geq 0)$ is a submartingale whenever $1 < p < \infty$ and $\mathbb{E}(|X(t)|^p) < \infty$ for all $t \geq 0$.

A vital estimate for much of our future work is the following:

Theorem 2.1.5 (Doob's martingale inequality) *If $(X(t), t \geq 0)$ is a positive submartingale then, for any $p > 1$,*

$$\mathbb{E}\left(\sup_{0 \leq s \leq t} X(s)^p\right) \leq q^p \, \mathbb{E}(X(t)^p),$$

where $1/p + 1/q = 1$.

See Williams [304], p. A143, for a nice proof in the discrete-time case and Dellacherie and Meyer [78], p. 18, or Revuz and Yor [260], Section 2.1, for the continuous case. Note that in the case $p = 2$ this inequality also holds for vector-valued martingales, and we will use this extensively below. More precisely, let $X = (X(t), t \geq 0)$ be a martingale taking values in $\mathbb{R}^d$. Then the component $(X_i(t)^2, t \geq 0)$ is a real-valued submartingale for each $1 \leq i \leq d$ and so, by Theorem 2.1.5, we have for each $t \geq 0$

$$\mathbb{E}\left(\sup_{0 \leq s \leq t} |X(s)|^2\right) \leq \sum_{i=1}^{d} \mathbb{E}\left(\sup_{0 \leq s \leq t} X_i(s)^2\right) \leq \sum_{i=1}^{d} q^2 \mathbb{E}(X_i(t)^2)$$
$$= q^2 \mathbb{E}(|X(t)|^2).$$

We will also need the following technical result.

Theorem 2.1.6 *Let $M = (M(t), t \geq 0)$ be a submartingale.*

(1) *For any countable dense subset D of $\mathbb{R}^+$, the following left and right limits exist and are almost surely finite for each $t > 0$:*

$$M(t-) = \lim_{s \in D, s \uparrow t} M(s); \qquad M(t+) = \lim_{s \in D, s \downarrow t} M(s).$$

(2) *If the filtration $(\mathcal{F}_t, t \geq 0)$ satisfies the usual hypotheses and if the map $t \to \mathbb{E}(M(t))$ is right-continuous, then M has a càdlàg modification.*

In fact (2) is a consequence of (1), and these results are both proved in Dellacherie and Meyer [78], pp. 73–6, and in Revuz and Yor [260], pp. 63–5.

The proofs of the next two results are based closely on the accounts of Bretagnolle [61] and of Protter [255], Chapter 1, Section 4.

Theorem 2.1.7 *Every Lévy process has a càdlàg modification that is itself a Lévy process.*

Proof Let X be a Lévy process that is adapted to its own augmented natural filtration. For each $u \in \mathbb{R}^d$ we recall the martingales M_u of Proposition 2.1.3. Let D be a countable, dense subset of $\mathbb{R}^+$. It follows immediately from

2.1 Martingales

Theorem 2.1.6(1) that at each $t > 0$ the left and right limits $M_u(t-)$ and $M_u(t+)$ exist along D almost surely. Now for each $u \in \mathbb{R}^d$, let $\mathcal{O}_u$ be that subset of Ω for which these limits fail to exist; then $\mathcal{O} = \bigcup_{u \in \mathbb{Q}^d} \mathcal{O}_u$ is also a set of P-measure zero.

Fix $\omega \in \mathcal{O}^c$ and for each $t \geq 0$ let $(s_n, n \in \mathbb{N})$ be a sequence in D increasing to t. Let $x^1(t)(\omega)$ and $x^2(t)(\omega)$ be two distinct accumulation points of the set $\{X(s_n)(\omega), n \in \mathbb{N}\}$, corresponding to limits along subsequences $(s_{n_i}, n_i \in \mathbb{N})$ and $(s_{n_j}, n_j \in \mathbb{N})$, respectively. We deduce from the existence of $M_u(t-)$ that $\lim_{s_n \uparrow t} e^{i(u, X(s_n)(\omega))}$ exists and hence that $x^1(t)(\omega)$ and $x^2(t)(\omega)$ are both finite.

Now choose $u \in \mathbb{Q}^d$ such that $(u, x_t^1(\omega) - x_t^2(\omega)) \neq 2n\pi$ for any $n \in \mathbb{Z}$. By continuity,

$$\lim_{s_{n_i} \uparrow t} e^{i(u, X(s_{n_i})(\omega))} = e^{i(u, x_t^1(\omega))} \text{ and } \lim_{s_{n_j} \uparrow t} e^{i(u, X(s_{n_j})(\omega))} = e^{i(u, x_t^2(\omega))},$$

and so we obtain a contradiction. Hence X always has a unique left limit along D, at every $t > 0$ on $\mathcal{O}^c$. A similar argument shows that it always has such right limits on $\mathcal{O}^c$. It then follows from elementary real analysis that the process Y is càdlàg, where for each $t \geq 0$

$$Y(t)(\omega) = \begin{cases} \lim_{s \in D, s \downarrow t} X(t)(\omega) & \text{if } \omega \in \mathcal{O}^c, \\ 0 & \text{if } \omega \in \mathcal{O}. \end{cases}$$

To see that Y is a modification of X, we use the dominated convergence theorem for each $t \geq 0$ to obtain

$$\mathbb{E}(e^{i(u, Y(t) - X(t))}) = \lim_{s \in D, s \downarrow t} \mathbb{E}(e^{i(u, X(s) - X(t))}) = 1,$$

by (L2) and (L3) in Section 1.3 and Lemma 1.3.2.

Hence $P(\{\omega, Y(t)(\omega) = X(t)(\omega)\}) = 1$ as required. That Y is a Lévy process now follows immediately from Lemma 1.4.7. $\square$

Example 2.1.8 It follows that the canonical Lévy process discussed in Subsection 1.4.1 has a càdlàg version that lives on the space of all càdlàg paths starting at zero. A classic result, originally due to Norbert Wiener (see [301, 302]), modifies the path-space construction to show that there is a Brownian motion that lives on the space of continuous paths starting at zero. We will see below that a general Lévy process has continuous sample paths if and only if it is Gaussian.

We can now complete our discussion of the usual hypotheses for Lévy processes.

Theorem 2.1.9 *If X is a Lévy process with càdlàg paths, then its augmented natural filtration is right-continuous.*

Proof For convenience, we will write $\mathcal{G}^X = \mathcal{G}$. First note that it is sufficient to prove that

$$\mathcal{G}_t = \bigcap_{n \in \mathbb{N}} \mathcal{G}_{t+1/n}$$

for each $t \geq 0$, so all limits as $w \downarrow t$ can be replaced by limits as $n \to \infty$. Fix $t, s_1, \ldots, s_m \geq 0$ and $u_1, \ldots, u_m \in \mathbb{R}^d$. Our first task is to establish that

$$\mathbb{E}\left(\exp\left[i \sum_{j=1}^m (u_j, X(s_j))\right] \Big| \mathcal{G}_t\right) = \mathbb{E}\left(\exp\left[i \sum_{j=1}^m (u_j, X(s_j))\right] \Big| \mathcal{G}_{t+}\right). \tag{2.1}$$

Now, (2.1) is clearly satisfied when $\max_{1 \leq j \leq m} s_j \leq t$ and the general case follows easily when it is established for $\min_{1 \leq j \leq m} s_j > t$, as follows. We take $m = 2$ for simplicity and consider $s_2 > s_1 > t$. Our strategy makes repeated use of the martingales described in Proposition 2.1.3. We begin by applying Proposition 1.1.5 to obtain

$$\mathbb{E}\big(\exp\{i[(u_1, X(s_1)) + (u_2, X(s_2))]\} \big| \mathcal{G}_{t+}\big)$$
$$= \lim_{w \downarrow t} \mathbb{E}\big(\exp\{i[(u_1, X(s_1)) + (u_2, X(s_2))]\} \big| \mathcal{G}_w\big)$$
$$= \exp\big[s_2 \eta(u_2)\big] \lim_{w \downarrow t} \mathbb{E}\big(\exp\big[i(u_1, X(s_1))\big] M_{u_2}(s_2) \big| \mathcal{G}_w\big)$$
$$= \exp\big[s_2 \eta(u_2)\big] \lim_{w \downarrow t} \mathbb{E}\big(\exp\big[i(u_1, X(s_1))\big] M_{u_2}(s_1) \big| \mathcal{G}_w\big)$$
$$= \exp\big[(s_2 - s_1)\eta(u_2)\big] \lim_{w \downarrow t} \mathbb{E}\big(\exp\big[i(u_1 + u_2, X(s_1))\big] \big| \mathcal{G}_w\big)$$
$$= \exp\big[(s_2 - s_1)\eta(u_2) + s_1 \eta(u_1 + u_2)\big] \lim_{w \downarrow t} \mathbb{E}(M_{u_1+u_2}(s_1) | \mathcal{G}_w)$$
$$= \exp\big[(s_2 - s_1)\eta(u_2) + s_1 \eta(u_1 + u_2)\big] \lim_{w \downarrow t} M_{u_1+u_2}(w)$$
$$= \lim_{w \downarrow t} \exp\big[i(u_1 + u_2, X(w))\big]$$
$$\quad \times \exp\big[(s_2 - s_1)\eta(u_2) + (s_1 - w)\eta(u_1 + u_2)\big]$$
$$= \exp\big[i(u_1 + u_2, X(t))\big] \exp\big[(s_2 - s_1)\eta(u_2) + (s_1 - t)\eta(u_1 + u_2)\big]$$
$$= \mathbb{E}\big(\exp\{i[(u_1, X(s_1)) + (u_2, X(s_2))]\} \big| \mathcal{G}_t\big),$$

where, in the penultimate step, we have used the fact that X is càdlàg.

2.1 Martingales

$P(A|\mathcal{G}) = E(\mathbb{1}_A | \mathcal{G})$

Now let $X^{(m)} = (X(s_1), \ldots, X(s_m))$; then by the unique correspondence between characteristic functions and probability measures we deduce that

$$P(X^{(m)}|\mathcal{G}_{t+}) = P(X^{(m)}|\mathcal{G}_t) \quad \text{a.s.}$$

$P(X^{(m)}$

and hence, by equation (1.1), we have

$$\mathbb{E}\big(g(X(s_1), \ldots, X(s_m)))|\mathcal{G}_{t+}\big) = \mathbb{E}\big(g(X(s_1), \ldots, X(s_m)))|\mathcal{G}_t\big)$$

for all $g : \mathbb{R}^{dm} \to \mathbb{R}$ with $\mathbb{E}(|g(X(s_1), \ldots, X(s_m))|) < \infty$. In particular, if we vary t, m and $s_1, \ldots, s_m$ we can deduce that

$$P(A|\mathcal{G}_{t+}) = P(A|\mathcal{G}_t)$$

for all $A \in \mathcal{G}_\infty$. Now, suppose that $A \in \mathcal{G}_{t+}$; then we have

$$\chi_A = P(A|\mathcal{G}_{t+}) = P(A|\mathcal{G}_t) = \mathbb{E}(\chi_A|\mathcal{G}_t) \quad \text{a.s.}$$

Hence, since $\mathcal{G}_t$ is augmented, we deduce that $\mathcal{G}_{t+} \subseteq \mathcal{G}_t$ and the result follows. □

Some readers may feel that using the augmented natural filtration is an unnecessary restriction. After all, nature may present us with a practical situation wherein the filtration is much larger. To deal with such circumstances we will, for the remainder of this book, always make the following assumptions:

- $(\Omega, \mathcal{F}, P)$ *is a fixed probability space equipped with a filtration* $(\mathcal{F}_t, t \geq 0)$ *that satisfies the usual hypotheses;*
- *every Lévy process* $X = (X(t), t \geq 0)$ *is assumed to be* $\mathcal{F}_t$*-adapted and to have càdlàg sample paths;*
- $X(t) - X(s)$ *is independent of* $\mathcal{F}_s$ *for all* $0 \leq s < t < \infty$.

Theorems 2.1.7 and 2.1.9 confirm that these are quite reasonable assumptions.

2.1.3 Martingale spaces

We can define an equivalence relation on the set of all martingales on a probability space by the prescription that $M_1 \sim M_2$ if and only if M_1 is a modification of M_2. Note that by Theorem 2.1.6 each equivalence class contains a càdlàg member.

Let $\mathcal{M}$ be the linear space of equivalence classes of $\mathcal{F}_t$-adapted L^2-martingales and define a (separating) family of seminorms $(\|\cdot\|_t, t \geq 0)$ by the prescription

$$||M||_t = \mathbb{E}(|X(t)|^2)^{1/2};$$

then $\mathcal{M}$ becomes a locally convex space with the topology induced by these seminorms (see Chapter 1 of Rudin [267]). We call $\mathcal{M}$ a *martingale space*.

For those unfamiliar with these notions, the key point is that a sequence $(M_n, n \in \mathbb{N})$ in $\mathcal{M}$ converges to $N \in \mathcal{M}$ if $||M_n - N||_t \to 0$ as $n \to \infty$ for all $t \geq 0$.

Lemma 2.1.10 $\mathcal{M}$ *is complete.*

Proof By the completeness of L^2, any Cauchy sequence $(M_n, n \in \mathbb{N})$ in $\mathcal{M}$ has a limit N that is an $\mathcal{F}_t$-adapted process with $\mathbb{E}(|N(t)|^2) < \infty$ for all $t \geq 0$. We are done if we can show that N is a martingale. We use the facts that each M_n is a martingale and that the conditional expectation $\mathbb{E}_s = \mathbb{E}(\cdot|\mathcal{F}_s)$ is an L^2-projection (and therefore a contraction). Hence, for each $0 \leq s < t < \infty$,

$$\begin{aligned}\mathbb{E}\big(|N(s) - \mathbb{E}_s(N(t))|^2\big) &= \mathbb{E}\big(|N(s) - M_n(s) + M_n(s) - \mathbb{E}_s(N(t))|^2\big) \\ &\leq 2||N(s) - M_n(s)||^2 + 2||\mathbb{E}_s(M_n(t) - N(t))||^2 \\ &\leq 2||N(s) - M_n(s)||^2 + 2||M_n(t) - N(t)||^2 \\ &\to 0 \text{ as } n \to \infty,\end{aligned}$$

where $||\cdot||$ without a subscript is the usual L^2-norm; the required result follows. $\square$

Exercise 2.1.11 Define another family of seminorms on $\mathcal{M}$ by the prescription

$$||M||'_t = \left(\sup_{0 \leq s \leq t} \mathbb{E}(|M(s)|^2)\right)^{1/2}$$

for each $M \in \mathcal{M}$, $t \geq 0$. Show that $(||\cdot||_t, t \geq 0)$ and $(||\cdot||'_t, t \geq 0)$ induce equivalent topologies on $\mathcal{M}$. (Hint: Use Doob's inequality.)

In what follows, when we speak of a process $M \in \mathcal{M}$ we will always understand M to be the càdlàg member of its equivalence class.

2.2 Stopping times

A *stopping time* is a random variable $T : \Omega \to [0, \infty]$ for which the event $(T \leq t) \in \mathcal{F}_t$ for each $t \geq 0$.

2.2 Stopping times

Any ordinary deterministic time is clearly a stopping time. A more interesting example, which has many important applications, is the *first hitting time* T_A *of a process to a set*. This is defined as follows. Let X be an $\mathcal{F}_t$-adapted càdlàg process and $A \in \mathcal{B}(\mathbb{R}^d)$; then

$$T_A = \inf\{t \geq 0; X(t) \in A\},$$

where we adopt the convention that $\inf\{\emptyset\} = \infty$. It is fairly straightforward to prove that T_A really is a stopping time if A is open or closed (see e.g. Protter [255], p. 5). The general case is more problematic; see e.g. Rogers and Williams [261], Chapter II, section 76, and references therein.

If X is an adapted process and T is a stopping time (with respect to the same filtration) then the *stopped random variable* $X(T)$ is defined by

$$X(T)(\omega) = X(T(\omega))(\omega)$$

(with the convention that $X(\infty)(\omega) = \lim_{t \to \infty} X(t)(\omega)$ if the limit exists (a.s.) and $X(\infty)(\omega) = 0$ otherwise) and the *stopped σ-algebra* $\mathcal{F}_T$ by

$$\mathcal{F}_T = \{A \in \mathcal{F}; A \cap \{T \leq t\} \in \mathcal{F}_t, \forall t \geq 0\}.$$

If X is càdlàg, it can be shown that $X(T)$ is $\mathcal{F}_T$-measurable (see e.g. Kunita [182], p. 8).

A key application of these concepts is in providing the following 'random time' version of the martingale notion.

Theorem 2.2.1 (Doob's optional stopping theorem) *If X is a càdlàg martingale and S and T are bounded stopping times for which $S \leq T$ (a.s.), then $X(S)$ and $X(T)$ are both integrable, with*

$$\mathbb{E}(X(T)|\mathcal{F}_S) = X(S) \quad \text{a.s.}$$

See Williams [304], p. 100, for a proof in the discrete case and Dellacherie and Meyer [78], pp. 8–9, or Revuz and Yor [260], Section 2.3, for the continuous case. An immediate corollary is that

$$\mathbb{E}(X(T)) = \mathbb{E}(X(0))$$

for each bounded stopping time T.

Exercise 2.2.2 If S and T are stopping times and $\alpha \geq 1$, show that $S + T$, αT, $S \wedge T$ and $S \vee T$ are also stopping times.

If T is an unbounded stopping time and one wants to employ Theorem 2.2.1, a useful trick is to replace T by the bounded stopping times $T \wedge n$ (where $n \in \mathbb{N}$) and then take the limit as $n \to \infty$ to obtain the required result. This procedure is sometimes called *localisation*.

Another useful generalisation of the martingale concept that we will use extensively is the *local martingale*. This is an adapted process $M = (M(t), t \geq 0)$ for which there exists a sequence of stopping times $\tau_1 \leq \cdots \leq \tau_n \to \infty$ (a.s.) such that each of the processes $(M(t \wedge \tau_n), t \geq 0)$ is a martingale. Any martingale is clearly a local martingale. For an example of a local martingale that is not a martingale see Protter [255], p. 33.

2.2.1 The Doob–Meyer decomposition

Do not worry too much about the technical details in this section unless, of course, they appeal to you. The main reason for including this material is to introduce the Meyer angle bracket $\langle \cdot \rangle$, and you should concentrate on getting a sound intuition about how this works.

In Exercise 2.1.4 we saw that if B is a one-dimensional standard Brownian motion and $\tilde{N}$ is a compensated Poisson process of intensity λ then B and $\tilde{N}$ are both martingales and, furthermore, so are the processes defined by $B(t)^2 - t$ and $\tilde{N}^2 - \lambda t$, for each $t \geq 0$. It is natural to ask whether this behaviour extends to more general martingales. Before we can answer this question, we need some further definitions. We take $d = 1$ throughout this section.

Let $\mathcal{I}$ be some index set and $X = \{X_i, i \in \mathcal{I}\}$ be a family of random variables. We say X is *uniformly integrable* if

$$\lim_{n \to \infty} \sup_{i \in \mathcal{I}} \mathbb{E}\big(|X_i| \chi_{\{|X_i| > n\}}\big) = 0.$$

A sufficient condition for this to hold is that $\mathbb{E}(\sup_{i \in \mathcal{I}} |X_i|) < \infty$; see e.g. Klebaner [170], pp. 171–2, or Williams [304], p. 128. Let $M = (M(t), t \geq 0)$ be a closed martingale, so that $M(t) = \mathbb{E}(X|\mathcal{F}_t)$, for each $t \geq 0$, for some random variable X where $\mathbb{E}(|X|) \leq \infty$; then it is easy to see that M is uniformly integrable. Conversely, any uniformly integrable martingale is closed; see e.g. Dellacherie and Meyer [78], p. 79.

A process $X = (X(t), t \geq 0)$ is in the <u>Dirichlet class</u> or <u>class D</u> if $\{X(\tau), \tau \in \mathcal{T}\}$ is uniformly integrable, where $\mathcal{T}$ is the family of all finite stopping times on our filtered probability space.

The process X is *integrable* if $\mathbb{E}(|X(t)|) < \infty$, for each $t \geq 0$.

The process X is <u>predictable</u> if the mapping $X : \mathbb{R}^+ \times \Omega \to \mathbb{R}$ given by $X(t, \omega) = X(t)(\omega)$ is measurable with respect to the smallest σ-algebra generated by all adapted left-continuous mappings from $\mathbb{R}^+ \times \Omega \to \mathbb{R}$. The

2.2 Stopping times

idea of predictability is very important in the theory of stochastic integration and will be developed more extensively in Chapter 4.

Our required generalisation is then the following result.

Theorem 2.2.3 (Doob–Meyer 1) *Let Y be a submartingale of class D; then there exists a unique predictable, integrable, increasing process $A = (A(t), t \geq 0)$ with $A(0) = 0$ (a.s.) such that the process given by $Y(t) - Y(0) - A(t)$ for each $t \geq 0$ is a uniformly integrable martingale.* ↔ $Y(t) - Y(0) = M(t) + A(t)$

The discrete-time version of this result is due to Doob [83] and is rather easy to prove (see e.g. Williams [304], p. 121). Its extension to continuous time is much harder and was carried out by Meyer in [222], [223]. For more recent accounts see e.g. Karatzas and Shreve [167], pp. 24–5, or Rogers and Williams [262], Chapter 6, Section 6.

In the case where each $Y(t) = M(t)^2$ for a square-integrable martingale M, it is common to use the 'inner-product' notation $\langle M, M \rangle(t) = A(t)$ for each $t \geq 0$ and we call $\langle M, M \rangle$ *Meyer's angle-bracket process*. This notation was originally introduced by Motoo and Watanabe [232]. The logic behind it is as follows.

Let $M, N \in \mathcal{M}$; then we may use the polarisation identity to define

$$\langle M, N \rangle(t) = \tfrac{1}{4}\big[\langle M+N, M+N\rangle(t) - \langle M-N, M-N\rangle(t)\big].$$

Exercise 2.2.4 Show that

$$M(t)N(t) - \langle M, N \rangle(t) \text{ is a martingale.}$$

Exercise 2.2.5 Deduce that, for each $t \geq 0$:
(1) $\langle M, N \rangle(t) = \langle N, M \rangle(t)$;
(2) $\langle \alpha M_1 + \beta M_2, N \rangle(t) = \alpha \langle M_1, N \rangle(t) + \beta \langle M_2, N \rangle(t)$ for each $M_1, M_2 \in \mathcal{M}$ and $\alpha, \beta \in \mathbb{R}$;
(3) $\mathbb{E}(\langle M, N \rangle(t)^2) \leq \mathbb{E}(\langle M, M \rangle(t))\mathbb{E}(\langle N, N \rangle(t))$, the equality holding if and only if $M(t) = cN(t)$ (a.s.) for some $c \in \mathbb{R}$. (Hint: Mimic the proof of the usual Cauchy–Schwarz inequality.)

The Doob–Meyer theorem has been considerably generalised and, although we will not have need of it, we quote the following more powerful result, a proof of which can be found in Protter [255] p. 94.

Theorem 2.2.6 (Doob–Meyer 2) *Any submartingale Y has a decomposition $Y(t) = Y(0) + M(t) + A(t)$, where A is an increasing, adapted process and M is a local martingale.*

Note that the processes A that appear in Theorem 2.2.6 may not be predictable in general; however, there is a unique one that is 'locally natural', in a sense made precise in Protter [255], p. 89.

We close this section by quoting an important theorem – Lévy's martingale characterisation of Brownian motion – which will play an important role below.

Theorem 2.2.7 *Let $X = (X(t), t \geq 0)$ be an adapted process with continuous sample paths having mean 0 and covariance $\mathbb{E}(X_i(t)X_j(s)) = a_{ij}(s \wedge t)$ for $1 \leq i, j \leq d$, $s, t \geq 0$; where $a = (a_{ij})$ is a positive definite symmetric $d \times d$ matrix. Then the following are equivalent:*

(1) *X is a Brownian motion with covariance a;*
(2) *X is a martingale with $\langle X_i, X_j \rangle(t) = a_{ij}t$ for each $1 \leq i, j \leq d$, $t \geq 0$;*
(3) *$\left(\exp[i(u, X(t)) + \frac{t}{2}(u, au)], t \geq 0 \right)$ is a martingale for each $u \in \mathbb{R}^d$.*

We postpone a proof until Chapter 4, where we can utilise Itô's formula for Brownian motion. In fact it is the following consequence that we will need in this chapter.

Corollary 2.2.8 *If X is a Lévy process satisfying the hypotheses of Theorem 2.2.7 then X is a Brownian motion if and only if*

$$\mathbb{E}(e^{i(u, X(t))}) = e^{-t(u, au)/2}$$

for each $t \geq 0$, $u \in \mathbb{R}^d$.

Proof The result is an easy consequence of Proposition 2.1.3 and Theorem 2.2.7(3). □

2.2.2 Stopping times and Lévy processes

We now give three applications of stopping times to Lévy processes. We begin by again considering the Lévy subordinator (see Subsection 1.3.2).

Theorem 2.2.9 *Let $B = (B(t), t \geq 0)$ be a one-dimensional standard Brownian motion and for each $t \geq 0$ define*

$$T(t) = \inf \left\{ s > 0; B(s) = \frac{t}{\sqrt{2}} \right\};$$

then $T = (T(t), t \geq 0)$ is the Lévy subordinator.

Proof (cf. Rogers and Williams [261] p. 18). Clearly each $T(t)$ is a stopping time. By Exercise 2.1.4(3), the process given for each $\theta \in \mathbb{R}$ by $M_\theta(t) = \exp[\theta B(t) - \frac{1}{2}\theta^2 t]$ is a continuous martingale with respect to the

2.2 Stopping times

augmented natural filtration for Brownian motion. By Theorem 2.2.1, for each $t \geq 0, n \in \mathbb{N}, \theta \geq 0$, we have

$$1 = \mathbb{E}\big(\exp\big[\theta B(T(t) \wedge n) - \tfrac{1}{2}\theta^2(T(t) \wedge n)\big]\big).$$

Now for each $n \in \mathbb{N}, t \geq 0$, let $A_{n,t} = \{\omega \in \Omega; T(t)(\omega) \leq n\}$; then

$$\mathbb{E}\big(\exp\big[\theta B(T(t) \wedge n) - \tfrac{1}{2}\theta^2(T(t) \wedge n)\big]\big)$$
$$= \mathbb{E}\big(\exp\big\{[\theta B(T(t)) - \tfrac{1}{2}\theta^2 T(t)]\chi_{A_{n,t}}\big\}\big)$$
$$+ \exp\big(-\tfrac{1}{2}\theta^2 n\big)\mathbb{E}\big(\exp[\theta B(n)]\chi_{A^c_{n,t}}\big).$$

But, for each $\omega \in \Omega, T(t)(\omega) > n \Rightarrow B(n) < t/\sqrt{2}$; hence

$$\exp\big(-\tfrac{1}{2}\theta^2 n\big) \mathbb{E}\big(e^{\theta B(n)}\chi_{A^c_{n,t}}\big)$$
$$< \exp\big[-\tfrac{1}{2}\theta^2 n + (t\theta/\sqrt{2})\big] \to 0 \quad \text{as} \quad n \to \infty.$$

By the monotone convergence theorem,

$$1 = \mathbb{E}\big(\exp\big[\theta B(T(t)) - \tfrac{1}{2}\theta^2 T(t)\big]\big) = \exp(\theta t/\sqrt{2})\, \mathbb{E}\big(\exp\big[-\tfrac{1}{2}\theta^2 T(t)\big]\big).$$

On substituting $\theta = \sqrt{2u}$ we obtain

$$\mathbb{E}\big(\exp[-uT(t)]\big) = \exp(-t\sqrt{u}),$$

as required. □

Exercise 2.2.10 Generalise the proof given above to find the characteristic function for the inverse Gaussian subordinator, as defined in Example 1.3.21.

If X is an $\mathcal{F}_t$-adapted process and T is a stopping time then we may define a new process $X_T = (X_T(t), t \geq 0)$ by the procedure

$$X_T(t) = X(T+t) - X(T)$$

for each $t \geq 0$. The following result is called the strong Markov property for Lévy processes. For the proof, we again follow Protter [255], p. 23, and Bretagnolle [61].

Theorem 2.2.11 (Strong Markov property) *If X is a Lévy process and T is a stopping time, then, on $(T < \infty)$:*

(1) *X_T is a Lévy process that is independent of $\mathcal{F}_T$;*
(2) *for each $t \geq 0$, $X_T(t)$ has the same law as $X(t)$;*
(3) *X_T has càdlàg paths and is $\mathcal{F}_{T+t}$-adapted.*

Proof We assume, for simplicity, that T is a bounded stopping time. Let $A \in \mathcal{F}_T$ and for each $n \in \mathbb{N}$, $1 \leq j \leq n$, let $u_j \in \mathbb{R}^d$, $t_j \in \mathbb{R}^+$. Recall from Proposition 2.1.3 the martingales given by $M_{u_j}(t) = e^{i(u_j, X(t)) - t\eta(u_j)}$ for each $t \geq 0$. Now we have

$$\mathbb{E}\left(\chi_A \exp\left[i \sum_{j=1}^n (u_j, X(T+t_j) - X(T+t_{j-1}))\right]\right)$$

$$= \mathbb{E}\left(\chi_A \prod_{j=1}^n \frac{M_{u_j}(T+t_j)}{M_{u_j}(T+t_{j-1})} \prod_{j=1}^n \phi_{t_j - t_{j-1}}(u_j)\right),$$

where we use the notation $\phi_t(u) = \mathbb{E}(e^{i(u, X(t))})$, for each $t \geq 0$, $u \in \mathbb{R}^d$.

Hence by conditioning and Theorem 2.2.1, we find that for each $1 \leq j \leq n$, $0 < a < b < \infty$, we have

$$\mathbb{E}\left(\chi_A \frac{M_{u_j}(T+b)}{M_{u_j}(T+a)}\right) = \mathbb{E}\left(\chi_A \frac{1}{M_{u_j}(T+a)} \mathbb{E}(M_{u_j}(T+b)|\mathcal{F}_{T+a})\right)$$

$$= \mathbb{E}(\chi_A) = P(A).$$

Repeating this argument n times yields

$$\mathbb{E}\left(\chi_A \exp\left[i \sum_{j=1}^n (u_j, X(T+t_j) - X(T+t_{j-1}))\right]\right)$$

$$= P(A) \prod_{j=1}^n \phi_{t_j - t_{j-1}}(u_j) \qquad (2.2)$$

Take $A = \Omega$, $n = 1$, $u_1 = u$, $t_1 = t$ in (2.2) to obtain

$$\mathbb{E}(e^{i(u, X_T(t))}) = \mathbb{E}(e^{i(u, X(t))}),$$

from which (2) follows immediately.

To verify that X_T is a Lévy process, first note that (L1) is immediate. (L2) follows from (2.2) by taking $A = \Omega$ and n arbitrary. The stochastic continuity (L3) of X_T follows directly from that of X and the stationary-increment property. We now show that X_T and $\mathcal{F}_T$ are independent. It follows from (2.2) on choosing appropriate $u_1, \ldots, u_n$ and $t_1, \ldots, t_n$ that for all $A \in \mathcal{F}_T$

$$\mathbb{E}\left(\chi_A \exp\left[i \sum_{j=1}^n (u_j, X_T(t_j))\right]\right) = \mathbb{E}\left(\exp\left[i \sum_{j=1}^n (u_j, X_T(t_j))\right]\right) P(A),$$

2.2 Stopping times

so that

$$\mathbb{E}\left(\exp\left[i\sum_{j=1}^{n}(u_j, X_T(t_j))\right]\Big|\mathcal{F}_T\right) = \mathbb{E}\left(\exp\left[i\sum_{j=1}^{n}(u_j, X_T(t_j))\right]\right),$$

and the result follows from (1.1). Part (1) is now fully proved. To verify (3), we need only observe that X_T inherits càdlàg paths from X. □

Exercise 2.2.12 Use a localisation argument to extend Theorem 2.2.11 to the case of unbounded stopping times.

Before we look at our final application of stopping times, we introduce a very important process associated to a Lévy process X. The *jump process* $\Delta X = (\Delta X(t), t \geq 0)$ is defined by

$$\Delta X(t) = X(t) - X(t-)$$

for each $t \geq 0$ ($X(t-)$ is the left limit at the point t; see Section 2.8.)

Theorem 2.2.13 *If N is a Lévy process that is increasing (a.s.) and is such that $(\Delta N(t), t \geq 0)$ takes values in $\{0, 1\}$, then N is a Poisson process.*

Proof Define a sequence of stopping times recursively by $T_0 = 0$ and $T_n = \inf\{t > T_{n-1}; (N(t) - N(T_{n-1})) \neq 0\}$ for each $n \in \mathbb{N}$. It follows from Theorem 2.2.11 that the sequence $(T_1, T_2 - T_1, \ldots, T_n - T_{n-1}, \ldots)$ is i.i.d.
By (L2) again, we have for each $s, t \geq 0$

$$P(T_1 > s + t) = P(N(s) = 0, N(t + s) - N(s) = 0)$$
$$= P(T_1 > s)P(T_1 > t).$$

From the fact that N is increasing (a.s.), it follows easily that the map $t \to P(T_1 > t)$ is decreasing and, by (L3), we find that the map $t \to P(T_1 > t)$ is continuous at $t = 0$. So the solution to the above functional equation is continuous everywhere, hence there exists $\lambda > 0$ such that $P(T_1 > t) = e^{-\lambda t}$ for each $t \geq 0$ (see e.g. Bingham et al. [46], pp. 4–6). So T_1 has an exponential distribution with parameter λ and

$$P(N(t) = 0) = P(T_1 > t) = e^{-\lambda t}$$

for each $t \geq 0$.
Now assume as an inductive hypothesis that

$$P(N(t) = n) = e^{-\lambda t}\frac{(\lambda t)^n}{n!};$$

then

$$P(N(t) = n+1) = P(T_{n+2} > t, T_{n+1} \leq t)$$
$$= P(T_{n+2} > t) - P(T_{n+1} > t).$$

But

$$T_{n+1} = T_1 + (T_2 - T_1) + \cdots + (T_{n+1} - T_n)$$

is the sum of $n+1$ i.i.d. exponential random variables and so has a gamma distribution with density

$$f_{T_{n+1}}(s) = e^{-\lambda s}\frac{\lambda^{n+1}s^n}{n!} \qquad \text{for } s > 0;$$

see Exercise 1.2.5. The required result follows on integration. □

2.3 The jumps of a Lévy process – Poisson random measures

We have already introduced the jump process $\Delta X = (\Delta X(t), t \geq 0)$ associated with a Lévy process. Clearly ΔX is an adapted process but it is not, in general, a Lévy process, as the following exercise indicates.

Exercise 2.3.1 Let N be a Poisson process and choose $0 \leq t_1 < t_2 < \infty$. Show that

$$P\big(\Delta N(t_2) - \Delta N(t_1) = 0 \big| \Delta N(t_1) = 1\big) \neq P\big(\Delta N(t_2) - \Delta N(t_1) = 0\big),$$

so that ΔN cannot have independent increments.

The following result demonstrates that ΔX is not a straightforward process to analyse.

Lemma 2.3.2 *If X is a Lévy process, then, for fixed $t > 0$, $\Delta X(t) = 0$ (a.s.).*

Proof Let $(t(n), n \in \mathbb{N})$ be a sequence in $\mathbb{R}^+$ with $t(n) \uparrow t$ as $n \to \infty$; then, since X has càdlàg paths, $\lim_{n\to\infty} X(t(n)) = X(t-)$. However, by (L3) the sequence $(X(t(n)), n \in \mathbb{N})$ converges in probability to $X(t)$ and so has a subsequence that converges almost surely to $X(t)$. The result follows by uniqueness of limits. □

Warning! Do not be tempted to assume that we also have $\Delta X(T) = 0$ (a.s.) when T is a stopping time.

2.3 The jumps of a Lévy process – Poisson random measures

Much of the analytic difficulty in manipulating Lévy processes arises from the fact that it is possible for them to have

$$\sum_{0 \leq s \leq t} |\Delta X(s)| = \infty \quad \text{a.s.}$$

and the way these difficulties are overcome exploits the fact that we always have

$$\sum_{0 \leq s \leq t} |\Delta X(s)|^2 < \infty \quad \text{a.s.}$$

We will gain more insight into these ideas as the discussion progresses.

Exercise 2.3.3 Show that $\sum_{0 \leq s \leq t} |\Delta X(s)| < \infty$ (a.s.) if X is a compound Poisson process.

Rather than exploring ΔX itself further, we will find it more profitable to count jumps of specified size. More precisely, let $0 \leq t < \infty$ and $A \in \mathcal{B}(\mathbb{R}^d - \{0\})$. Define

$$N(t, A) = \#\{0 \leq s \leq t; \Delta X(s) \in A\} = \sum_{0 \leq s \leq t} \chi_A(\Delta X(s)).$$

Note that for each $\omega \in \Omega$, $t \geq 0$, the set function $A \to N(t, A)(\omega)$ is a counting measure on $\mathcal{B}(\mathbb{R}^d - \{0\})$ and hence

$$\mathbb{E}(N(t, A)) = \int N(t, A)(\omega) dP(\omega)$$

is a Borel measure on $\mathcal{B}(\mathbb{R}^d - \{0\})$. We write $\mu(\cdot) = \mathbb{E}(N(1, \cdot))$ and call it the *intensity measure*[1] associated with X. We say that $A \in \mathcal{B}(\mathbb{R}^d - \{0\})$ is *bounded below* if $0 \notin \bar{A}$.

Lemma 2.3.4 *If A is bounded below, then $N(t, A) < \infty$ (a.s.) for all $t \geq 0$.*

Proof Define a sequence of stopping times $(T_n^A, n \in \mathbb{N})$ by $T_1^A = \inf\{t > 0; \Delta X(t) \in A\}$ and, for $n > 1$, by $T_n^A = \inf\{t > T_{n-1}^A; \Delta X(t) \in A\}$. Since X has càdlàg paths, we have $T_1^A > 0$ (a.s.) and $\lim_{n \to \infty} T_n^A = \infty$ (a.s.). Indeed, if either of these were not the case then the set of all jumps in A would have an accumulation point, and this is not possible if X is càdlàg (see the proof of Theorem 2.8.1 in Appendix 2.8). Hence, for each $t \geq 0$,

$$N(t, A) = \sum_{n \in \mathbb{N}} \chi_{\{T_n^A \leq t\}} < \infty \quad \text{a.s.}$$

□

---

[1] Readers should be aware that many authors use the term 'intensity measure' to denote the product of μ with Lebesgue measure on $\mathbb{R}^+$.

Be aware that if A fails to be bounded below then Lemma 2.3.4 may no longer hold, because of the accumulation of infinite numbers of small jumps.

Theorem 2.3.5

(1) *If A is bounded below, then $(N(t, A), t \geq 0)$ is a Poisson process with intensity $\mu(A)$.*
(2) *If $A_1, \ldots, A_m \in \mathcal{B}(\mathbb{R}^d - \{0\})$ are disjoint, then the random variables $N(t, A_1), \ldots, N(t, A_m)$ are independent.*

Proof (1) We first need to show that $(N(t, A), t \geq 0)$ is a Lévy process, as we can then deduce immediately that it is a Poisson process by Theorem 2.2.13.

(L1) is obvious. To verify (L2), note first that for $0 \leq s < t < \infty, n \in \mathbb{N}$, we have $N(t, A) - N(s, A) \geq n$ if and only if there exists $s < t_1 < \cdots < t_n \leq t$ such that
$$\Delta X(t_j) \in A \ (1 \leq j \leq n). \tag{2.3}$$
Furthermore, $\Delta X(u) \in A$ if and only if there exists $a \in A$ for which, given any $\epsilon > 0$, there exists $\delta > 0$ such that
$$0 < u - w < \delta \Rightarrow |X(w) - X(u) - a| < \epsilon. \tag{2.4}$$

From (2.3) and (2.4), we deduce that $(N(t, A) - N(s, A) = n) \in \sigma\{X(v) - X(u); s \leq u < v \leq t\}$, and (L2) follows.

To establish (L3), note first that if $N(t, A) = 0$ for some $t > 0$ then $N(s, A) = 0$ for all $0 \leq s < t$. Hence, since (L2) holds we find that for all $n \in \mathbb{N}$

$P(N(t, A) = 0)$
$= P\left(N\left(\frac{t}{n}, A\right) = 0, N\left(\frac{2t}{n}, A\right) = 0, \ldots, N(t, A) = 0\right)$
$= P\left(N\left(\frac{t}{n}, A\right) = 0, N\left(\frac{2t}{n}, A\right) - N\left(\frac{t}{n}, A\right) = 0,\right.$
$\left.\ldots, N(t, A) - N\left(\frac{(n-1)t}{n}, A\right) = 0\right) = \left[P\left(N\left(\frac{t}{n}, A\right) = 0\right)\right]^n.$

From this we deduce that
$$\limsup_{t \to 0} P(N(t, A) = 0) = \lim_{n \to \infty} \limsup_{t \to 0} \left[P\left(N\left(\frac{t}{n}, A\right) = 0\right)\right]^n,$$
and, since we can herein replace $\limsup_{t \to 0}$ by $\liminf_{t \to 0}$, we see that either $\lim_{t \to 0} P(N(t, A) = 0)$ exists and is 0 or 1 or $\liminf_{t \to 0} P(N(t, A) = 0) = 0$ and $\limsup_{t \to 0} P(N(t, A) = 0) = 1$.

2.3 The jumps of a Lévy process – Poisson random measures

First suppose that $\liminf_{t \to 0} P(N(t, A) = 0) = 0$ and that $\limsup_{t \to 0} P(N(t, A) = 0) = 1$. Recall that if $N(t, A) = 0$ for some $t > 0$ then $N(s, A) = 0$ for all $0 \leq s \leq t$. From this we see that the map $t \to P(N(t, A) = 0)$ is monotonic decreasing. So if $P(N(t, A) = 0) = \epsilon > 0$ for some $t \geq 0$ we must have $\liminf_{t \to 0} P(N(t, A) = 0) \geq \epsilon$. Hence, if $\liminf_{t \to 0} P(N(t, A) = 0) = 0$ then $P(N(t, A) = 0) = 0$ for all $t \geq 0$ and so $\limsup_{t \to 0} P(N(t, A) = 0) = 0$, which yields our desired contradiction.

Now suppose that $\lim_{t \to 0} P(N(t, A) = 0) = 0$; then $\lim_{t \to 0} P(N(t, A) \neq 0) = 1$. Let A and B be bounded below and disjoint. Since $N(t, A \cup B) \neq 0$ if and only if $N(t, A) \neq 0$ or $N(t, B) \neq 0$, we find that $\lim_{t \to 0} P(N(t, A \cup B) \neq 0) = 2$, which is also a contradiction.

Hence we have deduced that $\lim_{t \to 0} P(N(t, A) = 0) = 1$ and so $\lim_{t \to 0} P(N(t, A) \neq 0) = 0$, as required.

(2) Using arguments similar to those that led up to (2.3) and (2.4), we deduce that the events

$$(N(t, A_1) = n_1), \ldots, (N(t, A_m) = n_m)$$

are members of independent σ-algebras. □

Remark 1 It follows immediately that $\mu(A) < \infty$ whenever A is bounded below, hence the measure μ is σ-finite.

Remark 2 By Theorem 2.1.7, N has a càdlàg modification that is also a Poisson process. We will identify N with this modification henceforth, in accordance with our usual philosophy.

2.3.1 Random measures

Let $(S, \mathcal{A})$ be a measurable space and $(\Omega, \mathcal{F}, P)$ be a probability space. A *random measure* M on $(S, \mathcal{A})$ is a collection of random variables $(M(B), B \in \mathcal{A})$ such that:

(1) $M(\emptyset) = 0$;
(2) (σ-*additivity*) given any sequence $(A_n, n \in \mathbb{N})$ of mutually disjoint sets in $\mathcal{A}$,

$$M\left(\bigcup_{n \in \mathbb{N}} A_n\right) = \sum_{n \in \mathbb{N}} M(A_n) \quad \text{a.s.;}$$

[replace with finite additivity]

(3) (*independently scattered property*) for each disjoint family $(B_1, \ldots, B_n)$ in $\mathcal{A}$, the random variables $M(B_1), \ldots, M(B_n)$ are independent.

We say that we have a *Poisson random measure* if each $M(B)$ has a Poisson distribution whenever $M(B) < \infty$. In many cases of interest, we obtain a σ-finite measure λ on $(S, \mathcal{A})$ by the prescription $\lambda(A) = \mathbb{E}(M(A))$ for all $A \in \mathcal{A}$. Conversely we have:

Theorem 2.3.6 *Given a σ-finite measure λ on a measurable space $(S, \mathcal{A})$, there exists a Poisson random measure M on a probability space $(\Omega, \mathcal{F}, P)$ such that $\lambda(A) = \mathbb{E}(M(A))$ for all $A \in \mathcal{A}$.*

Proof See e.g. Ikeda and Watanabe [140], p. 42, or Sato [274], p. 122.

Suppose that $S = \mathbb{R}^+ \times U$, where U is a measurable space equipped with a σ-algebra $\mathcal{C}$, and $\mathcal{A} = \mathcal{B}(\mathbb{R}^+) \otimes \mathcal{C}$. Let $p = (p(t), t \geq 0)$ be an adapted process taking values in U such that M is a Poisson random measure on S, where $M([0, t) \times A) = \#\{0 \leq s < t; p(s) \in A\}$ for each $t \geq 0$, $A \in \mathcal{C}$. In this case we say that p is a *Poisson point process* and M is its associated Poisson random measure.

The final concept we need is a merger of the two important ideas of the random measure and the martingale. Let U be a topological space and take $\mathcal{C}$ to be its Borel σ-algebra. Let M be a random measure on $S = \mathbb{R}^+ \times U$. For each $A \in \mathcal{C}$, define a process $M_A = (M_A(t), t \geq 0)$ by $M_A(t) = M([0, t) \times A)$. We say that M is a *martingale-valued measure* if M_A is a martingale whenever $\bar{A} \cap V = \emptyset$. We call V the associated *forbidden set* (which may of course itself be $\emptyset$).

The key example of these concepts for our work is as follows.

Example 2.3.7 Let $U = \mathbb{R}^d$ and $\mathcal{C}$ be its Borel σ-algebra. Let X be a Lévy process; then ΔX is a Poisson point process and N is its associated Poisson random measure. For each $t \geq 0$ and A bounded below, we define the *compensated Poisson random measure* by

$$\tilde{N}(t, A) = N(t, A) - t\mu(A).$$

By Exercise 2.1.4(4), $(\tilde{N}(t, A), t \geq 0)$ is a martingale and so $\tilde{N}$ extends to a martingale-valued measure with forbidden set $\{0\}$.

In case you are unfamiliar with Poisson random measures we summarise below the main properties of N. These will be used extensively later.

(1) For each $t > 0$, $\omega \in \Omega$, $N(t, \cdot)(\omega)$ is a counting measure on $\mathcal{B}(\mathbb{R}^d - \{0\})$.
(2) For each A bounded below, $(N(t, A), t \geq 0)$ is a Poisson process with intensity $\mu(A) = \mathbb{E}(N(1, A))$.

2.3 The jumps of a Lévy process – Poisson random measures

(3) $(\tilde{N}(t, A), t \geq 0)$ is a martingale-valued measure, where $\tilde{N}(t, A) = N(t, A) - t\mu(A)$, for A bounded below.

2.3.2 Poisson integration

Let f be a Borel measurable function from $\mathbb{R}^d$ to $\mathbb{R}^d$ and let A be bounded below; then for each $t > 0$, $\omega \in \Omega$, we may define the *Poisson integral* of f as a random finite sum by

$$\int_A f(x) N(t, dx)(\omega) = \sum_{x \in A} f(x) N(t, \{x\})(\omega).$$

Note that each $\int_A f(x) N(t, dx)$ is an $\mathbb{R}^d$-valued random variable and gives rise to a càdlàg stochastic process as we vary t.

Now, since $N(t, \{x\}) \neq 0 \Leftrightarrow \Delta X(u) = x$ for at least one $0 \leq u \leq t$, we have

$$\int_A f(x) N(t, dx) = \sum_{0 \leq u \leq t} f(\Delta X(u)) \chi_A(\Delta X(u)). \tag{2.5}$$

Let $(T_n^A, n \in \mathbb{N})$ be the arrival times for the Poisson process $(N(t, A), t \geq 0)$. Then another useful representation for Poisson integrals, which follows immediately from (2.5), is

$$\int_A f(x) N(t, dx) = \sum_{n \in \mathbb{N}} f(\Delta(X(T_n^A))) \chi_{[0,t]}(T_n^A). \tag{2.6}$$

Henceforth, we will sometimes use μ_A to denote the restriction to A of the measure μ.

Theorem 2.3.8 *Let A be bounded below. Then:*

(1) *for each $t \geq 0$, $\int_A f(x) N(t, dx)$ has a compound Poisson distribution such that, for each $u \in \mathbb{R}^d$,*

$$\mathbb{E}\left(\exp\left[i\left(u, \int_A f(x) N(t, dx)\right)\right]\right) = \exp\left[t \int_{\mathbb{R}^d} (e^{i(u,x)} - 1) \mu_{f,A}(dx)\right]$$

where $\mu_{f,A}(B) = \mu(A \cap f^{-1}(B))$;

(2) *if $f \in L^1(A, \mu_A)$, we have*

$$\mathbb{E}\left(\int_A f(x) N(t, dx)\right) = t \int_A f(x) \mu(dx);$$

(3) *if* $f \in L^2(A, \mu_A)$, *we have*

$$\mathrm{Var}\left(\left|\int_A f(x)N(t,dx)\right|\right) = t\int_A |f(x)|^2 \mu(dx).$$

Proof (1) For simplicity, we will prove this result in the case $f \in L^1(A, \mu_A)$. The general proof for arbitrary measurable f can be found in Sato [274], p. 124. First let f be a simple function and write $f = \sum_{j=1}^n c_j \chi_{A_j}$, where each $c_j \in \mathbb{R}^d$. We can assume, without loss of generality, that the A_j are disjoint Borel subsets of A. By Theorem 2.3.5 we find that

$$\mathbb{E}\left(\exp\left[i\left(u, \int_A f(x)N(t,dx)\right)\right]\right)$$

$$= \mathbb{E}\left(\exp\left[i\left(u, \sum_{j=1}^n c_j N(t, A_j)\right)\right]\right)$$

$$= \prod_{j=1}^n \mathbb{E}\left(\exp\left[i(u, c_j N(t, A_j))\right]\right)$$

$$= \prod_{j=1}^n \exp\{t[\exp(i(u, c_j)) - 1]\mu(A_j)\}$$

$$= \exp\left[t\int_A \{\exp[i(u, f(x))] - 1\}\mu(dx)\right].$$

Given an arbitrary $f \in L^1(A, \mu_A)$, we can find a sequence of simple functions converging to f in L^1 and hence a subsequence that converges to f almost surely. Passing to the limit along this subsequence in the above yields the required result, via dominated convergence.

Parts (2) and (3) follow from (1) by differentiation. □

It follows from Theorem 2.3.8(2) that a Poisson integral will fail to have a finite mean if $f \notin L^1(A, \mu)$.

Exercise 2.3.9 Show that if $f : \mathbb{R}^d \to \mathbb{R}^d$ is Borel measurable then

$$\sum_{0 \leq u \leq t} |f(\Delta X(u))|\chi_A(\Delta X(u)) < \infty \quad \text{a.s.}$$

Consider the sequence of *jump size* random variables $(Y_f^A(n), n \in \mathbb{N})$, where each

$$Y_f^A(n) = \int_A f(x)N(T_n^A, dx) - \int_A f(x)N(T_{n-1}^A, dx). \tag{2.7}$$

2.3 The jumps of a Lévy process – Poisson random measures

It follows from (2.6) and (2.7) that

$$Y_f^A(n) = f(\Delta(X(T_n^A)))$$

for each $n \in \mathbb{N}$.

Theorem 2.3.10

(1) $(Y_f^A(n), n \in \mathbb{N})$ *are i.i.d. with common law given by*

$$P(Y_f^A(n) \in B) = \frac{\mu(A \cap f^{-1}(B))}{\mu(A)} \quad (2.8)$$

for each $B \in \mathcal{B}(\mathbb{R}^d)$.

(2) $\left(\int_A f(x) N(t, dx), t \geq 0 \right)$ *is a compound Poisson process.*

Proof (1) We begin by establishing (2.8). Using Theorem 2.3.8(2) and (2.7), together with the fact that $(T_n^A - T_{n-1}^A, n \in \mathbb{N})$ are i.i.d. exponentially distributed random variables with common mean $1/\mu(A)$, we obtain

$$P(Y_f^A(n) \in B) = \mathbb{E}(\chi_B(Y_f^A(n))) = \mathbb{E}\left[\mathbb{E}_{T_n^A - T_{n-1}^A}(\chi_B(Y_f^A(n))) \right]$$

$$= \int_0^\infty s \int_A \chi_B(f(x)) \mu(dx) p_{T_n^A - T_{n-1}^A}(ds)$$

$$= \frac{\mu(A \cap f^{-1}(B))}{\mu(A)},$$

as required. Hence our random variables are identically distributed. To see that they are independent, we use a similar argument to that above to write, for any finite set of natural numbers $\{i_1, i_2, \ldots, i_m\}$ and $B_{i_1}, B_{i_2}, \ldots, B_{i_m} \in \mathcal{B}(\mathbb{R}^d)$,

$$P\left(Y_f^A(i_1) \in B_{i_1}, Y_f^A(i_2) \in B_{i_2}, \ldots, Y_f^A(i_m) \in B_{i_m} \right)$$

$$= \mathbb{E}\left[\mathbb{E}_{T_1^A, T_2^A - T_1^A, \ldots, T_n^A - T_{n-1}^A} \prod_{j=1}^m \chi_{B_{i_j}}\left(Y_f^A(i_j) \right) \right]$$

$$= \int_0^\infty \int_0^\infty \cdots \int_0^\infty s_{i_1} s_{i_2} \cdots s_{i_m} \prod_{j=1}^m \int_A \chi_{B_{i_j}}(f(x)) \mu(dx)$$

$$\times p_{T_{i_1}^A}(ds_{i_1}) p_{T_{i_2}^A - T_{i_1}^A}(ds_{i_2}) \cdots p_{T_{i_m}^A - T_{i_{m-1}}^A}(ds_{i_m})$$

$$= P(Y_f^A(i_1) \in B_{i_1}) P(Y_f^A(i_2) \in B_{i_2}) \cdots P(Y_f^A(i_m) \in B_{i_m}),$$

by (2.8).

(2) First we observe that $(Y_f^A(n), n \in \mathbb{N})$ and the Poisson process $(N(t, A), t \geq 0)$ are independent. Indeed, this follows from a slight extension of the following argument. For each $m \in \mathbb{N}, n \in \mathbb{N} \cup \{0\}, t \geq 0, B \in \mathcal{B}(\mathbb{R}^d)$, we have

$$P\left(Y_f^A(m) \in B \big| N(t, A) = n\right) = P\left(Y_f^A(m) \in B \big| T_n^A \leq t, T_{n+1}^A > t\right)$$
$$= P(Y_f^A(m) \in B),$$

by a calculation similar to that in (1). For each $t \geq 0$, we have

$$\int_A f(x) N(t, dx) = Y_f^A(1) + Y_f^A(2) + \cdots + Y_f^A(N(t, A)).$$

The summands are i.i.d. by (1), and the result follows. □

For each $f \in L^1(A, \mu_A), t \geq 0$, we define the *compensated Poisson integral* by

$$\int_A f(x) \tilde{N}(t, dx) = \int_A f(x) N(t, dx) - t \int_A f(x) \mu(dx).$$

A straightforward argument, as in Exercise 2.1.4(4), shows that

$$\left(\int_A f(x) \tilde{N}(t, dx), t \geq 0\right)$$

is a martingale, and we will use this fact extensively later. By Theorem 2.3.8(1), (3) we can easily deduce the following two important facts:

$$\mathbb{E}\left(\exp\left[i\left(u, \int_A f(x) \tilde{N}(t, dx)\right)\right]\right)$$
$$= \exp\left\{t \int_{\mathbb{R}^d} [e^{i(u,x)} - 1 - i(u, x)] \mu_f(dx)\right\} \quad (2.9)$$

for each $u \in \mathbb{R}^d$ and, for $f \in L^2(A, \mu_A)$,

$$\mathbb{E}\left(\left|\int_A f(x) \tilde{N}(t, dx)\right|^2\right) = t \int_A |f(x)|^2 \mu(dx). \quad (2.10)$$

Exercise 2.3.11 For A, B bounded below and $f \in L^2(A, \mu_A), g \in L^2(B, \mu_B)$, show that

$$\left\langle \int_A f(x) \tilde{N}(t, dx), \int_B g(x) \tilde{N}(t, dx) \right\rangle = t \int_{A \cap B} f(x) g(x) \mu(dx).$$

Exercise 2.3.12 For each A bounded below define
$$\mathcal{M}_A = \left\{ \int_A f(x)\tilde{N}(t,dx),\ f \in L^2(A,\mu_A) \right\}.$$
Show that $\mathcal{M}_A$ is a closed subspace of the martingale space $\mathcal{M}$.

Exercise 2.3.13 Deduce that $\lim_{n\to\infty} T_n^A = \infty$ (a.s.) whenever A is bounded below.

2.3.3 Processes of finite variation

We begin by introducing a useful class of functions. Let $\mathcal{P} = \{a = t_1 < t_2 < \cdots < t_n < t_{n+1} = b\}$ be a partition of the interval $[a,b]$ in $\mathbb{R}$, and define its mesh to be $\delta = \max_{1 \le i \le n} |t_{i+1} - t_i|$. We define the *variation* $\operatorname{var}_{\mathcal{P}}(g)$ of a càdlàg mapping $g:[a,b] \to \mathbb{R}^d$ over the partition $\mathcal{P}$ by the prescription

$$\operatorname{var}_{\mathcal{P}}(g) = \sum_{i=1}^n |g(t_{i+1}) - g(t_i)|.$$

If $V(g) = \sup_{\mathcal{P}} \operatorname{var}_{\mathcal{P}}(g) < \infty$, we say that g has <u>*finite variation on*</u> $[a,b]$. If g is defined on the whole of $\mathbb{R}$ (or $\mathbb{R}^+$), it is said to have *finite variation* if it has finite variation on each compact interval.

It is a trivial observation that every non-decreasing g is of finite variation. Conversely, if g is of finite variation then it can always be written as the difference of two non-decreasing functions; to see this, just write

$$g = \frac{V(g) + g}{2} - \frac{V(g) - g}{2},$$

where $V(g)(t)$ is the variation of g on $[a,t]$. Functions of finite variation are important in integration: suppose that we are given a function g that we are proposing as an integrator, then as a minimum we will want to be able to define the Stieltjes integral $\int_I f\,dg$ for all continuous functions f, where I is some finite interval. It is shown on pp. 40–41 of Protter [255] that a necessary and sufficient condition for obtaining such an integral as a limit of Riemann sums is that g has finite variation (see also the discussion and references in Mikosch [228], pp. 88–92).

Exercise 2.3.14 Show that all the functions of finite variation on $[a,b]$ (or on $\mathbb{R}$) form a vector space.

A stochastic process $(X(t), t \ge 0)$ is of *finite variation* if the paths $(X(t)(\omega), t \ge 0)$ are of finite variation for almost all $\omega \in \Omega$. The following is an important example for us.

Example 2.3.15 (Poisson integrals) Let N be a Poisson random measure, with intensity measure μ, that counts the jumps of a Lévy process X and let $f : \mathbb{R}^d \to \mathbb{R}^d$ be Borel measurable. For A bounded below, let $Y = (Y(t), t \geq 0)$ be given by $Y(t) = \int_A f(x) N(t, dx)$; then Y is of finite variation on $[0, t]$ for each $t \geq 0$. To see this, we observe that, for all partitions $\mathcal{P}$ of $[0, t]$, by Exercise 2.3.9 we have

$$\operatorname{var}_{\mathcal{P}}(Y) \leq \sum_{0 \leq s \leq t} |f(\Delta X(s))| \chi_A(\Delta X(s)) < \infty \qquad \text{a.s.} \qquad (2.11)$$

Exercise 2.3.16 Let Y be a Poisson integral as above and let η be its Lévy symbol. For each $u \in \mathbb{R}^d$ consider the martingales $M_u = (M_u(t), t \geq 0)$ where each

$$M_u(t) = e^{i(u, Y(t)) - t\eta(u)}.$$

Show that M_u is of finite variation. (Hint: Use the mean value theorem.)

Exercise 2.3.17 Show that every subordinator is of finite variation.

In fact, a necessary and sufficient condition for a Lévy process to be of finite variation is that there is no Brownian part (i.e. $A = 0$ in the Lévy–Khinchine formula) and that $\int_{|x|<1} |x| \nu(dx) < \infty$; see e.g. Bertoin [36] p. 15, or Bretagnolle [60].

2.4 The Lévy–Itô decomposition

Here we will give a proof of one of the key results in the elementary theory of Lévy processes, namely the celebrated Lévy–Itô decomposition of the sample paths into continuous and jump parts. Our approach closely follows that of Bretagnolle [61]. First we will need a number of preliminary results.

Proposition 2.4.1 *Let M_j, $j = 1, 2$, be two càdlàg-centred martingales. Suppose that, for some j, M_j is L^2 and that for each $t \geq 0$ $\mathbb{E}(|V(M_k(t))|^2) < \infty$ where $k \neq j$; then*

$$\mathbb{E}[(M_1(t), M_2(t))] = \mathbb{E}\left(\sum_{0 \leq s \leq t} (\Delta M_1(s), \Delta M_2(s)) \right).$$

Proof For convenience, we work in the case $d = 1$. We suppose throughout that M_1 is L^2 and so M_2 has square-integrable variation. Let $\mathcal{P} = \{0 = t_0 < t_1 < t_2 < \cdots < t_m = t\}$ be a partition of $[0, t]$; then by the martingale

2.4 The Lévy–Itô decomposition

property we have

$$\mathbb{E}(M_1(t)M_2(t))$$
$$= \sum_{i=0}^{m-1}\sum_{j=0}^{m-1} \mathbb{E}\big([M_1(t_{i+1}) - M_1(t_i)][M_2(t_{j+1}) - M_2(t_j)]\big)$$
$$= \sum_{i=0}^{m-1} \mathbb{E}\big([M_1(t_{i+1}) - M_1(t_i)][M_2(t_{i+1}) - M_2(t_i)]\big).$$

Now let $(\mathcal{P}^{(n)}, n \in \mathbb{N})$ be a sequence of such partitions with

$$\lim_{n\to\infty} \max_{0 \le i(n) \le m(n)-1} \big|t_{i+1}^{(n)} - t_i^{(n)}\big| = 0.$$

Then, with probability 1, we claim that

$$\lim_{n\to\infty} \sum_{i(n)=0}^{m(n)-1} \big[M_1(t_{i(n)+1}) - M_1(t_{i(n)})\big]\big[M_2(t_{i(n)+1}) - M_2(t_{i(n)})\big]$$
$$= \sum_{0 \le s \le t} \Delta M_1(s) \Delta M_2(s).$$

To establish this claim, fix $\omega \in \Omega$ and assume (without loss of generality) that $(M_1(t)(\omega), t \ge 0)$ and $(M_2(t)(\omega), t \ge 0)$ have common points of discontinuity $A = (t_n, n \in \mathbb{N})$.

We first consider the set A^c. Let $(\mathcal{P}_n, n \in \mathbb{N})$ be a sequence of partitions of $[0, t]$ such that, for each $n \in \mathbb{N}$, $A \cap [t_j^{(n)}, t_{j+1}^{(n)}] = \emptyset$ for all $0 \le j \le m(n) - 1$. Dropping ω for notational convenience, we find that

$$\sum_{i=0}^{m(n)-1} \left\|\big[M_1(t_{i(n)+1}) - M_1(t_{i(n)})\big]\big[M_2(t_{i(n)+1}) - M_2(t_{i(n)})\big]\right\|$$
$$\le \max_{0 \le i \le m(n)-1} \big|M_1(t_{i(n)+1}) - M_1(t_{i(n)})\big| \operatorname{Var}_{\mathcal{P}_n}(M_2)$$
$$\to 0 \quad \text{as} \quad n \to \infty.$$

Turning our attention to A, we fix $\epsilon > 0$ and choose $\delta = (\delta_n, n \in \mathbb{N})$ to be such that

$$\max\big\{|M_1(t_n) - M_1(t_n - \delta_n) - \Delta M_1(t_n)|,$$
$$|M_2(t_n) - M_2(t_n - \delta_n) - \Delta M_2(t_n)|\big\} < \frac{\epsilon}{K 2^n},$$

where
$$K = 2 \sup_{0 \leq s \leq t} |M_1(s)| + 2 \sup_{0 \leq s \leq t} |M_2(s)|.$$

To establish the claim in this case, we consider
$$S(\delta) = \sum_{n=1}^{\infty} \{[M_1(t_n) - M_1(t_n - \delta_n)][M_2(t_n) - M_2(t_n - \delta_n)] - \Delta M_1(t_n)\Delta M_2(t_n)\}.$$

We then find that
$$|S(\delta)|$$
$$\leq \sum_{n=1}^{\infty} |(M_1(t_n) - M_1(t_n - \delta_n) - \Delta M_1(t_n)||M_2(t_n) - M_2(t_n - \delta_n)|$$
$$+ \sum_{n=1}^{\infty} |M_2(t_n) - M_2(t_n - \delta_n) - \Delta M_2(t_n)||\Delta M_1(t_n)|$$
$$\leq 2 \left(\sup_{0 \leq s \leq t} |M_1(s)| + \sup_{0 \leq s \leq t} |M_2(s)| \right) \sum_{n=1}^{\infty} \frac{\epsilon}{K 2^n} < \epsilon,$$

and the claim is thus established.

The result of the theorem follows by dominated convergence, using the fact that for each $n \in \mathbb{N}$
$$\left| \sum_{i(n)=0}^{m(n)-1} [M_1(t_{i(n)+1}) - M_1(t_{i(n)})][M_2(t_{i(n)+1}) - M_2(t_{i(n)})] \right|$$
$$\leq 2 \sup_{0 \leq s \leq t} |M_1(s)| V(M_2(t)),$$

and, on using Doob's martingale inequality,
$$\mathbb{E}\left(\sup_{0 \leq s \leq t} |M_1(s)| V(M_2(t)) \right) \leq \mathbb{E}\left(\sup_{0 \leq s \leq t} |M_1(s)|^2 \right) + \mathbb{E}(|V(M_2(t))|^2)$$
$$\leq 4 \mathbb{E}(|M_1(t)|^2) + \mathbb{E}(|V(M_2(t))|^2) < \infty.$$
□

The following special case of Proposition 2.4.1 plays a major role below.

Example 2.4.2 Let A and B be bounded below and suppose that $f \in L^2(A, \mu_A)$, $g \in L^2(B, \mu_B)$. For each $t \geq 0$, let $M_1(t) = \int_A f(x) \tilde{N}(t, dx)$

and $M_2(t) = \int_B g(x)\tilde{N}(t, dx)$; then, by (2.11),

$$V(M_1(t)) \leq V\left(\int_A f(x)N(t, dx)\right) + V\left(t\int_A f(x)\nu(dx)\right)$$
$$\leq \int_A |f(x)|N(t, dx) + t\int_A f(x)\nu(dx).$$

From this and the Cauchy–Schwarz inequality we have $\mathbb{E}(|V(M_1(t)|^2) < \infty$, and so we can apply Proposition 2.4.1 in this case. Note the important fact that $\mathbb{E}(M_1(t)M_2(t)) = 0$ for each $t \geq 0$ if $A \cap B = \emptyset$.

Exercise 2.4.3 Show that Proposition 2.4.1 fails to hold when $M_1 = M_2 = B$, where B is a standard Brownian motion.

Exercise 2.4.4 Let $N = (N(t), t \geq 0)$ be a Poisson process with arrival times $(T_n, n \in \mathbb{N})$ and let M be a centred càdlàg L^2-martingale. Show that, for each $t \geq 0$,

$$\mathbb{E}(M(t)N(t)) = \mathbb{E}\left(\sum_{n \in \mathbb{N}} \Delta M(T_n)\chi_{\{T_n \leq t\}}\right).$$

Exercise 2.4.5 Let A be bounded below and M be a centred càdlàg L^2-martingale that is continuous at the arrival times of $(N(t, A), t \geq 0)$. Show that M is orthogonal to every process in $\mathcal{M}_A$ (as defined in Exercise 2.3.12).

For A bounded below note that, for each $t \geq 0$,

$$\int_A xN(t, dx) = \sum_{0 \leq u \leq t} \Delta X(u)\chi_A(\Delta X(u))$$

is the sum of all the jumps taking values in the set A up to the time t. Since the paths of X are càdlàg, this is clearly a finite random sum.

Theorem 2.4.6 *If A_p, $p = 1, 2$, are disjoint and bounded below, then $\left(\int_{A_1} xN(t, dx), t \geq 0\right)$ and $\left(\int_{A_2} xN(t, dx), t \geq 0\right)$ are independent stochastic processes.*

Proof For each $p = 1, 2$, $t \geq 0$, write $X(t, A_p) = \int_{A_p} xN(t, dx)$ and let η_{A_p} be the Lévy symbol of each of these compound Poisson processes (recall Theorem 2.3.8). We will also have need of the centred càdlàg L^2-martingales

$(M_p(t), t \geq 0)$ for $p = 1, 2$, given by

$$M_p(t) = \exp\left[i(u_p, X(t, A_p)) - t\eta_{A_p}\right] - 1$$

for each $t \geq 0$, where $u_1, u_2 \in \mathbb{R}^d$. We will need the fact below that at least one M_p has square-integrable variation on finite intervals. This follows easily after using the mean value theorem to establish that, for each $t \geq 0$, there exists a complex-valued random variable $\rho(t)$ with $0 \leq |\rho(t)| \leq 1$ for which

$$M_p(t) = \rho(t)\left[i(u_p, X(t, A_p)) - t\eta_{A_p}\right].$$

Now for $0 \leq s \leq t < \infty$ we have

$$\mathbb{E}\big(M_1(t)M_2(s)\big) = \mathbb{E}\big(M_1(s)M_2(s)\big) + \mathbb{E}\big([M_1(t) - M_1(s)]M_2(s)\big).$$

Since A_1 and A_2 are disjoint, M_1 and M_2 have their jumps at distinct times and so $\mathbb{E}(M_1(s)M_2(s)) = 0$ by Proposition 2.4.1. However, M_1 is a martingale and so a straightforward conditioning argument yields $\mathbb{E}([M_1(t) - M_1(s)]M_2(s)) = 0$. Hence we have that, for all $u_1, u_2 \in \mathbb{R}^d$,

$$\mathbb{E}\big(e^{i(u_1, X(t, A_1))} e^{i(u_2, X(s, A_2))}\big) = \mathbb{E}\big(e^{i(u_1, X(t, A_1))}\big) \mathbb{E}\big(e^{i(u_2, X(s, A_2))}\big),$$

and so the random variables $X(t, A_1)$ and $X(s, A_2)$ are independent by Kac's theorem.

Now we need to show that the processes are independent. To this end, fix $n_1, n_2 \in \mathbb{N}$, choose $0 = t_0^p < t_1^p < \cdots < t_{n_p}^p < \infty$ and $u_j^p \in \mathbb{R}^d$, $0 \leq j \leq n_p$ and write $v_j^p = u_j^p + u_{j+1}^p + \cdots + u_n^p$, for $p = 1, 2$. By (L2) we obtain

$$\mathbb{E}\left(\exp\left[i\sum_{j=1}^{n_1}(u_j^1, X(t_j^1, A_1))\right]\right) \mathbb{E}\left(\exp\left[i\sum_{k=1}^{n_2}(u_k^2, X(t_k^2, A_2))\right]\right)$$

$$= \mathbb{E}\left(\exp\left[i\sum_{j=1}^{n_1}(v_j^1, X(t_j^1, A_1) - X(t_{j-1}^1, A_1))\right]\right)$$

$$\times \mathbb{E}\left(\exp\left[i\sum_{k=1}^{n_2}(v_k^2, X(t_k^2, A_2) - X(t_{k-1}^2, A_2))\right]\right)$$

$$= \prod_{j=1}^{n_1} \mathbb{E}\left(\exp\left[i(v_j^1, X(t_j^1 - t_{j-1}^1, A_1))\right]\right)$$

$$\times \prod_{k=1}^{n_2} \mathbb{E}\left(\exp\left[i(v_k^2, X(t_k^2 - t_{k-1}^2, A_2))\right]\right)$$

$$= \prod_{j=1}^{n_1} \prod_{k=1}^{n_2} \mathbb{E}\left(\exp\{i[(v_j^1, X(t_j^1 - t_{j-1}^1, A_1)) + (v_k^2, X(t_k^2 - t_{k-1}^2, A_2))]\}\right)$$

$$= \prod_{j=1}^{n_1} \prod_{k=1}^{n_2} \mathbb{E}\left(\exp\{i[(v_j^1, X(t_j^1, A_1) - X(t_{j-1}^1, A_1))\right.$$
$$\left. + (v_k^2, X(t_k^2, A_2) - X(t_{k-1}^2, A_2))]\}\right)$$
$$= \mathbb{E}\left(\exp\left[i \sum_{j=1}^{n_1}(v_j^1, X(t_j^1, A_1) - X(t_{j-1}^1, A_1))\right.\right.$$
$$\left.\left. + i \sum_{k=1}^{n_2}(v_k^2, X(t_k^2, A_2) - X(t_{k-1}^2, A_2))\right]\right)$$
$$= \mathbb{E}\left(\exp\left[i \sum_{j=1}^{n_1}(u_j^1, X(t_j^1, A_1)) + i \sum_{k=1}^{n_2}(u_k^2, X(t_k^2, A_2))\right]\right),$$

and again the result follows by Kac's theorem. □

We say that a Lévy processes X has *bounded jumps* if there exists $C > 0$ with
$$\sup_{0 \leq t < \infty} |\Delta X(t)| < C.$$

Theorem 2.4.7 *If X is a Lévy process with bounded jumps then we have $\mathbb{E}(|X(t)|^m) < \infty$ for all $m \in \mathbb{N}$.*

Proof Let $C > 0$ be as above and define a sequence of stopping times $(T_n, n \in \mathbb{N})$ by $T_1 = \inf\{t \geq 0, |X(t)| > C\}$ and, for $n > 1$, $T_n = \inf\{t > T_{n-1}, |X(t) - X(T_{n-1})| > C\}$. We note that $|\Delta X(T_n)| \leq C$ and that $T_{n+1} - T_n = \inf\{t \geq 0; |X(t + T_n) - X(T_n)| > C\}$, for all $n \in \mathbb{N}$.

Our first goal will be to establish that, for all $n \in \mathbb{N}$,
$$\sup_{0 \leq s < \infty} |X(s \wedge T_n)| \leq 2nC \tag{2.12}$$

and we will prove this by induction. To see that this holds for $n = 1$ observe that
$$\sup_{0 \leq s \leq \infty} |X(s \wedge T_1)| = |X(T_1)|$$
$$\leq |\Delta X(T_1)| + |X(T_1-)| \leq 2C.$$

Now suppose that the inequality (2.12) holds for some $n > 1$. We fix $\omega \in \Omega$ and consider the left-hand side of (2.12) when n is replaced by $n + 1$. Now the supremum of $|X(s \wedge T_{n+1})|$ is attained over the interval $[0, T_n(\omega))$ or over the interval $[T_n(\omega), T_{n+1}(\omega)]$. In the former case we are done, and in the latter case we have

$$\sup_{0 \leq s < \infty} |X(s \wedge T_{n+1})(\omega)|$$

$$= \sup_{T_n(\omega) \leq s \leq T_{n+1}(\omega)} |X(s)(\omega)|$$

$$\leq \sup_{T_n(\omega) \leq s \leq T_{n+1}(\omega)} |X(s)(\omega) - X(T_n)(\omega)| + |X(T_n)(\omega)|$$

$$\leq |X(T_{n+1})(\omega) - X(T_n)(\omega)| + 2nC$$

$$\leq |X(T_{n+1})(\omega) - X(T_{n+1}-)(\omega)|$$
$$+ |X(T_{n+1}-)(\omega) - X(T_n)(\omega)| + 2nC$$

$$\leq 2(n+1)C,$$

as required.

By the strong Markov property (Theorem 2.2.11), we deduce that, for each $n \geq 2$, the random variables $T_n - T_{n-1}$ are independent of $\mathcal{F}_{T_{n-1}}$ and have the same law as T_1. Hence by repeated use of Doob's optional stopping theorem, we find that there exists $0 < a < 1$ for which

$$\mathbb{E}(e^{-T_n}) = \mathbb{E}\big(e^{-T_1} e^{-(T_2-T_1)} \cdots e^{-(T_n-T_{n-1})}\big) = \big[\mathbb{E}(e^{-T_1})\big]^n = a^n. \quad (2.13)$$

Now combining (2.12) and (2.13) and using the Chebyshev–Markov inequality we see that for each $n \in \mathbb{N}, t \geq 0$,

$$P(|X(t)| \geq 2nC) \leq P(T_n < t) \leq e^t \, \mathbb{E}(e^{-T_n}) \leq e^t a^n. \quad (2.14)$$

Finally, to verify that each $\mathbb{E}(|X(t)|^m) < \infty$, observe that by (2.14) we have

$$\int_{|x| \geq 2nC} |x|^m p_{X(t)}(dx) = \sum_{r=n}^{\infty} \int_{2rC}^{2(r+1)C} |x|^m p_{X(t)}(dx)$$

$$\leq (2C)^m e^t \sum_{r=n}^{\infty} (r+1)^m a^r < \infty.$$

$\square$

For each $a > 0$, consider the compound Poisson process

$$\left(\int_{|x| \geq a} x N(t, dx), t \geq 0 \right)$$

and define a new stochastic process $Y_a = (Y_a(t), t \geq 0)$ by the prescription

$$Y_a(t) = X(t) - \int_{|x| \geq a} x N(t, dx).$$

2.4 The Lévy–Itô decomposition

Intuitively, Y_a is what remains of the Lévy process X when all the jumps of size greater than a have been removed. We can get more insight into its paths by considering the impact of removing each jump. Let $(T_n, n \in \mathbb{N})$ be the arrival times for the Poisson process $(N(t, B_a(0)^c), t \geq 0)$. Then we have

$$Y_a(t) = \begin{cases} X(t) & \text{for } 0 \leq t < T_1, \\ X(T_1-) & \text{for } t = T_1, \\ X(t) - X(T_1) + X(T_1-) & \text{for } T_1 < t < T_2, \\ Y_a(T_2-) & \text{for } t = T_2, \end{cases}$$

and so on recursively.

Theorem 2.4.8 Y_a *is a Lévy process.*

Proof (L1) is immediate. For (L2) we argue as in the proof of Theorem 2.3.5 and deduce that, for each $0 \leq s < t < \infty$, $Y_a(t) - Y_a(s)$ is $\mathcal{F}_{s,t}$-measurable where $\mathcal{F}_{s,t} = \sigma\{X(u) - X(v); s \leq v \leq u < t\}$. To establish (L3), use the fact that for each $b > 0, t \geq 0$,

$$P\big(|Y_a(t)| > b\big) \leq P\left(|X(t)| > \frac{b}{2}\right) + P\left(\left|\int_{|x|\geq a} x N(t, dx)\right| > \frac{b}{2}\right).$$

□

We then immediately deduce the following:

Corollary 2.4.9 *A Lévy process has bounded jumps if and only if it is of the form Y_a for some $a > 0$.*

The proof is left as a (straightforward) exercise for the reader.
For each $a > 0$, we define a Lévy process $\hat{Y}_a = (\hat{Y}_a(t), t \geq 0)$ by

$$\hat{Y}_a = Y_a(t) - \mathbb{E}(Y_a(t)).$$

It is then easy to verify that $\hat{Y}_a$ is a càdlàg centred L^2-martingale.

Exercise 2.4.10 Show that $\mathbb{E}(Y_a(t)) = t \, \mathbb{E}(Y_a(1))$ for each $t \geq 0$.

In the following, we will find it convenient to take $a = 1$ and write the processes $Y_1, \hat{Y}_1$ simply as $Y, \hat{Y}$, respectively. So Y is what remains of our Lévy process when all jumps whose magnitude is larger than 1 have been removed, and $\hat{Y}$ is the centred version of Y. We also introduce the notation $M(t, A) = \int_A x \tilde{N}(t, dx)$ for $t \geq 0$ and A bounded below.

The following is a key step towards our required result.

Theorem 2.4.11 *For each $t \geq 0$,*

$$\hat{Y}(t) = Y_c(t) + Y_d(t),$$

where Y_c and Y_d are independent Lévy processes, Y_c has continuous sample paths and

$$Y_d(t) = \int_{|x|<1} x \tilde{N}(t, dx).$$

Proof Let $(\epsilon_n, n \in \mathbb{N})$ be a sequence that decreases monotonically to zero, wherein $\epsilon_1 < 1$. For each $m \in \mathbb{N}$ let

$$B_m = \{x \in \mathbb{R}^d, \epsilon_{m+1} \leq |x| \leq \epsilon_m\}$$

and for each $n \in \mathbb{N}$ let $A_n = \bigcup_{m=1}^n B_m$. Our first task is to show that the sequence $(M(\cdot, A_n), n \in \mathbb{N})$ converges in martingale space to Y_d. First note that for each $t \geq 0$ the $M(t, B_m)$ are mutually orthogonal by Proposition 2.4.1. So, for each $n \geq 0$,

$$\mathbb{E}(|M(t, A_n)|^2) = \sum_{m=1}^n \mathbb{E}(|M(t, B_m)|^2). \qquad (2.15)$$

By Proposition 2.1.3, the argument in the proof of Theorem 2.4.6 and Exercise 2.4.5, we find that the Lévy processes $\hat{Y} - M(\cdot, A_n)$ and $M(\cdot, A_n)$ are independent, and so

$$\text{Var}(|\hat{Y}(t)|) = \text{Var}(|\hat{Y}(t) - M(t, A_n)|) + \text{Var}(|M(t, A_n)|).$$

Hence

$$\mathbb{E}(|M(t, A_n)|^2) = \text{Var}(|M(t, A_n)|) \leq \text{Var}(|\hat{Y}(t)|). \qquad (2.16)$$

By (2.15) and (2.16) we see that, for each $t \geq 0$, the sequence $(\mathbb{E}(M(t, A_n)^2), n \in \mathbb{N})$ is increasing and bounded above and hence convergent. Furthermore by (2.15), for each $n_1 \leq n_2$,

$$\mathbb{E}(|M(t, A_{n_2}) - M(t, A_{n_1})|^2) = \mathbb{E}(|M(t, A_{n_2})|^2) - \mathbb{E}(|M(t, A_{n_1})|^2).$$

Hence we deduce that

$$Y_d(t) = \int_{|x|<1} x \tilde{N}(t, dx) = L^2 - \lim_{n \to \infty} M(t, A_n),$$

and that Y_d lives in martingale space. Furthermore it follows from Theorem 1.3.7 that Y_d is a Lévy process, where we have used Chebyshev's inequality to deduce that, for each $b > 0$, $\lim_{t \downarrow 0} P(|Y_d(t) - M(t, A_n)| > b) = 0$ for all $n \in \mathbb{N}$. A similar argument shows that Y_c is a Lévy process in martingale space, where

$$Y_c(t) = L^2 - \lim_{n \to \infty} \left[\hat{Y}(t) - M(t, A_n) \right].$$

The fact that Y_c and Y_d are independent follows by a straightforward limiting argument applied to characteristic functions.

Now we need to show that Y_c has continuous sample paths. If $Y_c(t) = 0$ (a.s.) for all $t \geq 0$ we are finished, so suppose that this is not the case. We seek a proof by contradiction. Let $N \subseteq \Omega$ be the set on which the paths of Y_c fail to be continuous. If $P(N) = 0$, we can replace Y_c by a modification that has continuous sample paths, so we will assume that $P(N) > 0$. Then there exists some $b > 0$ and a stopping time T such that $P(|\Delta X(T)| > b) > 0$. Let $A = \{x \in \mathbb{R}^d; |x| > b\}$; then by Proposition 2.4.1 we have for each $t \geq 0$, $f \in L^2(A, \mu_A)$,

$$0 \neq \mathbb{E}\left(\left(Y_c(t), \int_{|x|>b} f(x) \tilde{N}(t, dx) \right) \right)$$
$$= \lim_{n \to \infty} \mathbb{E}\left(\left(\hat{Y}(t) - M(t, A_n), \int_{|x|>b} f(x) \tilde{N}(t, dx) \right) \right) = 0,$$

and we have obtained our desired contradiction. □

We recall that μ is the intensity measure of the Poisson random measure N.

Corollary 2.4.12 *μ is a Lévy measure.*

Proof We have already shown that $\mu((-1, 1)^c) < \infty$ (see Remark 1 after Theorem 2.3.5). We also have

$$\int_{|x| \leq 1} |x|^2 \mu(dx) = \lim_{n \to \infty} \int_{A_n} |x|^2 \mu(dx) = \lim_{n \to \infty} \mathbb{E}(|M(t, A_n)|^2)$$
$$= \mathbb{E}(|Y_d|^2) < \infty,$$

and the result is established. □

Corollary 2.4.13 *For each $t \geq 0$, $u \in \mathbb{R}^d$,*

$$\mathbb{E}(e^{i(u, Y_d(t))}) = \exp\left\{ t \int_{|x|<1} \left[e^{i(u,x)} - 1 - i(u, x) \right] \mu(dx) \right\}.$$

Proof Take limits in equation (2.9). □

Exercise 2.4.14 Deduce that for each $t \geq 0$, $1 \leq i \leq d$,

$$\langle Y_d^i, Y_d^i \rangle(t) = \int_{|x|<1} x_i^2 \mu(dx).$$

Theorem 2.4.15 Y_c *is a Brownian motion.*

Proof Our strategy is to prove that for all $u \in \mathbb{R}^d$, $t \geq 0$,

$$\mathbb{E}(e^{i(u, Y_c(t))}) = e^{-t(u, Au)/2}, \qquad (2.17)$$

where A is a positive definite symmetric $d \times d$ matrix. The result then follows from Corollary 2.2.8, the corollary to Lévy's martingale characterisation of Brownian motion.

For convenience we take $d = 1$. Note that, since Y_c has no jumps, all its moments exist by Theorem 2.4.7 and since Y_c is a centred Lévy process we must have

$$\phi_t(u) = \mathbb{E}(e^{i(u, Y_c(t))}) = e^{t\eta(u)}, \qquad (2.18)$$

where $\eta \in C^\infty(\mathbb{R})$ and $\eta'(0) = 0$. Repeated differentiation yields for each $t \geq 0$, $m \geq 2$,

$$\mathbb{E}(Y_c(t)^m) = a_1 t + a_2 t^2 + \cdots + a_{m-1} t^{m-1} \qquad (2.19)$$

where $a_1, a_2 \ldots, a_{m-1} \in \mathbb{R}$.

Let $\mathcal{P} = \{0 = t_0 < t_1 < \cdots < t_n = t\}$ be a partition of $[0, t]$ and, for the purposes of this proof, write $\Delta Y_c(t_j) = Y_c(t_{j+1}) - Y_c(t_j)$ for each $0 \leq j \leq n - 1$. Now by Taylor's theorem we find

$$\mathbb{E}(e^{iuY_c(t)} - 1) = \mathbb{E}\left(\sum_{j=0}^{n-1} \left(e^{iuY_c(t_{j+1})} - e^{iuY_c(t_j)}\right)\right)$$

$$= \mathbb{E}(I_1(t)) + \mathbb{E}(I_2(t)) + \mathbb{E}(I_3(t)),$$

where

$$I_1(t) = iu \sum_{j=0}^{n-1} e^{iuY_c(t_j)} \Delta Y_c(t_j),$$

$$I_2(t) = -\frac{u^2}{2} \sum_{j=0}^{n-1} e^{iuY_c(t_j)} [\Delta Y_c(t_j)]^2,$$

$$I_3(t) = -\frac{u^2}{2} \sum_{j=0}^{n-1} \left(e^{iu[Y_c(t_j) + \theta_j \Delta Y_c(t_j)]} - e^{iuY_c(t_j)}\right) [\Delta Y_c(t_j)]^2,$$

with each $0 < \theta_j < 1$.

2.4 The Lévy–Itô decomposition

Now by independent increments we find immediately that

$$\mathbb{E}(I_1(t)) = iu \sum_{j=0}^{n-1} \mathbb{E}(e^{iuY_c(t_j)}) \mathbb{E}(\Delta Y_c(t_j)) = 0$$

and

$$\mathbb{E}(I_2(t)) = -\frac{u^2}{2} \sum_{j=0}^{n-1} \mathbb{E}(e^{iuY_c(t_j)}) \mathbb{E}((\Delta Y_c(t_j))^2)$$

$$= -\frac{au^2}{2} \sum_{j=0}^{n-1} \phi_{t_j}(u)(t_{j+1} - t_j), \qquad (2.20)$$

where we have used (2.18) and (2.19) and written $a_1 = a \geq 0$.

The analysis of $I_3(t)$ is more delicate, and we will need to introduce, for each $\alpha > 0$, the event

$$B_\alpha = \max_{0 \leq j \leq n-1} \sup_{t_j \leq u, v \leq t_{j+1}} |Y_c(v) - Y_c(u)| \leq \alpha$$

and write

$$\mathbb{E}(I_3(t)) = \mathbb{E}(I_3(t)\chi_{B_\alpha}) + \mathbb{E}(I_3(t)\chi_{B_\alpha^c}).$$

Now on using the elementary inequality $|e^{iy} - 1| \leq 2$, for any $y \in \mathbb{R}$, we deduce that

$$|\mathbb{E}(I_3(t))\chi_{B_\alpha^c}| \leq u^2 \sum_{j=0}^{n-1} \int_{B_\alpha^c} [\Delta Y_c(t_j)(\omega)]^2 dP(\omega)$$

$$\leq u^2 (P(B_\alpha^c))^{1/2} \left[\mathbb{E}\left(\sum_{j=0}^{n-1} \Delta Y_c(t_j)^2 \right)^2 \right]^{1/2}$$

$$\leq u^2 (P(B_\alpha^c))^{1/2} O(t^2 + t^3)^{1/2} \qquad (2.21)$$

where we have used the Cauchy–Schwarz inequality and (2.19).

On using the mean value theorem and (2.19) again, we obtain

$$|\mathbb{E}(I_3(t)\chi_{B_\alpha})| \leq \frac{|u|^3}{2} \int_{B_\alpha} \sum_{j=0}^{n-1} |\Delta Y_c(t_j)(\omega)|^3 dP(\omega) \leq \frac{\alpha at|u|^3}{2}. \qquad (2.22)$$

Now let $(\mathcal{P}^{(n)}, n \in \mathbb{N})$ be a sequence of partitions with $\lim_{n \to \infty} \delta_n = 0$, where the mesh of each partition $\delta_n = \max_{1 \leq j^{(n)} \leq m^{(n)}} |t_{j+1}^{(n)} - t_j^{(n)}|$, and for each $n \in \mathbb{N}$ write the corresponding $I_k(t)$ as $I_k^{(n)}(t)$ for $j = 1, 2, 3$, and write each B_α as $B_\alpha^{(n)}$. Now

$$\max_{1 \leq j \leq m^{(n)}} \sup_{t_j^{(n)} \leq u, v \leq t_{j+1}^{(n)}} |Y_c(v) - Y_c(u)| \leq \sup_{0 \leq u, v \leq t, |u-v| \leq \delta_n} |Y_c(v) - Y_c(u)|$$

$$\to 0 \quad \text{as} \quad n \to \infty,$$

by sample-path continuity, and so it follows (e.g. by dominated convergence) that $\lim_{n \to \infty} P((B_\alpha^{(n)})^c) = 0$. Hence we obtain, by (2.21) and (2.22), that

$$\limsup_{n \to \infty} \mathbb{E}(I_3^{(n)}(t)) \leq \frac{\alpha a t |u|^3}{2}.$$

But α can be made arbitrarily small, and so we deduce that

$$\lim_{n \to \infty} \mathbb{E}(I_3^{(n)}(t)) = 0. \tag{2.23}$$

Taking limits in (2.20) yields

$$\lim_{n \to \infty} \mathbb{E}(I_2^{(n)}(t)) = -\frac{au^2}{2} \int_0^t \phi_s(u) ds. \tag{2.24}$$

Combining the results of (2.23) and (2.24), we find that

$$\phi_t(u) - 1 = -\frac{au^2}{2} \int_0^t \phi_s(u) ds.$$

Hence $\phi_t(u) = e^{-at|u|^2/2}$, as required. □

At last we are ready for the main result of this section.

Theorem 2.4.16 (The Lévy–Itô decomposition) *If X is a Lévy process, then there exists $b \in \mathbb{R}^d$, a Brownian motion B_A with covariance matrix A and an independent Poisson random measure N on $\mathbb{R}^+ \times (\mathbb{R}^d - \{0\})$ such that, for each $t \geq 0$,*

$$X(t) = bt + B_A(t) + \int_{|x|<1} x \tilde{N}(t, dx) + \int_{|x| \geq 1} x N(t, dx). \tag{2.25}$$

Proof This follows from Theorems 2.4.11 and 2.4.15 with

$$b = \mathbb{E}\left(X(1) - \int_{|x| \geq 1} x N(1, dx)\right).$$

The fact that B_A and N are independent follows from the argument of Theorem 2.4.6 via Exercise 2.4.4. □

2.4 The Lévy–Itô decomposition

Note We will sometimes find it convenient for each $t \geq 0$, to write
$$B_A(t) = \left(B_A^1(t), \ldots, B_A^d(t)\right)$$
in the form
$$B_A^i(t) = \sum_{j=1}^m \sigma_j^i B^j(t)$$
where $B^1, \ldots, B^m$ are standard one-dimensional Brownian motions and σ is a $d \times m$ real-valued matrix for which $\sigma \sigma^T = A$.

Exercise 2.4.17 Write down the Lévy–Itô decompositions for the cases where X is (a) α-stable, (b) a subordinator, (c) a subordinated process.

Exercise 2.4.18 Show that an α-stable Lévy process has finite mean if $1 < \alpha \leq 2$ and infinite mean otherwise.

Exercise 2.4.19 Deduce that if X is a Lévy process then, for each $t \geq 0$, $\sum_{0 \leq s \leq t} [\Delta X(s)]^2 < \infty$ (a.s.).

An important by-product of the Lévy–Itô decomposition is the Lévy–Khintchine formula.

Corollary 2.4.20 *If X is a Lévy process then for each $u \in \mathbb{R}^d$, $t \geq 0$,*
$$\mathbb{E}(e^{i(u, X(t))}) = \exp\left(t\left\{i(b, u) - \tfrac{1}{2}(u, Au)\right.\right.$$
$$\left.\left. + \int_{\mathbb{R}^d - \{0\}} [e^{i(u, y)} - 1 - i(u, y)\chi_B(y)] \mu(dy) \right\}\right). \quad (2.26)$$

Proof By independence we have
$$\mathbb{E}(e^{i(u, X(t))}) = \mathbb{E}(e^{i(u, Y_c(t))}) \mathbb{E}(e^{i(u, Y_d(t))}) \mathbb{E}\left(e^{i\left(u, \int_{|x| \geq 1} x N(t, dx)\right)}\right),$$
and the result follows by using equation (2.17) and the results of Corollary 2.4.13 and Theorem 2.3.8. □

Now let ρ be an arbitrary infinitely divisible probability measure; then by Corollary 1.4.6 we can construct a canonical Lévy process X for which ρ appears as the law of $X(1)$. Note that X is adapted to its augmented natural filtration and thus we obtain a proof of the first part of the Lévy–Khinchine formula (Theorem 1.2.14).

Note 1 We emphasise that the above argument is not circular, in that we have at no time used the Lévy–Khinchine formula in the proof of the Lévy–Itô decomposition. We have used extensively, however, the weaker result $\mathbb{E}(e^{i(u,X(t))}) = e^{t\eta(u)}$, where $u \in \mathbb{R}^d$, $t \geq 0$, with $\eta(u) = \log\left[\mathbb{E}(e^{i(u,X(1))})\right]$. This is a consequence of the definition of a Lévy process (see Theorem 1.3.3).

Note 2 The process $\left(\int_{|x|<1} x \tilde{N}(t, dx), t \geq 0\right)$ in (2.25) is the *compensated sum of small jumps*. The compensation takes care of the analytic complications in the Lévy–Khinchine formula in a probabilistically pleasing way – since it is an L^2-martingale.

The process $\left(\int_{|x|\geq 1} x N(t, dx), t \geq 0\right)$ describing the 'large jumps' in (2.25) is a compound Poisson process by Theorem 2.3.10.

Note 3 In the light of (2.25), it is worth revisiting the result of Theorem 2.4.7. If X is a Lévy process then the Lévy process whose value at time $t \geq 0$ is $X(t) - \int_{|x|\geq 1} x N(t, dx)$ has finite moments to all orders. However, $\left(\int_{|x|\geq 1} x N(t, dx), t \geq 0\right)$ may have no finite moments, e.g. consider the case where X is α-stable with $0 < \alpha \leq 1$.

Corollary 2.4.21 *The characteristics (b, A, ν) of a Lévy process are uniquely determined by the process.*

Proof This follows from the construction that led to Theorem 2.4.16. □

Corollary 2.4.22 *Let G be a group of matrices acting on $\mathbb{R}^d$. A Lévy process is G-invariant if and only if, for each $g \in G$,*

$$b = gb, \qquad A = gAg^{\mathrm{T}} \qquad \text{and} \qquad \nu \text{ is } G\text{-invariant.}$$

Proof This follows immediately from the above corollary and the Lévy–Khintchine formula. □

Exercise 2.4.23 Show that a Lévy process is $O(d)$-invariant if and only if it has characteristics $(0, aI, \nu)$ where $a \geq 0$ and ν is $O(d)$-invariant. Show that a Lévy process is symmetric if and only if it has characteristics $(0, A, \nu)$ where A is an arbitrary positive definite symmetric matrix and ν is symmetric, i.e. $\nu(B) = \nu(-B)$ for all $B \in \mathcal{B}(\mathbb{R}^d - \{0\})$.

2.5 The interlacing construction

Exercise 2.4.24 Let X be a Lévy process for which

$$\int_{|x|\geq 1} |x|^n v(dx) < \infty$$

for all $n \geq 2$. For each $n \geq 2$, $t \geq 0$, define

$$X^{(n)}(t) = \sum_{0 \leq s \leq t} [\Delta X(s)]^n \quad \text{and} \quad Y^{(n)}(t) = X^{(n)}(t) - \mathbb{E}(X^{(n)}(t)).$$

Show that each $(Y^{(n)}(t), t \geq 0)$ is a martingale.

Note that these processes were introduced by Nualart and Schoutens [237] and called *Teugels martingales* therein.

Jump and creep

Suppose that X is a Lévy process with Lévy–Itô decomposition of the form

$$X(t) = \int_{|x|<1} x \tilde{N}(t, dx),$$

for all $t \geq 0$. Subtle behaviour can take place in the case $v(\hat{B} - \{0\}) = \infty$. Intuitively, the resulting path can be seen as the outcome of a competition between an infinite number of jumps of small size and an infinite drift. A deep analysis of this has been carried out by Millar [230], in the case where $d = 1$ and $v((0, 1)) > 0$. For each $x > 0$, let $T_x = \inf\{t \geq 0; X(t) > x\}$; then

$$P(X(T_x-) = x < X(T_x)) = P(X(T_x-) < x = X(T_x)) = 0,$$

so that either paths jump across x or they hit x continuously. Furthermore, either $P(X(T_x) = x) > 0$ for all $x > 0$ or $P(X(T_x) = x) = 0$ for all $x > 0$. In the first case, every positive point can be hit continuously in X and this phenomena is called *creep* in Bertoin [36], pp. 174–5. In the second case, only jumps can occur (almost surely). Millar [230] classified completely the conditions for creep or jump for general one-dimensional Lévy processes, in terms of their characteristics. For example, a sufficient condition for creep is $A = 0$ and $\int_{-1}^{0} |x|v(dx) = \infty$, $\int_{0}^{1} xv(dx) < \infty$. This is satisfied by 'spectrally negative' α-stable Lévy processes ($0 < \alpha < 2$), i.e. those for which $c_1 = 0$ in Theorem 1.2.20(2).

2.5 The interlacing construction

In this section we are going to use the interlacing technique to gain greater insight into the Lévy–Itô decomposition. First we need some preliminaries.

2.5.1 Limit events – a review

Let $(A(n), n \in \mathbb{N})$ be a sequence of events in $\mathcal{F}$. We define the tail events

$$\liminf_{n \to \infty} A(n) = \bigcup_{n=1}^{\infty} \bigcap_{k=n}^{\infty} A(k) \quad \text{and} \quad \limsup_{n \to \infty} A(n) = \bigcap_{n=1}^{\infty} \bigcup_{k=n}^{\infty} A(k).$$

Elementary manipulations yield

$$P\left(\liminf_{n \to \infty} A(n)^c\right) = 1 - P\left(\limsup_{n \to \infty} A(n)\right). \tag{2.27}$$

The following is a straightforward consequence of the continuity of probability:

$$P\left(\liminf_{n \to \infty} A(n)\right) \leq \liminf_{n \to \infty} P(A(n)) \leq \limsup_{n \to \infty} P(A(n))$$
$$\leq P\left(\limsup_{n \to \infty} A(n)\right). \tag{2.28}$$

For a proof see e.g. Rosenthal [264], p. 26.

We will need Borel's lemma (sometimes called the first Borel–Cantelli lemma), which is proved in many textbooks on elementary probability. The proof is simple, but we include it here for completeness.

Lemma 2.5.1 (Borel's lemma) *If $(A(n), n \in \mathbb{N})$ is a sequence of events for which $\sum_{n=1}^{\infty} P(A(n)) < \infty$, then $P(\limsup_{n \to \infty} A(n)) = 0$.*

Proof Given any $\epsilon > 0$ we can find $n_0 \in \mathbb{N}$ such that $m > n_0 \Rightarrow \sum_{n=m}^{\infty} P(A(n)) < \epsilon$, hence we find

$$P\left(\limsup_{n \to \infty} A(n)\right) \leq P\left(\bigcup_{n=m}^{\infty} A(n)\right) \leq \sum_{n=m}^{\infty} P(A(n)) < \epsilon,$$

and the result follows. □

For the second Borel–Cantelli lemma, which we will not use in this book, see e.g. Rosenthal [264], p. 26, or Grimmett and Stirzaker [123], p. 288.

2.5.2 Interlacing

Let $Y = (Y(t), t \geq 0)$ be a Lévy process with jumps bounded by 1, so that we have the Lévy–Itô decomposition

$$Y(t) = bt + B_A(t) + \int_{|x|<1} x \tilde{N}(t, dx)$$

for each $t \geq 0$. For the following construction to be non-trivial we will find it convenient to assume that Y may have jumps of arbitrarily small size, i.e. that there exists no $0 < a < 1$ such that $\nu((-a, a)) = 0$.

Now define a sequence $(\epsilon_n, n \in \mathbb{N})$ that decreases monotonically to zero by

$$\epsilon_n = \sup\left\{y \geq 0, \int_{0 < |x| < y} x^2 \nu(dx) \leq \frac{1}{8^n}\right\},$$

where ν is the Lévy measure of Y. We define an associated sequence of Lévy processes $Y_n = (Y_n(t), t \geq 0)$, wherein the size of each jump is bounded below by ϵ_n and above by 1, as follows:

$$Y_n(t) = bt + B_A(t) + \int_{\epsilon_n \leq |x| < 1} x \tilde{N}(t, dx)$$

$$= C_n(t) + \int_{\epsilon_n \leq |x| < 1} x N(t, dx),$$

where, for each $n \in \mathbb{N}$, C_n is the Brownian motion with drift given by

$$C_n(t) = B_A(t) + t\left[b - \int_{\epsilon_n \leq |x| < 1} x \nu(dx)\right],$$

for each $t \geq 0$.

Now $\int_{\epsilon_n \leq |x| < 1} x N(t, dx)$ is a compound Poisson process with jumps $\Delta Y(t)$ taking place at times $(T_n^m, m \in \mathbb{N})$. We can thus build the process Y_n by interlacing, as in Example 1.3.13:

$$Y_n(t) = \begin{cases} C_n(t) & \text{for } 0 \leq t < T_n^1, \\ C_n(T_n^1) + \Delta Y(T_n^1) & \text{for } t = T_n^1, \\ Y_n(T_n^1) + C_n(t) - C_n(T_n^1) & \text{for } T_n^1 < t < T_n^2, \\ Y_n(T_n^2-) + \Delta Y(T_n^2) & \text{for } t = T_2, \end{cases}$$

and so on recursively.

Our main result is the following (cf. Fristedt and Gray [106], Theorem 4, p. 608):

Theorem 2.5.2 *For each $t \geq 0$,*

$$\lim_{n \to \infty} Y_n(t) = Y(t) \quad \text{a.s.}$$

and the convergence is uniform on compact intervals of $\mathbb{R}^+$.

Proof Fix $T \geq 0$ then, for each $0 \leq t \leq T$, $n \in \mathbb{N}$ we have

$$Y_{n+1}(t) - Y_n(t) = \int_{\epsilon_{n+1} < |x| < \epsilon_n} x \tilde{N}(t, dx),$$

which is an L^2-martingale. Hence by Doob's martingale inequality we obtain

$$\mathbb{E}\left(\sup_{0 \leq t \leq T} |Y_{n+1}(t) - Y_n(t)|^2\right) \leq 4\mathbb{E}(|Y_{n+1}(T) - Y_n(T)|^2)$$

$$= 4T \int_{\epsilon_{n+1} < |x| < \epsilon_n} |x|^2 \nu(dx) \leq \frac{4T}{8^n},$$

where we have used (2.10). By Chebyshev's inequality

$$P\left(\sup_{0 \leq t \leq T} |Y_{n+1}(t) - Y_n(t)| \geq \frac{1}{2^n}\right) \leq \frac{4T}{2^n}$$

and by Borel's lemma (Lemma 2.5.1), we deduce that

$$P\left(\limsup_{n \to \infty} \sup_{0 \leq t \leq T} |Y_{n+1}(t) - Y_n(t)| \geq \frac{1}{2^n}\right) = 0;$$

so, by (2.27),

$$P\left(\liminf_{n \to \infty} \sup_{0 \leq t \leq T} |Y_{n+1}(t) - Y_n(t)| < \frac{1}{2^n}\right) = 1.$$

Hence given any $\delta > 0$ there exists $n_0 \in \mathbb{N}$ such that, for $m, n > n_0$, we have

$$\sup_{0 \leq t \leq T} |Y_n(t) - Y_m(t)| \leq \sum_{r=m}^{n-1} \sup_{0 \leq t \leq T} |Y_{r+1}(t) - Y_r(t)| < \sum_{r=m}^{n-1} \frac{1}{2^r} < \delta$$

with probability 1, from which we see that $(Y_n(t), n \in \mathbb{N})$ is almost surely uniformly Cauchy on compact intervals and hence is almost surely uniformly convergent on compact intervals. □

Now let X be an arbitrary Lévy process; then by the Lévy–Itô decomposition, for each $t \geq 0$,

$$X(t) = Y(t) + \int_{|x| \geq 1} x N(t, dx).$$

But $\int_{|x|\geq 1} xN(t,dx)$ is a compound Poisson process and so the paths of X can be obtained by a further interlacing with jumps of size bigger than 1, as in Example 1.3.13.

2.6 Semimartingales

A key aim of stochastic calculus is to make sense of $\int_0^t F(s)dX(s)$ for a suitable class of adapted processes and integrators X. It turns out, as we will see, that the processes we will now define are ideally suited for the role of integrators. We say that X is a *semimartingale* if it is an adapted process such that, for each $t \geq 0$,

$$X(t) = X(0) + M(t) + C(t),$$

where $M = (M(t), t \geq 0)$ is a local martingale and $C = (C(t), t \geq 0)$ is an adapted process of finite variation. In many cases of interest to us the process M will be a martingale.

The Doob–Meyer decomposition (Theorem 2.2.3) implies that $(M(t)^2, t \geq 0)$ is a semimartingale whenever M is square-integrable. Another important class of semimartingales is given by the following result.

Proposition 2.6.1 *Every Lévy process is a semimartingale.*

Proof By the Lévy–Itô decomposition we have, for each $t \geq 0$,

$$X(t) = M(t) + C(t),$$

where

$$M(t) = B_A(t) + \int_{|x|<1} x\tilde{N}(t,dx), \qquad C(t) = bt + \int_{|x|\geq 1} xN(t,dx).$$

We saw above that $M = (M(t), t \geq 0)$ is a martingale.

But $Y(t) = \int_{|x|\geq 1} xN(t,dx)$ is a compound Poisson process and thus for any partition $\mathcal{P}$ of $[0,t]$ we find that

$$\mathrm{var}_\mathcal{P}(Y) \leq \sum_{0\leq s\leq t} |\Delta X(s)|\chi_{[1,\infty)}(\Delta X(s)) < \infty \qquad \text{a.s.},$$

and the required result follows. □

In Chapter 4, we will explore the problem of defining

$$\int_0^t F(s)dX(s) = \int_0^t F(s)dM(s) + \int_0^t F(s)dC(s),$$

for a class of semimartingales. Observe that if F is say locally bounded and measurable and $\mathcal{N}$ is the set on which C fails to be of finite variation then we can define

$$\int_0^t F(s)dC(s)(\omega) = \begin{cases} \int_0^t F(s)(\omega)dC(s)(\omega) & \text{if } \omega \in \Omega - \mathcal{N}, \\ 0 & \text{if } \omega \in \mathcal{N}. \end{cases}$$

In general $\int_0^t F(s)dM(s)$ cannot be defined as a Stieltjes integral; indeed the only continuous martingales that are of finite variation are constants (see Revuz and Yor [260], p. 114). We will learn how to get around this problem in Chapter 4.

The Lévy–Itô decomposition admits an interesting generalisation to arbitrary semimartingales. We sketch this very vaguely – full details can be found in Jacod and Shiryaev [151], p. 84. We define a random measure M_X on $\mathbb{R}^+ \times \mathbb{R}^d$ in the usual way:

$$M_X(t, A) = \#\{0 \leq s \leq t; \Delta X(s) \in A\}$$

for each $t \geq 0$, $A \in \mathcal{B}(\mathbb{R}^d)$. It can be shown that a *compensator* ν_X always exists, this being a random measure on $\mathbb{R}^+ \times \mathbb{R}^d$ for which (in particular) $\int_{\mathbb{R}^d} f(x)[M_X(t, dx) - \nu_X(t, dx)]$ is a martingale for all measurable f such that the integral exists.

For all $t \geq 0$ we then have the decomposition

$$X(t) = B(t) + X^c(t) + \int_{\mathbb{R}^d} h(x)\big[M_X(t, dx) - \nu_X(t, dx)\big]$$
$$+ \int_{\mathbb{R}^d} [x - h(x)]M_X(t, dx),$$

where X^c is a continuous local martingale and B is an adapted process. The mapping h appearing here is a fixed *truncation function*, so that h is bounded and has compact support and $h(x) = x$ in a neighbourhood of the origin. Write $C_{ij} = \langle X_i^c, X_j^c \rangle$; then the *characteristics* of the semimartingale X are (B, C, ν_X). Note that B depends upon the choice of h.

Resources for general material about semimartingales include Jacod and Shiryaev [151], Métivier [221] and Protter [255].

2.7 Notes and further reading

Martingales were first developed by Doob [83] in discrete time and many of their properties were extended to continuous time by P.A. Meyer. His work and that of his collaborators is summarised in Dellacherie and Meyer [78]. Readers should also consult early editions of *Séminaire de Probabilités*: for reviews of some of these articles, consult the database at http://www-irma.u-strasbg.fr/irma/semproba/e_index.shtml. See also the collection of articles edited by Emery and Yor [96].

Brémaud [59] contains a systematic martingale-based approach to point processes, with a number of applications including queues, filtering and control.

The Lévy–Itô decomposition is implicit in work of Lévy [192] and was rigorously established by Itô in [142]. The interlacing construction also appears, at least implicitly, for the first time in this paper. Bretagnolle [61] was responsible for the martingale-based approach used in the present text. Note that he credits this to unpublished work of Marie Duflo. An alternative proof that is closer in spirit to that of Itô can be found in Chapter 4 of Sato [274].

The objects which we have called 'martingale-valued measures' were called 'martingale measures' by Walsh [299]; however, the latter terminology has now become established in mathematical finance to denote probability measures under which the discounted stock price is a martingale (see Chapter 5).

2.8 Appendix: càdlàg functions

Let $I = [a, b]$ be an interval in $\mathbb{R}^+$. A mapping $f : I \to \mathbb{R}^d$ is said to be *càdlàg* (from the French *continue à droite et limité à gauche*) if, for all $t \in (a, b]$, f has a left limit at t and f is right-continuous at t, i.e.

- for all sequences $(t_n, n \in \mathbb{N})$ in I with each $t_n < t$ and $\lim_{n\to\infty} t_n = t$ we have that $\lim_{n\to\infty} f(t_n)$ exists;
- for all sequences $(t_n, n \in \mathbb{N})$ in I with each $t_n \geq t$ and $\lim_{n\to\infty} t_n = t$ we have that $\lim_{n\to\infty} f(t_n) = f(t)$.

A càglàd function (i.e. one that is left-continuous with right limits) is defined similarly.

Clearly any continuous function is càdlàg but there are plenty of other examples, e.g. take $d = 1$ and consider the indicator functions $f(t) = \chi_{[a,b)}(t)$ where $a < b$.

If f is a càdlàg function we will denote the left limit at each point $t \in (a, b]$ as $f(t-) = \lim_{s \uparrow t} f(s)$, and we stress that $f(t-) = f(t)$ if and only if f is continuous at t. We define the *jump at t* by

$$\Delta f(t) = f(t) - f(t-).$$

Clearly a càdlàg function can only have jump discontinuities.

The following result is of great importance for stochastic calculus.

Theorem 2.8.1 *If f is a càdlàg function then the set $S = \{t, \Delta f(t) \neq 0\}$ is at most countable.*

Proof For each $k > 0$, define
$$S_k = \{t, |\Delta f(t)| > k\}.$$

Suppose that S_k has at least one accumulation point x and choose a sequence $(x_n, n \in \mathbb{N})$ in S_k that converges to x. We assume, without loss of generality, that the convergence is from the left and that the sequence is increasing (we can, of course, always achieve this by passing to a subsequence if necessary).

Now, given any $n \in \mathbb{N}$, since $x_n \in S_k$ it follows that f has a left limit at x_n and so, given $\epsilon > 0$, we can find $\delta > 0$ such that, for all y with $y < x_n$ satisfying $x_n - y < \delta$, we have $f(x_n-) - f(y) = \epsilon_0(y)$ where $|\epsilon_0(y)| < \epsilon$.

Now fix $n_0 \in \mathbb{N}$ such that, for all $m > n > n_0$, $x_m - x_n < \delta$; then we have

$$f(x_n) - f(x_m) = \underbrace{f(x_n) - f(x_n-)} + f(x_n-) - f(x_m) = k_0 + \epsilon_0(m),$$

where $|k_0| > k$.

From this it is clear that $(f(x_n), n \in \mathbb{N})$ cannot be Cauchy and so f does not have a left limit at x. Hence S_k has no accumulation points and so is at most countable. However,

$$S = \bigcup_{n \in \mathbb{N}} S_{1/n},$$

from which we deduce that S is countable, as required. □

Note that a more general theorem, which establishes the countability of the set of discontinuities of the first kind for arbitrary real-valued functions, can be found in Hobson [131], p. 304 (see also Klebaner [170], p. 3).

Many useful properties of continuous functions continue to hold for càdlàg functions and we list some of these below.

(1) Let $D(a, b)$ denote the set of all càdlàg functions on $[a, b]$; then $D(a, b)$ is a linear space with respect to pointwise addition and scalar multiplication.
(2) If $f, g \in D(a, b)$ then $fg \in D(a, b)$. Furthermore, if $f(x) \neq 0$ for all $x \in [a, b]$ then $1/f \in D(a, b)$.
(3) If $f \in C(\mathbb{R}^d, \mathbb{R}^d)$ and $g \in D(a, b)$ then the composition $f \circ g \in D(a, b)$.
(4) Every càdlàg function is bounded on finite closed intervals and attains its bounds there.
(5) Every càdlàg function is uniformly right-continuous on finite closed intervals.

2.8 Appendix: càdlàg functions

(6) The uniform limit of a sequence of càdlàg functions on $[a, b]$ is itself càdlàg.
(7) Any càdlàg function can be uniformly approximated on finite intervals by a sequence of step functions.
(8) Every càdlàg function is Borel measurable.

All the above can be proved by tweaking the technique used for establishing the corresponding result for continuous functions. Note that by symmetry these results also hold for càglàd functions.

If $f \in D(a, b)$ we will sometimes find it convenient to consider the associated mapping $\tilde{f} : (a, b] \to \mathbb{R}$ defined by $\tilde{f}(x) = f(x-)$ whenever $x \in (a, b]$. Note that f and $\tilde{f}$ differ at most on a countable number of points and $\tilde{f}$ is càglàd on $(a, b]$. It is not difficult to verify that

$$\sup_{a < x \leq b} |f(x-)| \leq \sup_{a \leq x \leq b} |f(x)|.$$

Using (4), we can define seminorms on $D(\mathbb{R}^+) = D((0, \infty))$ by taking the supremum, i.e. $\|f\|_{a,b} = \sup_{a \leq t \leq b} |f(t)|$ for all $0 \leq a \leq b < \infty$; then the $\{\|\cdot\|_{0,n}, n \in \mathbb{N}\}$ form a separating family and so we obtain a complete metric on $D(\mathbb{R}^+)$ by the prescription

$$d(f, g) = \max_{n \in \mathbb{N}} \frac{\|f - g\|_{0,n}}{2^n(1 + \|f - g\|_{0,n})},$$

(see e.g. Rudin [267] p. 33). Note however that d is not separable. In order to turn $D(\mathbb{R}^+)$ into a Polish space (i.e. a separable topological space that is metrisable by a complete metric), we need to use a topology different from that induced by d. Such a topology exists and is usually called the *Skorohod topology*. We will not have need of it herein and refer the interested reader to Chapter 6 of Jacod and Shiryaev [151] or Chapter 3 of Billingsley [45] for details.

3
Markov processes, semigroups and generators

Summary Markov processes and the important subclass of Feller processes are introduced and shown to be determined by the associated semigroups. We take an analytic diversion into semigroup theory and investigate the important concepts of generator and resolvent. Returning to Lévy processes, we obtain two key representations for the generator: first, as a pseudo-differential operator; second, in 'Lévy–Khintchine form', which is the sum of a second-order elliptic differential operator and a (compensated) integral of difference operators. We also study the subordination of such semigroups and their action in L^p-spaces.

The structure of Lévy generators, but with variable coefficients, extends to a general class of Feller processes, via Courrège's theorems, and also to Hunt processes associated with symmetric Dirichlet forms, where the Lévy–Khintchine-type structure is apparent within the Beurling–Deny formula.

3.1 Markov processes, evolutions and semigroups
3.1.1 Markov processes and transition functions

Intuitively, a stochastic process is Markovian (or, a Markov process) if using the whole past history of the process to predict its future behaviour is no more effective than a prediction based only on a knowledge of the present. This translates into precise mathematics as follows.

Let $(\Omega, \mathcal{F}, P)$ be a probability space equipped with a filtration $(\mathcal{F}_t, t \geq 0)$. Let $X = (X(t), t \geq 0)$ be an adapted process. We say that X is a *Markov process* if, for all $f \in B_b(\mathbb{R}^d)$, $0 \leq s \leq t < \infty$,

$$\mathbb{E}(f(X(t))|\mathcal{F}_s) = \mathbb{E}(f(X(t))|X(s)) \qquad \text{a.s.} \qquad (3.1)$$

Notes

(1) The defining equation (3.1) is sometimes called the 'Markov property'.

3.1 Markov processes, evolutions and semigroups

(2) $\mathbb{R}^d$ may be replaced here by any Polish space, i.e. a separable topological space that is metrisable by a complete metric.
(3) In discrete time, we obtain the well-known notion of the Markov chain.

Example 3.1.1 (Lévy processes) Let X be a Lévy process; then it follows by Exercise 2.1.2 that X is a Markov process.

Recall that $B_b(\mathbb{R}^d)$ is a Banach space with respect to the norm

$$\|f\| = \sup\{|f(x)|, x \in \mathbb{R}^d\}$$

for each $f \in B_b(\mathbb{R}^d)$.

With each Markov process X, we associate a family of operators $(T_{s,t}, 0 \leq s \leq t < \infty)$ from $B_b(\mathbb{R}^d)$ to the Banach space (under the supremum norm) of all bounded functions on $\mathbb{R}^d$ by the prescription

$$(T_{s,t}f)(x) = \mathbb{E}(f(X(t))|X(s) = x)$$

for each $f \in B_b(\mathbb{R}^d)$, $x \in \mathbb{R}^d$. We recall that I is the identity operator, $If = f$, for each $f \in B_b(\mathbb{R}^d)$. We say that the Markov process X is _normal_ if $T_{s,t}(B_b(\mathbb{R}^d)) \subseteq B_b(\mathbb{R}^d)$, for each $0 \leq s \leq t < \infty$.

Theorem 3.1.2 *If X is a normal Markov process, then*

(1) $T_{s,t}$ *is a linear operator on $B_b(\mathbb{R}^d)$ for each $0 \leq s \leq t < \infty$.*
(2) $T_{s,s} = I$ *for each $s \geq 0$.*
(3) $T_{r,s}T_{s,t} = T_{r,t}$ *whenever $0 \leq r \leq s \leq t < \infty$.*
(4) $f \geq 0 \Rightarrow T_{s,t}f \geq 0$ *for all $0 \leq s \leq t < \infty$, $f \in B_b(\mathbb{R}^d)$.*
(5) $T_{s,t}$ *is a contraction, i.e. $\|T_{s,t}\| \leq 1$ for each $0 \leq s \leq t < \infty$.*
(6) $T_{s,t}(1) = 1$ *for all $t \geq 0$.*

Proof Parts (1), (2), (6) and (4) are obvious.

For (3) let $f \in B_b(\mathbb{R}^d)$, $x \in \mathbb{R}^d$; then, for each $0 \leq r \leq s \leq t < \infty$, applying conditioning and the Markov property (3.1) yields

$$(T_{r,t}f)(x) = \mathbb{E}(f(X(t))|X(r) = x) = \mathbb{E}(\mathbb{E}(f(X(t))|\mathcal{F}_s)|X(r) = x)$$
$$= \mathbb{E}(\mathbb{E}(f(X(t))|X(s))|X(r) = x) = \mathbb{E}(T_{s,t}f(X(s))|X(r) = x)$$
$$= (T_{r,s}(T_{s,t}f))(x).$$

(5) For each $f \in B_b(\mathbb{R}^d)$, $0 \leq s \leq t < \infty$,

$$\|T_{s,t}f\| = \sup_{x \in \mathbb{R}^d} \left|\mathbb{E}(f(X(t))|X(s) = x)\right| \leq \sup_{x \in \mathbb{R}^d} \mathbb{E}(|f(X(t))| \big| X(s) = x)$$
$$\leq \sup_{x \in \mathbb{R}^d} |f(x)| \sup_{x \in \mathbb{R}^d} \mathbb{E}(1|X(s) = x)$$
$$= \|f\|.$$

Hence each $T_{s,t}$ is a bounded operator and

$$||T_{s,t}|| = \sup\{||T_{s,t}(g)||, ||g|| = 1\} \leq 1.$$

□

Any family satisfying (1) to (6) of Theorem 3.1.2 is called a *Markov evolution*. Note that of all the six conditions, (3) is the most important, as this needs the Markov property for its proof. The other five properties all hold for arbitrary stochastic processes.

For each $0 \leq s \leq t < \infty$, $A \in \mathcal{B}(\mathbb{R}^d)$, $x \in \mathbb{R}^d$, define

$$p_{s,t}(x, A) = (T_{s,t}\chi_A)(x) = P\big(X(t) \in A | X(s) = x\big). \qquad (3.2)$$

By the properties of conditional probability, each $p_{s,t}(x, \cdot)$ is a probability measure on $\mathcal{B}(\mathbb{R}^d)$. We call the mappings $p_{s,t}$ *transition probabilities*, as they give the probabilities of 'transitions' of the process from the point x at time s to the set A at time t.

If X is an arbitrary Markov process, by equation (1.1) we have

$$(T_{s,t}f)(x) = \int_{\mathbb{R}^d} f(y) p_{s,t}(x, dy) \qquad (3.3)$$

for each $0 \leq s \leq t < \infty$, $f \in B_b(\mathbb{R}^d)$, $x \in \mathbb{R}^d$.

From (3.3) we see that a Markov process is normal if and only if the mappings $x \to p_{s,t}(x, A)$ are measurable for each $A \in \mathcal{B}(\mathbb{R}^d)$, $0 \leq s \leq t < \infty$.

Normal Markov processes are a natural class to deal with from both analytic and probabilistic perspectives, and from now on we will concentrate exclusively on these.

Exercise 3.1.3 Let X be a Lévy process and let q_t be the law of $X(t)$ for each $t \geq 0$. Show that

$$p_{s,t}(A, x) = q_{t-s}(A - x)$$

for each $0 \leq s \leq t < \infty$, $A \in \mathcal{B}(\mathbb{R}^d)$, $x \in \mathbb{R}^d$.

Exercise 3.1.4 A Markov process is said to have a *transition density* if for each $x \in \mathbb{R}^d$, $0 \leq s \leq t < \infty$, there exists a measurable function $y \to \rho_{s,t}(x, y)$ such that

$$p_{s,t}(x, A) = \int_A \rho_{s,t}(x, y) dy.$$

Deduce that a Lévy process $X = (X(t), t \geq 0)$ has a transition density if and only if q_t has a density f_t for each $t \geq 0$, and hence show that

$$p_{s,t}(x, y) = f_{t-s}(y - x)$$

for each $0 \leq s \leq t < \infty, x, y \in \mathbb{R}^d$.

Write down the transition densities for (a) standard Brownian motion, (b) the Cauchy process.

The following result will be familiar to students of Markov chains in its discrete form.

Theorem 3.1.5 (The Chapman–Kolmogorov equations) *If X is a normal Markov process then for each $0 \leq r \leq s \leq t < \infty$, $x \in \mathbb{R}^d$; $A \in \mathcal{B}(\mathbb{R}^d)$,*

$$p_{r,t}(x, A) = \int_{\mathbb{R}^d} p_{s,t}(y, A) p_{r,s}(x, dy). \tag{3.4}$$

Proof Note that since X is normal, the mappings $y \to p_{s,t}(y, A)$ are integrable. Now applying Theorem 3.1.2 and (3.3), we obtain

$$p_{r,t}(x, A) = (T_{r,t}\chi_A)(x) = (T_{r,s}(T_{s,t}\chi_A))(x)$$

$$= \int_{\mathbb{R}^d} (T_{s,t}\chi_A)(y) p_{r,s}(x, dy) = \int_{\mathbb{R}^d} p_{s,t}(y, A) p_{r,s}(x, dy).$$

$\square$

Exercise 3.1.6 Suppose that the Markov process X has a transition density as in Exercise 3.1.4. Deduce that

$$\rho_{r,t}(x, z) = \int_{\mathbb{R}^d} \rho_{r,s}(x, y) \rho_{s,t}(y, z) dy$$

for each $0 \leq r \leq s \leq t < \infty$ and $x, z \in \mathbb{R}^d$.

We have started with a Markov process X and then obtained the Chapman–Kolmogorov equations for the transition probabilities. There is a partial converse to this, which we will now develop. First we need a definition.

Let $\{p_{s,t}; 0 \leq s \leq t < \infty\}$ be a family of mappings from $\mathbb{R}^d \times \mathcal{B}(\mathbb{R}^d) \to [0, 1]$. We say that they are a *normal transition family* if, for each $0 \leq s \leq t < \infty$:

(1) the maps $x \to p_{s,t}(x, A)$ are measurable for each $A \in \mathcal{B}(\mathbb{R}^d)$;
(2) $p_{s,t}(x, \cdot)$ is a probability measure on $\mathcal{B}(\mathbb{R}^d)$ for each $x \in \mathbb{R}^d$;
(3) the Chapman–Kolmogorov equations (3.4) are satisfied.

Theorem 3.1.7 *If $\{p_{s,t}; 0 \leq s \leq t < \infty\}$ is a normal transition family and μ is a fixed probability measure on $\mathbb{R}^d$, then there exists a probability space $(\Omega, \mathcal{F}, P_\mu)$, a filtration $(\mathcal{F}_t, t \geq 0)$ and a Markov process $(X(t), t \geq 0)$ on that space such that:*

(1) $P(X(t) \in A | X(s) = x) = p_{s,t}(x, A)$ *(a.s.) for each $0 \leq s \leq t < \infty$, $x \in \mathbb{R}^d$; $A \in \mathcal{B}(\mathbb{R}^d)$;*
(2) $X(0)$ *has law μ.*

Proof We remind readers of the Kolmogorov existence theorem (Theorem 1.1.16). We take Ω to be the set of all mappings from $\mathbb{R}^+$ to $\mathbb{R}^d$ and $\mathcal{F}$ to be the σ-algebra generated by cylinder sets $I_{t_0,t_1,\ldots,t_n}^{A_0 \times A_1 \times \cdots \times A_n}$, where $0 \leq t_0 \leq t_1 \leq \cdots \leq t_n < \infty$ and $A_0, A_1, \ldots, A_n \in \mathcal{B}(\mathbb{R}^d)$. Define

$$p_{t_0,t_1,\ldots,t_n}(A_0 \times A_1 \times \cdots \times A_n)$$
$$= \int_{A_0} \mu(dx_0) \int_{A_1} p_{0,t_1}(x_0, dx_1) \int_{A_2} p_{t_1,t_2}(x_1, dx_2) \cdots$$
$$\times \int_{A_n} p_{t_{n-1},t_n}(x_{n-1}, dx_n).$$

By the Chapman–Kolmogorov equations (3.4), we can easily verify that $(p_{t_0,t_1,\ldots,t_n}, 0 \leq t_1 \leq t_2 \leq \cdots \leq t_n < \infty)$ satisfy Kolmogorov's consistency criteria and so, by Kolmogorov's existence theorem, Theorem 1.1.16, there exists a probability measure P_μ and a process $X = (X(t), t \geq 0)$ on $(\Omega, \mathcal{F}, P_\mu)$ having the $p_{t_0,t_1,\ldots,t_n}$ as finite-dimensional distributions. X is adapted to its natural filtration.

(2) is now immediate. To establish (1) observe that by the above construction, for each $0 \leq s \leq t < \infty$, $A \in \mathcal{B}(\mathbb{R}^d)$,

$$P(X(t) \in A) = \int_{\mathbb{R}^d} \int_{\mathbb{R}^d} p_{s,t}(x, A) \mu(dx_0) p_{0,s}(x_0, dx).$$

However,

$$P(X(t) \in A) = \int_{\mathbb{R}^d} P(X(t) \in A | X(s) = x) p_{X(s)}(dx)$$
$$= \int_{\mathbb{R}^d} \int_{\mathbb{R}^d} P(X(t) \in A | X(s) = x) \mu(dx_0) p_{0,s}(x_0, dx),$$

and the result follows.

3.1 Markov processes, evolutions and semigroups

Finally we must show that X is Markov. Let $A \in \mathcal{F}_s$; then for each $f \in B_b(\mathbb{R}^d)$

$$\mathbb{E}\big(\chi_A \mathbb{E}(f(X(t))|\mathcal{F}_s)\big) = \mathbb{E}(f(X(t))\chi_A) = P(f(X(t)) \in A)$$
$$= \int_{\mathbb{R}^d} \int_{\mathbb{R}^d} \int_A f(y)\mu(dx_0) p_{0,s}(x_0, dx) p_{s,t}(x, dy).$$

However, by (3.3),

$$\mathbb{E}\big(\chi_A \mathbb{E}(f(X(t))|X(s))\big) = \mathbb{E}\left(\chi_A \int_{\mathbb{R}^d} f(y) p_{s,t}(X(s), dy)\right)$$
$$= \int_{\mathbb{R}^d} \int_{\mathbb{R}^d} \int_A f(y)\mu(dx_0) p_{0,s}(x_0, dx) p_{s,t}(x, dy).$$

We have shown that

$$\mathbb{E}\big(\chi_A \mathbb{E}(f(X(t)))|\mathcal{F}_s\big) = \mathbb{E}\big(\chi_A \mathbb{E}(f(X(t))|X(s))\big),$$

and the result follows since $\{\chi_A, A \in \mathcal{F}_s\}$ is total[1] in $L^2(\Omega, \mathcal{F}_s, P_\mu)$. □

We call the process X constructed in this way a *canonical Markov process*.

A great simplification in the study of Markov processes is made by reduction to the following important subclass. A Markov process is said to be *(time-) homogeneous* if

$$T_{s,t} = T_{0,t-s}$$

for all $0 \leq s \leq t < \infty$; using (3.3), it is easily verified that this holds if and only if

$$p_{s,t}(x, A) = p_{0,t-s}(x, A)$$

for each $0 \leq s \leq t < \infty$, $x \in \mathbb{R}^d$, $A \in \mathcal{B}(\mathbb{R}^d)$. If a Markov process is not homogeneous, it is often said to be *inhomogeneous*.

For homogeneous Markov processes, we will always write the operators $T_{0,t}$ as T_t and the transition probabilities $p_{0,t}$ as p_t.

The key evolution property Theorem 3.1.2(3) now takes the form

$$T_{s+t} = T_s T_t \tag{3.5}$$

---

[1] A set of vectors S is *total* in a Hilbert space H if the set of all finite linear combinations of vectors from S is dense in H.

for each $s, t \geq 0$. Theorem 3.1.2(2) now reads $T_0 = I$ and the Chapman–Kolmogorov equations can be written as

$$p_{t+s}(x, A) = \int_{\mathbb{R}^d} p_s(y, A) p_t(x, dy) \tag{3.6}$$

for each $s, t \geq 0$, $x \in \mathbb{R}^d$, $A \in \mathcal{B}(\mathbb{R}^d)$.

In general any family of linear operators on a Banach space that satisfies (3.5) is called a *semigroup*. By (3.3) and Theorem 3.1.7 we see that the semigroup effectively determines the process if the transition probabilities are normal. There is a deep and extensive analytical theory of semigroups, which we will begin to study in the next chapter. In order to be able to make more effective use of this and to deal with one of the most frequently encountered classes of Markov processes, we will make a further definition.

A homogeneous Markov process X is said to be a *Feller process* if

(1) $T_t : C_0(\mathbb{R}^d) \subseteq C_0(\mathbb{R}^d)$ for all $t \geq 0$,
(2) $\lim_{t \to 0} ||T_t f - f|| = 0$ for all $f \in C_0(\mathbb{R}^d)$.

In this case, the semigroup associated with X is called a *Feller semigroup*. More generally, we say that any semigroup defined on the Banach space $C_0(\mathbb{R}^d)$ is *Feller* if it satisfies (2) above and (one-parameter versions of) all the conditions of Theorem 3.1.2.

Note 1 Some authors prefer to use $C_b(\mathbb{R}^d)$ in the definition of a Feller process instead of $C_0(\mathbb{R}^d)$. Although this can make life easier, the space $C_0(\mathbb{R}^d)$ has nicer analytical properties than $C_b(\mathbb{R}^d)$ and this can allow the proof of important probabilistic theorems such as the one below. In particular, for most of the semigroups which we study in this book, condition (2) above fails when we replace $C_0(\mathbb{R}^d)$ with $C_b(\mathbb{R}^d)$. For more on this theme, see Schilling [276]. We will consider this point again in Chapter 6.

Note 2 There is also a notion of a *strong Feller semigroup*, for which it is required that $T_t(B_b(\mathbb{R}^d)) \subseteq C_b(\mathbb{R}^d)$ for each $t \geq 0$. We will not have need of this concept in this book.

Theorem 3.1.8 *If X is a Feller process, then its transition probabilities are normal.*

Proof See Revuz and Yor [260], p. 83. □

The class of all Feller processes is far from empty, as the following result shows.

Theorem 3.1.9 *Every Lévy process is a Feller process.*

3.1 Markov processes, evolutions and semigroups

Proof If X is a Lévy process then it is a homogeneous Markov process by Exercise 2.1.2. Let q_t be the law of $X(t)$; then, by Proposition 1.4.4, $(q_t, t \geq 0)$ is a weakly continuous convolution semigroup of probability measures and, using the result of Exercise 3.1.3 and (3.3), we see that for each $f \in B_b(\mathbb{R}^d)$, $x \in \mathbb{R}^d, t \geq 0$,

$$(T_t f)(x) = \int_{\mathbb{R}^d} f(x+y) q_t(dy).$$

Now let $f \in C_0(\mathbb{R}^d)$. We need to prove that $T_t f \in C_0(\mathbb{R}^d)$ for each $t \geq 0$. First observe that if $(x_n, n \in \mathbb{N})$ is any sequence converging to $x \in \mathbb{R}^d$ then, by the dominated convergence theorem,

$$\lim_{n \to \infty} (T_t f)(x_n) = \lim_{n \to \infty} \int_{\mathbb{R}^d} f(x_n + y) q_t(dy)$$

$$= \int_{\mathbb{R}^d} f(x+y) q_t(dy) = (T_t f)(x),$$

from which it follows that $T_t f$ is continuous. We can then apply dominated convergence again to deduce that

$$\lim_{|x| \to \infty} |(T_t f)(x)| \leq \lim_{|x| \to \infty} \int_{\mathbb{R}^d} |f(x+y)| q_t(dy)$$

$$= \int_{\mathbb{R}^d} \lim_{|x| \to \infty} |f(x+y)| q_t(dy) = 0.$$

To prove the second part of the Feller condition, observe that the result is trivial if $f = 0$, so assume that $f \neq 0$ and use the stochastic continuity of X to deduce that, for any $\epsilon > 0$ and any $r > 0$, there exists $t_0 > 0$ such that $0 \leq t < t_0 \Rightarrow q_t(B_r(0)^c) < \epsilon/(4\|f\|)$.

Since every $f \in C_0(\mathbb{R}^d)$ is uniformly continuous, we can find $\delta > 0$ such that $\sup_{x \in \mathbb{R}^d} |f(x+y) - f(x)| < \epsilon/2$ for all $y \in B_\delta(0)$.

Choosing $r = \delta$, we then find that, for all $0 \leq t \leq t_0$,

$$\|T_t f - f\| = \sup_{x \in \mathbb{R}^d} |T_t f(x) - f(x)|$$

$$\leq \int_{B_\delta(0)} \sup_{x \in \mathbb{R}^d} |f(x+y) - f(x)| q_t(dy)$$

$$+ \int_{B_\delta(0)^c} \sup_{x \in \mathbb{R}^d} |f(x+y) - f(x)| q_t(dy)$$

$$< \frac{\epsilon}{2} q_t(B_\delta(0)) + 2\|f\| q_t(B_\delta(0)^c) < \epsilon,$$

and the required result follows. □

3.1.2 Sub-Markov processes

Much of the material that we have discussed so far can be extended to a more general formalism. Suppose that we are given a family $\{p_{s,t}, 0 \leq s \leq t < \infty\}$ of mappings from $\mathbb{R}^d \times \mathcal{B}(\mathbb{R}^d) \to [0, 1]$ which satisfy (1) and (3) of the definition of a normal transition family, but (2) is weakened to

(2′) for each $0 \leq s \leq t < \infty$, $x \in \mathbb{R}^d$, $p_{s,t}(x, \cdot)$ is a finite measure on $\mathcal{B}(\mathbb{R}^d)$, with $p_{s,t}(x, \mathbb{R}^d) \leq 1$.

We can then extend the $p_{s,t}$ to give a normal transition family by using the following device. We introduce a new point Δ, called the *cemetery point*, and work in the one-point compactification $\mathbb{R}^d \cup \{\Delta\}$; then $\{\tilde{p}_{s,t}, 0 \leq s \leq t < \infty\}$ is a normal transition family, where we define

$$\tilde{p}_{s,t}(x, A) = p_{s,t}(x, A) \quad \text{whenever } x \in \mathbb{R}^d,\ A \in \mathcal{B}(\mathbb{R}^d),$$
$$\tilde{p}_{s,t}(x, \{\Delta\}) = 1 - p_{s,t}(x, \mathbb{R}^d) \quad \text{whenever } x \in \mathbb{R}^d,$$
$$\tilde{p}_{s,t}(\{\Delta\}, \mathbb{R}^d) = 0, \qquad \tilde{p}_{s,t}(\Delta, \{\Delta\}) = 1.$$

Exercise 3.1.10 Check that the members of the family $\{\tilde{p}_{s,t}, 0 \leq s \leq t < \infty\}$ satisfy the Chapman–Kolmogorov equations.

Given such a family, we can then apply Theorem 3.1.7 to construct a Markov process $X = (X(t), t \geq 0)$ on $\mathbb{R}^d \cup \{\Delta\}$. We emphasise that X is not, in general, a Markov process on $\mathbb{R}^d$ and we may introduce the *lifetime* of X as the random variable l_X, where

$$l_X(\omega) = \inf\{t > 0;\ X(t)(\omega) \notin \mathbb{R}^d\}$$

for each $\omega \in \Omega$.

We call X a *sub-Markov process*; it is homogeneous if $p_{s,t} = p_{t-s}$ for each $0 \leq s \leq t < \infty$. We may associate a semigroup $(\tilde{T}_t, t \geq 0)$ of linear operators on $B_b(\mathbb{R}^d \cup \{\infty\})$ with such a homogeneous Markov process X, but it is more interesting to consider the semigroup $(T_t, t \geq 0)$ of linear operators on $B_b(\mathbb{R}^d)$ given by

$$(T_t f)(x) = \int_{\mathbb{R}^d} f(y) p_t(x, dy)$$

for each $t \geq 0$, $f \in B_b(\mathbb{R}^d)$, $x \in \mathbb{R}^d$. This satisfies all the conditions of Theorem 3.1.2 except (6), which is weakened to $T_t(1) \leq 1$ for all $t \geq 0$.

If, also, each $T_t(C_0(\mathbb{R}^d)) \subseteq C_0(\mathbb{R}^d)$ and $\lim_{t \to 0} ||T_t f - f|| = 0$, we say that X is a *sub-Feller process* and $(T_t, t \geq 0)$ is a *sub-Feller semigroup*. Many results obtained in this chapter for Feller processes and Feller semigroups extend naturally to the sub-Feller case; see e.g. Berg and Forst [35] and Jacob [148].

3.2 Semigroups and their generators

In the last section, we saw that the theory of homogeneous Markov processes is very closely related to the properties of families of linear operators in Banach spaces called semigroups. In this section, we will develop some understanding of these from a purely analytical point of view, which we can then feed back into later probabilistic discussions. Readers who feel they lack the necessary background in functional analysis are recommended to study the appendix to this chapter section 3.8, where they can learn in particular about unbounded operators and related concepts used below such as domains, closure, graph norms, cores and resolvents.

Most of the material given below is standard. There are many good books on semigroup theory and we have followed Davies [76] very closely. Many books about Markov processes also contain introductory material of a similar type to that given below, and readers may consult e.g. Jacob [149], Ethier and Kurtz [99] or Ma and Röckner [204].

Let B be a real Banach space and $L(B)$ be the algebra of all bounded linear operators on B. A <u>one-parameter semigroup of contractions</u> on B is a family of bounded, linear operators $(T_t, t \geq 0)$ on B for which

(1) $T_{s+t} = T_s T_t$ for all $s, t \geq 0$,
(2) $T_0 = I$,
(3) $||T_t|| \leq 1$ for all $t \geq 0$,
(4) the map $t \to T_t$ from $\mathbb{R}^+$ to $L(B)$ is strongly continuous at zero, i.e. $\lim_{t \downarrow 0} ||T_t \psi - \psi|| = 0$ for all $\psi \in B$,

From now on we will say that $(T_t, t \geq 0)$ is a *semigroup* whenever it satisfies the above conditions.

Lemma 3.2.1 *If $(T_t, t \geq 0)$ is a semigroup in a Banach space B, then the map $t \to T_t$ is strongly continuous from $\mathbb{R}^+$ to $L(B)$, i.e. $\lim_{s \to t} ||T_t \psi - T_s \psi|| = 0$ for all $t \geq 0, \psi \in B$.*

Proof If $(T_t, t \geq 0)$ is a semigroup then it is strongly continuous at zero. Fix

$t \geq 0$, $\psi \in B$; then for all $h > 0$ we have

$$\|T_{t+h}\psi - T_t\psi\| = \|T_t(T_h - I)\psi\| \quad \text{by (1) and (2)}$$
$$\leq \|T_t\| \|(T_h - I)\psi\| \leq \|(T_h - I)\psi\| \quad \text{by (3)}.$$

A similar argument holds when $h < 0$, and the result follows. □

Note Semigroups satisfying just the conditions (1), (2) and (4) given above are studied in the literature, and these are sometimes called C_0-*semigroups*. Although they are no longer necessarily contractions, it can be shown that there exist $M > 0$ and $\beta \in \mathbb{R}$ such that $\|T_t\| \leq Me^{\beta t}$ for all $t \geq 0$. Although all the theory given below extends naturally to encompass this more general class, we will be content to study the more restrictive case as this is sufficient for our needs. Indeed, the reader should quickly confirm that every Feller semigroup is a (contraction) semigroup on $C_0(\mathbb{R}^d)$ in the above sense.

Example 3.2.2 Let $B = C_0(\mathbb{R})$ and consider the semigroup $(T_t, t \geq 0)$ defined by $(T_t f)(x) = f(x + t)$ for each $f \in C_0(\mathbb{R})$, $x \in \mathbb{R}^d$, $t \geq 0$. This is called the *translation semigroup*. Now if $f \in C_0^\infty(\mathbb{R})$ is real-analytic, so that it can be represented by a Taylor series, we have

$$(T_t f)(x) = \sum_{n=0}^\infty \frac{t^n}{n!} (D^n f)(x) = \text{`}e^{tD} f\text{'},$$

where $Df(x) = f'(x)$ defines the operator of differentiation.

Exercise 3.2.3 Check the semigroup conditions (1) to (4) for the translation semigroup.

Exercise 3.2.4 Let A be a bounded operator in a Banach space B and for each $t \geq 0$, $\psi \in B$, define

$$T_t \psi = \sum_{n=0}^\infty \frac{t^n}{n!} A^n \psi = \text{`}e^{tA}\psi\text{'}.$$

Show that $(T_t, t \geq 0)$ is a strongly continuous semigroup of bounded operators in B, that $(T_t, t \geq 0)$ is norm-continuous, in that $\lim_{t \downarrow 0} \|T_t - I\| = 0$.

These examples have a valuable moral. Given a semigroup $(T_t, t \geq 0)$, we should try to find a linear operator A for which $T_t = e^{tA}$ can be given meaning. In general, just as D is an unbounded operator that does not operate on the whole of $C_0(\mathbb{R}^d)$ so we would expect A to be unbounded in general.

3.2 Semigroups and their generators

Now let $(T_t, t \geq 0)$ be an arbitrary semigroup in a Banach space B. We define

$$D_A = \left\{ \psi \in B; \exists \phi_\psi \in B \text{ such that } \lim_{t \downarrow 0} \left\| \frac{T_t \psi - \psi}{t} - \phi_\psi \right\| = 0 \right\}.$$

It is easy to verify that D_A is a linear space and thus we may define a linear operator A in B, with domain D_A, by the prescription

$$A\psi = \phi_\psi,$$

so that, for each $\psi \in D_A$,

$$A\psi = \lim_{t \downarrow 0} \frac{T_t \psi - \psi}{t}.$$

We call A the *infinitesimal generator*, or sometimes just the *generator*, of the semigroup $(T_t, t \geq 0)$. In the case where $(T_t, t \geq 0)$ is the Feller semigroup associated with a Feller process $X = (X(t), t \geq 0)$, we sometimes call A the *generator of X*.

In the following, we will utilise the *Bochner integral* of measurable mappings $f : \mathbb{R}^+ \to B$, which we write in the usual way as $\int_0^t f(s) ds$. This is defined, in a similar way to the Lebesgue integral, as a limit of integrals of simple B-valued functions, and we will take for granted that standard results such as dominated convergence can be extended to this context. A nice introduction to this topic can be found in Appendix E of Cohn [73], pp. 350–7. For an alternative approach based on the Riemann integral, see Ethier and Kurtz [99], pp. 8–9.

Let $(T_t, t \geq 0)$ be a semigroup in B and let $\psi \in B$. Consider the family of vectors $(\psi(t), t \geq 0)$, where each $\psi(t)$ is defined as a Bochner integral

$$\psi(t) = \int_0^t T_u \psi \, du.$$

For $s > 0$, we will frequently apply the continuity of T_s together with the semigroup condition (1) to write

$$T_s \psi(t) = \int_0^t T_{s+u} \psi \, du. \tag{3.7}$$

Readers who are worried about moving T_s through the integral should read Cohn [73], p. 352.

The following technical lemma plays a key role later.

Lemma 3.2.5 $\psi(t) \in D_A$ for each $t \geq 0$, $\psi \in B$ and

$$A\psi(t) = T_t\psi - \psi.$$

Proof Using (3.7), the fundamental theorem of calculus, the fact that $T_0 = I$ and a standard change of variable, we find for each $t \geq 0$,

$$\lim_{h\downarrow 0}\frac{1}{h}[T_h\psi(t) - \psi(t)] = \lim_{h\downarrow 0}\left(\frac{1}{h}\int_0^t T_{h+u}\psi\,du - \frac{1}{h}\int_0^t T_u\psi\,du\right)$$

$$= \lim_{h\downarrow 0}\left(\frac{1}{h}\int_h^{t+h} T_u\psi\,du - \frac{1}{h}\int_0^t T_u\psi\,du\right)$$

$$= \lim_{h\downarrow 0}\left(\frac{1}{h}\int_t^{t+h} T_u\psi\,du - \frac{1}{h}\int_0^h T_u\psi\,du\right)$$

$$= T_t\psi - \psi,$$

and the required result follows. □

Theorem 3.2.6

(1) D_A is dense in B.
(2) $T_t D_A \subseteq D_A$ for each $t \geq 0$.
(3) $T_t A\psi = A T_t\psi$ for each $t \geq 0$, $\psi \in D_A$.

Proof (1) By Lemma 3.2.5, $\psi(t) \in D_A$ for each $t \geq 0$, $\psi \in B$, but, by the fundamental theorem of calculus, $\lim_{t\downarrow 0}(\psi(t)/t) = \psi$; hence D_A is dense in B as required.

For (2) and (3), suppose that $\psi \in D_A$ and $t \geq 0$; then, by the definition of A and the continuity of T_t, we have

$$AT_t\psi = \left[\lim_{h\downarrow 0}\frac{1}{h}(T_h - I)\right]T_t\psi = \lim_{h\downarrow 0}\frac{1}{h}(T_{t+h} - T_t)\psi$$

$$= T_t\left[\lim_{h\downarrow 0}\frac{1}{h}(T_h - I)\right]\psi = T_t A\psi.$$

□

The strong derivative in B of the mapping $t \to T_t\psi$, where $\psi \in D_A$, is given by

$$\frac{d}{dt}T_t\psi = \lim_{h\downarrow 0}\frac{T_{t+h}\psi - T_t\psi}{h}.$$

From the proof of Theorem 3.2.6, we deduce that

$$\frac{d}{dt}T_t\psi = AT_t\psi. \tag{3.8}$$

More generally, it can be shown that $t \to T_t\psi$ is the unique solution of the following initial-value problem in Banach space:

$$\frac{d}{dt}u(t) = Au(t), \qquad u(0) = \psi;$$

see e.g. Davies [76], p. 5. This justifies the notation $T_t = e^{tA}$.

Theorem 3.2.7 *A is closed.*

Proof Let $(\psi_n, n \in \mathbb{N}) \in D_A$ be such that $\lim_{n\to\infty} \psi_n = \psi \in B$ and $\lim_{n\to\infty} A\psi_n = \phi \in B$. We must prove that $\psi \in D_A$ and $\phi = A\psi$.

First observe that, for each $t \geq 0$, by continuity, equation (3.8) and Theorem 3.2.6(3),

$$T_t\psi - \psi = \lim_{n\to\infty}(T_t\psi_n - \psi_n) = \lim_{n\to\infty}\int_0^t T_s A\psi_n \, ds$$

$$= \int_0^t T_s\phi \, ds \tag{3.9}$$

where the passage to the limit in the last line is justified by the fact that

$$\lim_{n\to\infty}\left\|\int_0^t T_s A\psi_n \, ds - \int_0^t T_s\phi \, ds\right\| \leq \lim_{n\to\infty}\int_0^t \|T_s(A\psi_n - \phi)\| ds$$

$$\leq t \lim_{n\to\infty}\|(A\psi_n - \phi)\| = 0.$$

Now, by the fundamental theorem of calculus applied to (3.9), we have

$$\lim_{t\downarrow 0}\frac{1}{t}(T_t\psi - \psi) = \phi,$$

from which the required result follows. □

The next result is extremely useful in applications.

Theorem 3.2.8 *If $D \subseteq D_A$ is such that*

(1) *D is dense in B,*
(2) *$T_t(D) \subseteq D$ for all $t \geq 0$,*

then D is a core for A.

Proof Let $\overline{D}$ be the closure of D in D_A with respect to the graph norm $|||.|||$ (where we recall that $|||\psi||| = ||\psi|| + ||A\psi||$ for each $\psi \in D_A$).

Let $\psi \in D_A$; then by hypothesis (1), we know there exist $(\psi_n, n \in \mathbb{N})$ in D such that $\lim_{n\to\infty} \psi_n = \psi$. Now define the Bochner integrals $\psi(t) = \int_0^t T_s \psi \, ds$ and $\psi_n(t) = \int_0^t T_s \psi_n \, ds$ for each $n \in \mathbb{N}$ and $t \geq 0$. By hypothesis (2) and Lemma 3.2.5, we deduce that each $\psi_n(t) \in \overline{D_A}$. Using Lemma 3.2.5 again and the fact that T_t is a contraction, we obtain for each $t \geq 0$

$$\lim_{n\to\infty} |||\psi(t) - \psi_n(t)|||$$
$$= \lim_{n\to\infty} ||\psi(t) - \psi_n(t)|| + \lim_{n\to\infty} ||A\psi(t) - A\psi_n(t)||$$
$$\leq \lim_{n\to\infty} \int_0^t ||T_s(\psi - \psi_n)|| ds + \lim_{n\to\infty} ||(T_t\psi - \psi) - (T_t\psi_n - \psi_n)||$$
$$\leq (t+2) \lim_{n\to\infty} ||\psi - \psi_n|| = 0,$$

and so $\psi(t) \in \overline{D}$ for each $t \geq 0$. Furthermore, by Lemma 3.2.5 and the fundamental theorem of calculus, we find

$$\lim_{t\downarrow 0} \left|\left|\left|\frac{1}{t}\psi(t) - \psi\right|\right|\right|$$
$$= \lim_{t\downarrow 0} \left|\left|\frac{1}{t}\int_0^t T_s\psi \, ds - \psi\right|\right| + \lim_{t\downarrow 0} \left|\left|\frac{1}{t}A\psi(t) - A\psi\right|\right|$$
$$= \lim_{t\downarrow 0} \left|\left|\frac{1}{t}\int_0^t T_s\psi \, ds - \psi\right|\right| + \lim_{t\downarrow 0} \left|\left|\frac{1}{t}(T_t\psi - \psi) - A\psi\right|\right| = 0.$$

From this we can easily deduce that $D_A \subseteq \overline{D}$, from which it is clear that D is a core for A, as required. $\square$

We now turn our attention to the resolvent $R_\lambda(A) = (\lambda - A)^{-1}$, which is defined for all λ in the resolvent set $\rho(A)$. Of course, there is no a priori reason why $\rho(A)$ should be non-empty. Fortunately we have the following:

Theorem 3.2.9 *If A is the generator of a semigroup $(T_t, t \geq 0)$, then $(0, \infty) \subseteq \rho(A)$ and, for each $\lambda > 0$,*

$$R_\lambda(A) = \int_0^\infty e^{-\lambda t} T_t \, dt. \tag{3.10}$$

Proof For each $\lambda > 0$, we define a linear operator $S_\lambda(A)$ by the Laplace transform on the right-hand side of (3.10). Our goal is to prove that this really is the resolvent. Note first of all that $S_\lambda(A)$ is a bounded operator on B; indeed,

3.2 Semigroups and their generators

for each $\psi \in B$, $t \geq 0$, on using the contraction property of T_t we find that

$$||S_\lambda(A)\psi|| \leq \int_0^\infty e^{-\lambda t}||T_t\psi||dt \leq ||\psi|| \int_0^\infty e^{-\lambda t}dt = \frac{1}{\lambda}||\psi||.$$

Hence we have $||S_\lambda(A)|| \leq 1/\lambda$.

Now define $\psi_\lambda = S_\lambda(A)\psi$ for each $\psi \in B$. Then by continuity, change of variable and the fundamental theorem of calculus, we have

$$\lim_{h \downarrow 0} \frac{1}{h}(T_h \psi_\lambda - \psi_\lambda)$$

$$= \lim_{h \downarrow 0} \left(\frac{1}{h} \int_0^\infty e^{-\lambda t} T_{t+h}\psi\, dt - \frac{1}{h} \int_0^\infty e^{-\lambda t} T_t \psi\, dt \right)$$

$$= \lim_{h \downarrow 0} \left(\frac{1}{h} \int_h^\infty e^{-\lambda(t-h)} T_t \psi\, dt - \frac{1}{h} \int_0^\infty e^{-\lambda t} T_t \psi\, dt \right)$$

$$= -\lim_{h \downarrow 0} e^{\lambda h} \frac{1}{h} \int_0^h e^{-\lambda t} T_t \psi\, dt + \lim_{h \downarrow 0} \frac{1}{h}(e^{\lambda h} - 1) \int_0^\infty e^{-\lambda t} T_t \psi\, dt$$

$$= -\psi + \lambda S_\lambda(A)\psi.$$

Hence $\psi_\lambda \in D_A$ and $A\psi_\lambda = -\psi + \lambda S_\lambda(A)\psi$, i.e. for all $\psi \in B$

$$(\lambda - A)S_\lambda(A)\psi = \psi.$$

So $\lambda - A$ is surjective for all $\lambda > 0$ and its right inverse is $S_\lambda(A)$.

Our proof is complete if we can show that $\lambda - A$ is also injective. To establish this, assume that there exists $\psi \neq 0$ such that $(\lambda - A)\psi = 0$ and define $\psi_t = e^{\lambda t}\psi$ for each $t \geq 0$. Then differentiation yields the initial-value problem

$$\psi_t' = \lambda e^{\lambda t}\psi = A\psi_t$$

with initial condition $\psi_0 = \psi$. But, by the remarks following equation (3.8), we see that $\psi_t = T_t \psi$ for all $t \geq 0$. We then have

$$||T_t\psi|| = ||\psi_t|| = |e^{\lambda t}|\,||\psi||,$$

and so $||T_t|| \geq ||T_t\psi||/||\psi|| = |e^{\lambda t}| > 1$, since $\lambda > 0$. But we know that each T_t is a contraction, and so we have a contradiction. Hence we must have $\psi = 0$ and the proof is complete. □

The final question that we will consider in this section leads to a converse to the last theorem. Suppose that A is a given densely defined closed linear operator in a Banach space B. Under what conditions is it the generator of a semigroup? The answer to this is given by the celebrated Hille–Yosida theorem.

Theorem 3.2.10 (Hille–Yosida) *Let A be a densely defined closed linear operator in a Banach space B and let $R_\lambda(A) = (\lambda - A)^{-1}$ be its resolvent for $\lambda \in \rho(A) \subseteq \mathbb{C}$. A is the generator of a one-parameter contraction semigroup in B if and only if*

(1) $(0, \infty) \subseteq \rho(A)$,
(2) $\|R_\lambda(A)\| \leq 1/\lambda$ for all $\lambda > 0$.

Proof Necessity has already been established in the proof of Theorem 3.2.9. We will not prove sufficiency here but direct the reader to standard texts such as Davies [76], Ma and Röckner [204] and Jacob [149]. □

The Hille–Yosida theorem can be generalised to give necessary and sufficient conditions for the closure of a closable operator to generate a semigroup. This result can be found in e.g. Section 4.1 of Jacob [149] or Chapter 1 of Ethier and Kurtz [99].

3.3 Semigroups and generators of Lévy processes

Here we will investigate the application to Lévy processes of some of the analytical concepts introduced in the last section. To this end, we introduce a Lévy process $X = (X(t), t \geq 0)$ that is adapted to a given filtration $(\mathcal{F}_t, t \geq 0)$ in a probability space $(\Omega, \mathcal{F}, P)$. The mapping η is the Lévy symbol of X, so that

$$\mathbb{E}(e^{i(u, X(t))}) = e^{t\eta(u)}$$

for all $u \in \mathbb{R}^d$. From Theorem 1.2.17 we know that η is a continuous, hermitian, conditionally positive mapping from $\mathbb{R}^d$ to C that satisfies $\eta(0) = 0$ and whose precise form is given by the Lévy–Khintchine formula. For each $t \geq 0$, q_t will denote the law of $X(t)$. We have already seen in Theorem 3.1.9 that X is a Feller process and if $(T_t, t \geq 0)$ is the associated Feller semigroup then

$$(T_t f)(x) = \int_{\mathbb{R}^d} f(x + y) q_t(dy)$$

for each $f \in B_b(\mathbb{R}^d)$, $x \in \mathbb{R}^d$, $t \geq 0$, i.e.

$$(T_t f)(x) = \mathbb{E}(f(X(t) + x)). \tag{3.11}$$

3.3.1 Translation-invariant semigroups

Let $(\tau_a, a \in \mathbb{R}^d)$ be the translation group acting in $B_b(\mathbb{R}^d)$, so that $(\tau_a f)(x) = f(x - a)$ for each $a, x \in \mathbb{R}^d$, $f \in B_b(\mathbb{R}^d)$.

3.3 Semigroups and generators of Lévy processes

We then find that

$$(T_t(\tau_a f))(x) = \mathbb{E}\big((\tau_a f)(X(t)+x)\big) = \mathbb{E}\big(f(X(t)+x-a)\big)$$
$$= (T_t f)(x-a) = (\tau_a(T_t f))(x),$$

i.e.

$$T_t \tau_a = \tau_a T_t$$

for each $t \geq 0$, $a \in \mathbb{R}^d$. The semigroup $(T_t, t \geq 0)$ is then said to be *translation invariant*. This property gives us another way of characterising Lévy processes within the class of Markov processes.

In the following result, we will take $(\Omega, \mathcal{F}, P)$ to be the canonical triple given by the Kolmogorov existence theorem, as used in Theorem 3.1.7.

Theorem 3.3.1 *If $(T_t, t \geq 0)$ is the semigroup associated with a canonical Feller process X for which $X(0) = 0$ (a.s.), then this semigroup is translation invariant if and only if X is a Lévy process.*

Proof We have already seen that the semigroup associated with a Lévy process is translation invariant. Conversely, let $(T_t, t \geq 0)$ be a translation-invariant Feller semigroup associated with a Feller process X with transition probabilities $(p_t, t \geq 0)$. Then for each $a, x \in \mathbb{R}^d$, $t \geq 0$, $f \in C_0(\mathbb{R}^d)$, by (3.3) we have

$$(\tau_a(T_t f))(x) = \int_{\mathbb{R}^d} f(y) p_t(x-a, dy).$$

Moreover,

$$(T_t(\tau_a f))(x) = \int_{\mathbb{R}^d} (\tau_a f)(y) p_t(x, dy) = \int_{\mathbb{R}^d} f(y-a) p_t(x, dy)$$
$$= \int_{\mathbb{R}^d} f(y) p_t(x, dy+a).$$

So, by translation invariance, we have

$$\int_{\mathbb{R}^d} f(y) p_t(x-a, dy) = \int_{\mathbb{R}^d} f(y) p_t(x, dy+a).$$

Now we may apply the Riesz representation theorem for continuous linear functionals on $C_0(\mathbb{R}^d)$ (see e.g. Cohn [73], pp. 209–10) to deduce that

$$p_t(x-a, B) = p_t(x, B+a) \tag{3.12}$$

for all $t \geq 0$, $a, x \in \mathbb{R}^d$, $B \in \mathcal{B}(\mathbb{R}^d)$.

Let q_t be the law of $X(t)$ for each $t \geq 0$, so that $q_t(B) = p_t(0, B)$ for each $B \in \mathcal{B}(\mathbb{R}^d)$; then, by (3.12), we have $p_t(x, B) = q_t(B-x)$ for each

$x \in \mathbb{R}^d$. Now apply the Chapman–Kolmogorov equations to deduce that, for all $s, t \geq 0$,

$$q_{t+s}(B) = p_{t+s}(0, B) = \int_{\mathbb{R}^d} p_t(y, B) p_s(0, dy) = \int_{\mathbb{R}^d} q_t(B - y) q_s(dy),$$

so $(q_t, t \geq 0)$ is a convolution semigroup of probability measures. It is vaguely continuous, since $(T_t, t \geq 0)$ is a Feller semigroup and so

$$\lim_{t \downarrow 0} \int_{\mathbb{R}^d} f(y) q_t(dy) = \lim_{t \downarrow 0} (T_t f)(0) = f(0).$$

for all $f \in C_0(\mathbb{R}^d)$. Hence by Theorem 1.4.5 and the note at the end of Subsection 1.4.1, we deduce that the co-ordinate process on $(\Omega, \mathcal{F}, P)$ is a Lévy process. $\square$

Exercise 3.3.2 Let X be a Lévy process with infinitesimal generator A. Deduce that, for all $a \in \mathbb{R}^d$, $\tau_a(D_A) \subseteq D_A$ and that for all $f \in D_A$

$$\tau_a A f = A \tau_a f.$$

3.3.2 Representation of semigroups and generators by pseudo-differential operators

We now turn our attention to the infinitesimal generators of Lévy processes.[1] Here we will require a very superficial knowledge of pseudo-differential operators acting in the Schwartz space $S(\mathbb{R}^d)$ of rapidly decreasing functions. Those requiring some background in this may consult the final part of Section 3.8. There is also no harm (apart from a slight reduction in generality) in replacing $S(\mathbb{R}^d)$ by $C_c^\infty(\mathbb{R}^d)$ in what follows.

Let $f \in S(\mathbb{R}^d)$. We recall that its Fourier transform is $\hat{f} \in S(\mathbb{R}^d, \mathbb{C})$, where

$$\hat{f}(u) = (2\pi)^{-d/2} \int_{\mathbb{R}^d} e^{-i(u,x)} f(x) dx$$

for all $u \in \mathbb{R}^d$, and the Fourier inversion formula yields

$$f(x) = (2\pi)^{-d/2} \int_{\mathbb{R}^d} \hat{f}(u) e^{i(u,x)} du$$

for each $x \in \mathbb{R}^d$.

A number of useful results about the Fourier transform are collected in the appendix at the end of this chapter. The next theorem is of great importance in the analytic study of Lévy processes and of their generalisations.

---

[1] In order to continue denoting the infinitesimal generator as A, we will henceforth use a to denote the positive definite symmetric matrix appearing in the Lévy–Khinchine formula.

3.3 Semigroups and generators of Lévy processes

Theorem 3.3.3 *Let X be a Lévy process with Lévy symbol η and characteristics (b, a, v). Let $(T_t, t \geq 0)$ be the associated Feller semigroup and A be its infinitesimal generator.*

(1) *For each $t \geq 0$, $f \in S(\mathbb{R}^d)$, $x \in \mathbb{R}^d$,*

$$(T_t f)(x) = (2\pi)^{-d/2} \int_{\mathbb{R}^d} e^{i(u,x)} e^{t\eta(u)} \hat{f}(u) du,$$

so that T_t is a pseudo-differential operator with symbol $e^{t\eta}$.

(2) *For each $f \in S(\mathbb{R}^d)$, $x \in \mathbb{R}^d$,*

$$(Af)(x) = (2\pi)^{-d/2} \int_{\mathbb{R}^d} e^{i(u,x)} \eta(u) \hat{f}(u) du,$$

so that A is a pseudo-differential operator with symbol η.

(3) *For each $f \in S(\mathbb{R}^d)$, $x \in \mathbb{R}^d$,*

$$(Af)(x) = b^i \partial_i f(x) + \tfrac{1}{2} a^{ij} \partial_i \partial_j f(x)$$
$$+ \int_{\mathbb{R}^d - \{0\}} [f(x+y) - f(x) - y^i \partial_i f(x) \chi_{\hat{B}}(y)] v(dy). \tag{3.13}$$

Proof (1) We apply Fourier inversion within (3.11) to find for all $t \geq 0$, $f \in S(\mathbb{R}^d)$, $x \in \mathbb{R}^d$,

$$(T_t f)(x) = \mathbb{E}(f(X(t) + x)) = (2\pi)^{-d/2} \mathbb{E}\left(\int_{\mathbb{R}^d} e^{i(u, x + X(t))} \hat{f}(u) du \right).$$

Since $\hat{f} \in S(\mathbb{R}^d) \subset L^1(\mathbb{R}^d)$, we have

$$\left| \int_{\mathbb{R}^d} e^{i(u,x)} \mathbb{E}(e^{i(u,X(t))}) \hat{f}(u) du \right| \leq \int_{\mathbb{R}^d} \left| e^{i(u,x)} \mathbb{E}(e^{i(u,X(t))}) \right| |\hat{f}(u)| du$$
$$\leq \int_{\mathbb{R}^d} |\hat{f}(u)| du < \infty,$$

so we can apply Fubini's theorem to obtain

$$(T_t f)(x) = (2\pi)^{-d/2} \int_{\mathbb{R}^d} e^{i(u,x)} \mathbb{E}(e^{i(u,X(t))}) \hat{f}(u) du$$
$$= (2\pi)^{-d/2} \int_{\mathbb{R}^d} e^{i(u,x)} e^{t\eta(u)} \hat{f}(u) du.$$

(2) For each $f \in S(\mathbb{R}^d)$, $x \in \mathbb{R}^d$, we have by result (1),

$$(Af)(x) = \lim_{t \downarrow 0} \frac{1}{t}[(T_t f)(x) - f(x)]$$

$$= (2\pi)^{-d/2} \lim_{t \downarrow 0} \int_{\mathbb{R}^d} e^{i(u,x)} \frac{e^{t\eta(u)} - 1}{t} \hat{f}(u) du.$$

Now, by the mean value theorem and Exercise 1.2.16, there exists $K > 0$ such that

$$\int_{\mathbb{R}^d} \left| e^{i(u,x)} \frac{e^{t\eta(u)} - 1}{t} \hat{f}(u) \right| du \leq \int_{\mathbb{R}^d} |\eta(u) \hat{f}(u)| du$$

$$\leq K \int_{\mathbb{R}^d} (1 + |u|^2|) |\hat{f}(u)| du < \infty,$$

since $(1 + |u|^2) \hat{f}(u) \in S(\mathbb{R}^d, \mathbb{C})$.

We can now use dominated convergence to deduce the required result.

(3) Applying the Lévy–Khinchine formula to result (2), we obtain for each $f \in S(\mathbb{R}^d)$, $x \in \mathbb{R}^d$,

$$(Af)(x) = (2\pi)^{-d/2} \int_{\mathbb{R}^d} e^{i(x,u)} \left\{ i(b,u) - \tfrac{1}{2}(au,u) \right.$$

$$\left. + \int_{\mathbb{R}^d - \{0\}} [e^{i(u,y)} - 1 - i(u,y) \chi_{\hat{B}}(y)] \nu(dy) \right\} \hat{f}(u) du.$$

The required result now follows immediately from elementary properties of the Fourier transform, all of which can be found in Section 3.8. Of course an interchange of integrals is required, but this is justified by Fubini's theorem in a similar way to the arguments given above. □

Note 1 The alert reader will have noticed that we have appeared to have cheated in our proof of (2), in that we have computed the generator using the pointwise limit instead of the uniform one. In fact the operators defined by both limits coincide in this context; see Sato [274], Lemma 31.7, p. 209.

Note 2 An alternative derivation of the important formula (3.13), which does not employ the calculus of pseudo-differential operators or Schwartz space, can be found in Sato [274], pp. 205–12. It is also shown therein that $C_c^\infty(\mathbb{R}^d)$ is a core for A and that $C_0^2(\mathbb{R}^d) \subseteq D_A$.

Note that $C_0^2(\mathbb{R}^d)$ is dense in $C_0(\mathbb{R}^d)$. We will establish the result $C_0^2(\mathbb{R}^d) \subseteq D_A$ later on, using stochastic calculus. An alternative analytic approach to these ideas may be found in Courrège [74].

3.3 Semigroups and generators of Lévy processes

Note 3 The results of Theorem 3.3.3 can be written in the convenient shorthand form

$$(\widehat{T(t)f})(u) = e^{t\eta(u)}\hat{f}(u), \qquad \widehat{Af}(u) = \eta(u)\hat{f}(u)$$

for each $t \geq 0$, $f \in S(\mathbb{R}^d)$, $u \in \mathbb{R}^d$.

We will now consider a number of examples of specific forms of (3.13) corresponding to important examples of Lévy processes.

Example 3.3.4 (Standard Brownian motion) Let X be a standard Brownian motion in $\mathbb{R}^d$. Then X has characteristics $(0, I, 0)$, and so we see from (3.13) that

$$A = \frac{1}{2}\sum_{i=1}^{d}\partial_i^2 = \frac{1}{2}\Delta,$$

where Δ is the usual Laplacian operator.

Example 3.3.5 (Brownian motion with drift) Let X be a Brownian motion with drift in $\mathbb{R}^d$. Then X has characteristics $(b, a, 0)$ and A is a diffusion operator of the form

$$A = b^i \partial_i + \frac{1}{2} a^{ij} \partial_i \partial_j.$$

Of course, we can construct far more general diffusions in which each b^i and a^{ij} is a function of x, and we will discuss this later in the chapter. The rationale behind the use of the term 'diffusion' will be explained in Chapter 6.

Example 3.3.6 (The Poisson process) Let X be a Poisson process with intensity $\lambda > 0$. Then X has characteristics $(0, 0, \lambda\delta_1)$ and A is a difference operator,

$$(Af)(x) = \lambda(f(x+1) - f(x)),$$

for all $f \in S(\mathbb{R}^d)$, $x \in \mathbb{R}^d$. Note that $||Af|| \leq 2\lambda ||f||$, so that A extends to a bounded operator on the whole of $C_0(\mathbb{R}^d)$.

Example 3.3.7 (The compound Poisson process) We leave it as an exercise for the reader to verify that

$$(Af)(x) = \int_{\mathbb{R}^d} [f(x+y) - f(x)] \nu(dy)$$

for all $f \in S(\mathbb{R}^d)$, $x \in \mathbb{R}^d$, where ν is a finite measure. The operator A again extends to a bounded operator on the whole of $C_0(\mathbb{R}^d)$.

Example 3.3.8 (Rotationally invariant stable processes) Let X be a rotationally invariant stable process of index α, where $0 < \alpha < 2$. Its symbol is given by $\eta(u) = -|u|^\alpha$ for all $u \in \mathbb{R}^d$ (see Subsection 1.2.5), where we have taken $\sigma = 1$ for convenience. It is instructive to pretend that η is the symbol for a legitimate differential operator; then, using the usual correspondence $u_j \to -i\partial_j$ for $1 \leq j \leq d$, we would write

$$A = \eta(D) = -\left(\sqrt{-\partial_1^2 - \partial_2^2 - \cdots - \partial_d^2}\right)^\alpha = -(-\Delta)^{\alpha/2}.$$

In fact, it is very useful to interpret $\eta(D)$ as a fractional power of the Laplacian. We will consider fractional powers of more general generators in the next section.

Example 3.3.9 (Relativistic Schrödinger operators) Fix $m, c > 0$ and recall from Subsection 1.2.6 the Lévy symbol $-E_{m,c}$, which represents (minus) the free energy of a particle of mass m moving at relativistic speeds (when $d = 3$):

$$E_{m,c}(u) = \sqrt{m^2 c^4 + c^2 |u|^2} - mc^2.$$

Arguing as above, we make the correspondence $u_j \to -i\partial_j$, for $1 \leq j \leq d$. Readers with a background in physics will recognise that this is precisely the prescription for quantisation of the free energy, and the corresponding generator is then given by

$$A = -\left(\sqrt{m^2 c^4 - c^2 \Delta} - mc^2\right).$$

Physicists call $-A$ a *relativistic Schrödinger operator*. Of course, it is more natural from the point of view of quantum mechanics to consider this as an operator in $L^2(\mathbb{R}^d)$, and we will address such considerations later in this chapter. For more on relativistic Schrödinger operators from both a probabilistic and physical point of view, see Carmona et al. [64] and references therein.

Note Readers trained in physics should note that we are employing a system of units wherein $\hbar = 1$.

Exercise 3.3.10 Show that Schwartz space is a core for the Laplacian. (Hint: Use Theorem 3.2.8.)

3.3 Semigroups and generators of Lévy processes

We will now examine the resolvent of a Lévy process from the Fourier-analytic point of view and show that it is always a convolution operator.

Theorem 3.3.11 *If X is a Lévy process, with associated Feller semigroup $(T_t, t \geq 0)$ and resolvent R_λ for each $\lambda > 0$, then there exists a finite measure μ_λ on $\mathbb{R}^d$ such that*

$$R_\lambda(f) = f * \mu_\lambda$$

for each $f \in S(\mathbb{R}^d)$.

Proof Fix $\lambda > 0$, let η be the Lévy symbol of X and define $r_\lambda : \mathbb{R}^d \to \mathbb{C}$ by $r_\lambda(u) = 1/[\lambda - \eta(u)]$. Since $\Re(\eta(u)) \leq 0$ for all $u \in \mathbb{R}^d$, it is clear that r_λ is well defined, and we have

$$r_\lambda(u) = \int_0^\infty e^{-\lambda t} e^{t\eta(u)} dt$$

for each $u \in \mathbb{R}^d$. We will now show that r_λ is positive definite. For each $c_1, \ldots, c_n \in \mathbb{C}$ and $u_1, \ldots, u_n \in \mathbb{R}^d$,

$$\sum_{i,j=1}^d c_i \overline{c_j} r_\lambda(u_i - u_j) = \int_0^\infty e^{-\lambda t} \sum_{i,j=1}^d c_i \overline{c_j} e^{t\eta(u_i - u_j)} dt \geq 0,$$

as $u \to e^{t\eta(u)}$, is positive definite. Since $u \to \eta(u)$ is continuous, so also is $u \to r_\lambda(u)$ and hence, by a slight variant on Bochner's theorem, there exists a finite measure μ_λ on $\mathcal{B}(\mathbb{R}^d)$ for which

$$r_\lambda(u) = \widehat{\mu_\lambda}(u) = (2\pi)^{-d/2} \int_{\mathbb{R}^d} e^{-i(u,x)} \mu_\lambda(dx)$$

for all $u \in \mathbb{R}^d$.

Now we can apply Theorem 3.2.9, Theorem 3.3.3(2), Fubini's theorem and known results on the Fourier transform of a convolution (see Section 3.8) to find that for all $f \in S(\mathbb{R}^d), x \in \mathbb{R}^d$,

$$(R_\lambda f)(x) = \int_0^\infty e^{-\lambda t} (T_t f)(x) dt$$

$$= (2\pi)^{-d/2} \int_0^\infty e^{-\lambda t} \left(\int_{\mathbb{R}^d} e^{i(u,x)} e^{t\eta(u)} \hat{f}(u) du \right) dt$$

$$= (2\pi)^{-d/2} \int_{\mathbb{R}^d} e^{i(u,x)} \hat{f}(u) \left(\int_0^\infty e^{-\lambda t} e^{t\eta(u)} dt \right) du$$

$$= (2\pi)^{-d/2} \int_{\mathbb{R}^d} e^{i(u,x)} \hat{f}(u) r_\lambda(u) \, du$$

$$= (2\pi)^{-d/2} \int_{\mathbb{R}^d} e^{i(u,x)} \hat{f}(u) \widehat{\mu_\lambda}(u) \, du$$

$$= (2\pi)^{-d/2} \int_{\mathbb{R}^d} e^{i(u,x)} \widehat{f * \mu_\lambda}(u) \, du$$

$$= (f * \mu_\lambda)(x). \qquad \square$$

Exercise 3.3.12 Show that, for all $B \in \mathcal{B}(\mathbb{R}^d)$,

$$\mu_\lambda(B) = \int_0^\infty e^{-\lambda t} p_{X(t)}(-B) \, dt;$$

see Bertoin [36], p. 23.

Just like the semigroup and its generator, the resolvent can also be represented as a pseudo-differential operator. In fact, for each $\lambda > 0$, R_λ has symbol $[\lambda - \eta(\cdot)]^{-1}$. The following makes this precise:

Corollary 3.3.13 *For each* $\lambda > 0$, $f \in S(\mathbb{R}^d)$, $x \in \mathbb{R}^d$,

$$(R_\lambda f)(x) = (2\pi)^{-d/2} \int_{\mathbb{R}^d} e^{i(x,u)} \frac{\hat{f}(u)}{\lambda - \eta(u)} \, du$$

Proof This is implicit in the proof of Theorem 3.3.11. $\qquad \square$

We remark that an interesting partial converse to Theorem 3.3.3(1) is established in Reed and Simon [258], as follows.

Let $F : \mathbb{R}^d \to \mathbf{C}$ be such that there exists $k \in \mathbb{R}$ with $\Re(F(x)) \geq k$ for all $x \in \mathbb{R}^d$.

Theorem 3.3.14 $(T_t, t \geq 0)$ *is a positivity-preserving semigroup in* $L^2(\mathbb{R}^d)$ *with*

$$\widehat{T_t f}(u) = e^{-tF(u)} \hat{f}(u)$$

for all $f \in S(\mathbb{R}^d)$, $u \in \mathbb{R}^d$, $t \geq 0$, *if and only if* $F = -\eta$, *where* η *is a Lévy symbol.*

The proof can be found in pages 215–22 of Reed and Simon [258].

3.3.3 Subordination of semigroups

We now apply some of the ideas developed above to the subordination of semigroups. It is recommended that readers recall the basic properties of subordinators as described in Subsection 1.3.2.

In the following, X will always denote a Lévy process in $\mathbb{R}^d$ with symbol η_X, Feller semigroup $(T_t^X, t \geq 0)$ and generator A^X.

Let $S = (S(t), t \geq 0)$ be a subordinator, so that S is a one-dimensional, non-decreasing Lévy process and, for each $u, t > 0$,

$$\mathbb{E}(e^{-uS(t)}) = e^{-t\psi(u)},$$

where ψ is the Bernstein function given by

$$\psi(u) = bu + \int_0^\infty (1 - e^{-uy})\lambda(dy)$$

with $b \geq 0$ and $\int_0^\infty (y \wedge 1)\lambda(dy) < \infty$.

Recall from Theorem 1.3.25 and Proposition 1.3.27 that $Z = (Z(t), t \geq 0)$ is also a Lévy process, where we define each $Z(t) = X(T(t))$ and the symbol of Z is $\eta^Z = -\psi \circ (-\eta^X)$. We write $(T_t^Z, t \geq 0)$ and A^Z for the semigroup and generator associated with Z, respectively.

Theorem 3.3.15

(1) For all $t \geq 0$, $f \in B_b(\mathbb{R}^d)$, $x \in \mathbb{R}^d$,

$$(T_t^Z f)(x) = \int_0^\infty (T_s^X f)(x) p_{S(t)}(ds).$$

(2) For all $f \in S(\mathbb{R}^d)$,

$$A^Z f = b A^X f + \int_0^\infty (T_s^X f - f)\lambda(ds).$$

Proof (1) In Exercise 1.3.26, we established that for each $t \geq 0$, $B \in \mathcal{B}(\mathbb{R}^d)$, $p_{Z(t)}(B) = \int_0^\infty p_{X(s)}(B) p_{S(t)}(ds)$. Hence for each $t \geq 0$, $f \in B_b(\mathbb{R}^d)$, $x \in \mathbb{R}^d$, we obtain

$$(T_t^Z f)(x) = \mathbb{E}(f(Z(t) + x)) = \int_{\mathbb{R}^d} f(x+y) p_{Z(t)}(dy)$$

$$= \int_0^\infty \left(\int_{\mathbb{R}^d} f(x+y) p_{X(s)}(dy) \right) p_{S(t)}(ds)$$

$$= \int_0^\infty (T_s^X f)(x) p_{S(t)}(ds).$$

(2) From the first equation in the proof of Theorem 1.3.33, we obtain for each $u \in \mathbb{R}^d$,

$$\eta^Z(u) = b\eta_X(u) + \int_0^\infty \{\exp[s\eta^X(u)] - 1\}\lambda(ds), \qquad (3.14)$$

but by Theorem 3.3.3(2) we have

$$(A^Z f)(x) = (2\pi)^{-d/2} \int_{\mathbb{R}^d} e^{i(u,x)} \eta_Z(u) \hat{f}(u) du. \qquad (3.15)$$

The required result now follows from substitution of (3.14) into (3.15), a straightforward application of Fubini's theorem and a further application of Theorem 3.3.3(1), (2). The details are left as an exercise for the reader. □

The formula $\eta_Z = -\psi \circ (-\eta_X)$ suggests a natural functional calculus wherein we define $A^Z = -\psi(-A^X)$ for any Bernstein function ψ. As an example, we may generalise the fractional power of the Laplacian, discussed in the last section, to define $(-A^X)^\alpha$ for any Lévy process X and any $0 < \alpha < 1$. To carry this out, we employ the α-stable subordinator (see Example 1.3.18). This has characteristics $(0, \lambda)$ where

$$\lambda(dx) = \frac{\alpha}{\Gamma(1 - \alpha)} \frac{dx}{x^{1+\alpha}}.$$

Theorem 3.3.15(2) then yields the beautiful formula

$$-(-A^X)^\alpha f = \frac{\alpha}{\Gamma(1 - \alpha)} \int_0^\infty (T_s^X f - f) \frac{ds}{s^{1+\alpha}} \qquad (3.16)$$

for all $f \in S(\mathbb{R}^d)$.

Theorem 3.3.15 has a far-reaching generalisation, which we will now quote without proof.

Theorem 3.3.16 (Phillips) *Let $(T_t, t \geq 0)$ be a strongly continuous contraction semigroup of linear operators on a Banach space B with infinitesimal generator A and let $(S(t), t \geq 0)$ be a subordinator with characteristics (b, λ).*

- *The prescription*

$$T_t^S \phi = \int_0^\infty (T_s \phi) p_{S(t)}(ds),$$

for each $t \geq 0$, $\phi \in B$, defines a strongly continuous contraction semigroup $(T_t^S, t \geq 0)$ in B.

3.3 Semigroups and generators of Lévy processes

- If A^S is the infinitesimal generator of $(T_t^S, t \geq 0)$, then D_A is a core for A^S and, for each $\phi \in D_A$,

$$A^S \phi = bA\phi + \int_0^\infty (T_s^X \phi - \phi)\lambda(ds).$$

- If $B = C_0(\mathbb{R}^d)$ and $(T_t, t \geq 0)$ is a Feller semigroup, then $(T_t^S, t \geq 0)$ is also a Feller semigroup.

For a proof of this result see e.g. Sato [274], pp. 212–5, or Section 5.3 in Jacob [149].

This powerful theorem enables the extension of (3.16) to define fractional powers of a large class of infinitesimal generators of semigroups (see also Schilling [275]).

To give Theorem 3.3.16 a probabilistic flavour, let $X = (X(t), t \geq 0)$ be a homogeneous Markov process and $S = (S(t), t \geq 0)$ be an independent subordinator; then we can form the process $Y = (Y(t), t \geq 0)$, where $Y(t) = X(T(t))$ for each $t \geq 0$. For each $t \geq 0$, $f \in B_b(\mathbb{R}^d)$, $x \in \mathbb{R}^d$, define

$$(T_t^Y f)(x) = \mathbb{E}(f(Y(t))|Y(0) = x).$$

Then by direct computation (or appealing to Phillips' theorem, Theorem 3.3.16), we have that $(T_t^Y, t \geq 0)$ is a semigroup and

$$(T_t^Y f)(x) = \int_0^\infty (T_s^X f)(x) p_{S(t)}(ds),$$

where $(T_t^X, t \geq 0)$ is the semigroup associated with X.

Exercise 3.3.17 Deduce that $(T_t^Y, t \geq 0)$ is a Feller semigroup whenever $(T_t^X, t \geq 0)$ is and that Y is a Feller process in this case.

Exercise 3.3.18 Show that for all $t \geq 0$, $B \in \mathcal{B}(\mathbb{R}^d)$,

$$p_{Y(t)}(B) = \int_0^\infty p_{X(s)}(B) p_{T(t)}(ds),$$

and hence deduce that, for all $x \in \mathbb{R}^d$,

$$P(Y(t) \in B | Y(0) = x) = \int_0^\infty P(X(s) \in B | X(0) = x) p_{T(t)}(ds)$$

(a.s. with respect to $p_{X(0)}$).

3.4 L^p-Markov semigroups

We have seen above how Feller processes naturally give rise to associated Feller semigroups acting in the Banach space $C_0(\mathbb{R}^d)$. Sometimes, it is more appropriate to examine the process via semigroups induced in $L^p(\mathbb{R}^d)$, where $1 \leq p < \infty$, and the present section is devoted to this topic.

3.4.1 L^p-Markov semigroups and Lévy processes

We fix $1 \leq p < \infty$ and let $(T_t, t \geq 0)$ be a strongly continuous contraction semigroup of operators in $L^p(\mathbb{R}^d)$. We say that it is *sub-Markovian* if $f \in L^p(\mathbb{R}^d)$ and

$$0 \leq f \leq 1 \quad \text{a.e.} \quad \Rightarrow \quad 0 \leq T_t f \leq 1 \quad \text{a.e.}$$

for all $t \geq 0$.

Any semigroup on $L^p(\mathbb{R}^d)$ can be restricted to the dense subspace $C_c(\mathbb{R}^d)$. If this restriction can then be extended to a semigroup on $B_b(\mathbb{R}^d)$ that satisfies $T_t(1) = 1$ then the semigroup is said to be *conservative*.

A semigroup that is both sub-Markovian and conservative is said to be L^p-*Markov*.

Notes

(1) Readers should be mindful that the phrases 'strongly continuous' and 'contraction' in the above definition are now with respect to the L^p-norm, given by $||g||_p = \left(\int_{\mathbb{R}^d} |g(x)|^p dx\right)^{1/p}$ for each $g \in L^p(\mathbb{R}^d)$.

(2) If $(T_t, t \geq 0)$ is sub-Markovian then it is L^p-*positivity preserving*, in that $f \in L^p(\mathbb{R}^d)$ and $f \geq 0$ (a.e.) $\Rightarrow T_t f \geq 0$ (a.e.) for all $t \geq 0$; see Jacob [149], p. 365, for a proof.

Example 3.4.1 Let $X = (X(t), t \geq 0)$ be a Markov process on $\mathbb{R}^d$ and define the usual stochastic evolution

$$(T_t f)(x) = \mathbb{E}\big(f(X(t))|X(0) = x\big)$$

for each $f \in B_b(\mathbb{R}^d)$, $x \in \mathbb{R}^d$, $t \geq 0$. Suppose that $(T_t, t \geq 0)$ also yields a strongly continuous contraction semigroup on $L^p(\mathbb{R}^d)$; then it is clearly L^p-Markov.

Our good friends the Lévy processes provide a natural class for which the conditions of the last example hold, as the next theorem demonstrates.

3.4 L^p-Markov semigroups

Theorem 3.4.2 *If $X = (X(t), t \geq 0)$ is a Lévy process then, for each $1 \leq p < \infty$, the prescription $(T_t f)(x) = \mathbb{E}(f(X(t) + x))$ where $f \in L^p(\mathbb{R}^d)$, $x \in \mathbb{R}^d$, $t \geq 0$ gives rise to an L^p-Markov semigroup $(T_t \geq 0)$.*

Proof Let q_t be the law of $X(t)$ for each $t \geq 0$. We must show that each $T_t : L^p(\mathbb{R}^d) \to L^p(\mathbb{R}^d)$. In fact, for all $f \in L^p(\mathbb{R}^d)$, $t \geq 0$, by Jensen's inequality (or Hölder's inequality if you prefer) and Fubini's theorem, we obtain

$$\|T_t f\|_p^p = \int_{\mathbb{R}^d} \left| \int_{\mathbb{R}^d} f(x+y) q_t(dy) \right|^p dx$$

$$\leq \int_{\mathbb{R}^d} \int_{\mathbb{R}^d} |f(x+y)|^p q_t(dy) dx$$

$$= \int_{\mathbb{R}^d} \left(\int_{\mathbb{R}^d} |f(x+y)|^p dx \right) q_t(dy)$$

$$= \int_{\mathbb{R}^d} \left(\int_{\mathbb{R}^d} |f(x)|^p dx \right) q_t(dy) = \|f\|_p^p,$$

and we have proved that each T_t is a contraction in $L^p(\mathbb{R}^d)$.

Now we need to establish the semigroup property $T_{s+t} f = T_s T_t f$ for all $s, t \geq 0$. By Theorem 3.1.9 and the above, we see that this holds for all $f \in C_0(\mathbb{R}^d) \cap L^p(\mathbb{R}^d)$. However, this space is dense in $L^p(\mathbb{R}^d)$, which follows from that fact that $C_c(\mathbb{R}^d) \subset C_0(\mathbb{R}^d) \cap L^p(\mathbb{R}^d)$, and the result follows by continuity since each T_t is bounded.

Finally, we must prove strong continuity. First we let $f \in C_c(\mathbb{R}^d)$ and choose a ball B centred on the origin in $\mathbb{R}^d$. Then, using Jensen's inequality and Fubini's theorem as above, we obtain for each $t \geq 0$

$$\|T_t f - f\|_p^p = \int_{\mathbb{R}^d} \left| \int_{\mathbb{R}^d} [f(x+y) - f(x)] q_t(dy) \right|^p dx$$

$$\leq \int_{\mathbb{R}^d} \left(\int_B |f(x+y) - f(x)|^p dx \right) q_t(dy)$$

$$+ \int_{\mathbb{R}^d} \left(\int_{B^c} |f(x+y) - f(x)|^p dx \right) q_t(dy)$$

$$\leq \int_{\mathbb{R}^d} \left(\int_B |f(x+y) - f(x)|^p dx \right) q_t(dy)$$

$$+ \int_{\mathbb{R}^d} \left(\int_{B^c} 2^p \max\{|f(x+y)|^p, |f(x)|^p\} dx \right) q_t(dy)$$

$$\leq \sup_{y \in B} \sup_{x \in \mathbb{R}^d} |f(x+y) - f(x)|^p |B| + 2^p \|f\|_p^p q_t(B^c),$$

where $|B|$ denotes the Lebesgue measure of B. By choosing B to have sufficiently small radius, we obtain $\lim_{t \downarrow 0} \|T_t f - f\|_p = 0$ from the uniform continuity of f and the weak continuity of $(q_t, t \geq 0)$, just as in the proof of Theorem 3.1.9.

Now let $f \in L^p(\mathbb{R}^d)$ be arbitrary and choose a sequence $(f_n, n \in \mathbb{N})$ in $C_c(\mathbb{R}^d)$ that converges to f. Using the triangle inequality and the fact that each T_t is a contraction we obtain, for each $t \geq 0$,

$$\|T_t f - f\| \leq \|T_t f_n - f_n\| + \|T_t (f - f_n)\| + \|f - f_n\|$$
$$\leq \|T_t f_n - f_n\| + 2\|f - f_n\|,$$

from which the required result follows. □

For the case $p = 2$ we can compute explicitly the domain of the infinitesimal generator of a Lévy process. To establish this result, let X be a Lévy process with Lévy symbol η and let A be the infinitesimal generator of the associated L^2-Markov semigroup.

Exercise 3.4.3 Using the fact that the Fourier transform is a unitary isomorphism of $L^2(\mathbb{R}^d, \mathbb{C})$ (see Subsection 3.8.4), show that

$$(T_t f)(x) = (2\pi)^{-d/2} \int_{\mathbb{R}^d} e^{i(u,x)} e^{t\eta(u)} \hat{f}(u) du$$

for all $t \geq 0, x \in \mathbb{R}^d, f \in L^2(\mathbb{R}^d)$.

Define $\mathcal{H}_\eta(\mathbb{R}^d) = \left\{ f \in L^2(\mathbb{R}^d); \int_{\mathbb{R}^d} |\eta(u)|^2 |\hat{f}(u)|^2 du < \infty \right\}$. Then we have

Theorem 3.4.4 $D_A = \mathcal{H}_\eta(\mathbb{R}^d)$.

Proof We follow Berg and Forst [35], p. 92. Let $f \in D_A$; then $Af = \lim_{t \downarrow 0} [(1/t)(T_t f - f)]$ in $L^2(\mathbb{R}^d)$. We take Fourier transforms and use the continuity of $\mathcal{F}$ to obtain

$$\widehat{Af} = \lim_{t \downarrow 0} \frac{1}{t} (\widehat{T_t f} - \hat{f}).$$

By the result of Exercise 3.4.3, we have

$$\widehat{Af} = \lim_{t \downarrow 0} \frac{1}{t} (e^{t\eta} \hat{f} - \hat{f});$$

hence, for any sequence $(t_n, n \in \mathbb{N})$ in $\mathbb{R}^+$ for which $\lim_{n \to \infty} t_n = 0$, we get

$$\widehat{Af} = \lim_{n \to \infty} \frac{1}{t_n}(e^{t_n \eta} \hat{f} - \hat{f}) \quad \text{a.e.}$$

However, $\lim_{n \to \infty} [(1/t_n)(e^{t_n \eta} - 1)] = \eta$ and so $\widehat{Af} = \eta \hat{f}$ (a.e.). But then $\eta \hat{f} \in L^2(\mathbb{R}^d)$, i.e. $f \in \mathcal{H}_\eta(\mathbb{R}^d)$.

So we have established that $D_A \subseteq \mathcal{H}_\eta(\mathbb{R}^d)$.

Conversely, let $f \in \mathcal{H}_\eta(\mathbb{R}^d)$; then by Exercise 3.4.3 again,

$$\lim_{t \to 0} \frac{1}{t}(\widehat{T_t f} - \hat{f}) = \lim_{t \to 0} \frac{1}{t}(e^{t\eta} \hat{f} - \hat{f}) = \eta \hat{f} \in L^2(\mathbb{R}^d).$$

Hence, by the unitarity and continuity of the Fourier transform, $\lim_{t \downarrow 0} [(1/t)(T_t f - f)] \in L^2(\mathbb{R}^d)$ and so $f \in D_A$. □

Readers should note that the proof has also established the pseudo-differential operator representation

$$Af = (2\pi)^{-d/2} \int_{\mathbb{R}^d} e^{i(u,x)} \eta(u) \hat{f}(u) du$$

for all $f \in \mathcal{H}_\eta(\mathbb{R}^d)$.

The space $\mathcal{H}_\eta(\mathbb{R}^d)$ is called an *anisotropic Sobolev space* by Jacob [148]. Note that if we take X to be a standard Brownian motion then $\eta(u) = -\frac{1}{2}|u|^2$ for all $u \in \mathbb{R}^d$ and

$$\mathcal{H}_\eta(\mathbb{R}^d) = \left\{ f \in L^2(\mathbb{R}^d); \int_{\mathbb{R}^d} |u|^4 |\hat{f}(u)|^2 du < \infty \right\}.$$

This is precisely the Sobolev space, which is usually denoted $\mathcal{H}_2(\mathbb{R}^d)$ and which can be defined equivalently as the completion of $C_c^\infty(\mathbb{R}^d)$ with respect to the norm

$$\|f\|_2 = \left(\int_{\mathbb{R}^d} (1 + |u|^2)^2 |\hat{f}(u)|^2 du \right)^{1/2}$$

for each $f \in C_c^\infty(\mathbb{R}^d)$. By Theorem 3.4.4, $\mathcal{H}_2(\mathbb{R}^d)$ is the domain of the Laplacian Δ acting in $L^2(\mathbb{R}^d)$.

Exercise 3.4.5 Write down the domains of the fractional powers of the Laplacian $(-\Delta)^{\alpha/2}$, where $0 < \alpha < 2$.

For more on this topic, including interpolation between L^p and L^q sub-Markovian semigroups ($p < q < \infty$) and between L^p sub-Markovian semi-groups and Feller semigroups, see Farkas *et al.* [100].

3.4.2 Self-adjoint semigroups

We begin with some general considerations.

Let H be a Hilbert space and $(T_t, t \geq 0)$ be a strongly continuous contraction semigroup in H. We say that $(T_t, t \geq 0)$ is *self-adjoint* if $T_t = T_t^*$ for each $t \geq 0$.

Theorem 3.4.6 *There is a one-to-one correspondence between the generators of self-adjoint semigroups in H and linear operators A in H such that $-A$ is positive and self-adjoint.*

Proof We follow Davies [76], pp. 99–100. In fact we will prove only that half of the theorem which we will use, and we commend [76] to the reader for the remainder.

Suppose that $(T_t, t \geq 0)$ is a self-adjoint semigroup with generator A, and consider the Bochner integral

$$X\psi = \int_0^\infty T_t e^{-t} \psi \, dt$$

for each $\psi \in H$; then it is easily verified that X is a bounded self-adjoint operator (in fact, X is a contraction). Furthermore, by Theorem 3.2.9, we have $X = (I + A)^{-1}$, hence $(I + A)^{-1}$ is self-adjoint. We now invoke the spectral theorem (Theorem 3.8.8) to deduce that there exists a projection-valued measure P in H such that $(I + A)^{-1} = \int_\mathbb{R} \lambda P(d\lambda)$. If we define $f : \mathbb{R} \to \mathbb{R}$ by $f(\lambda) = (1/\lambda) - 1$ then $A = \int_\mathbb{R} f(\lambda) P(d\lambda)$ is self-adjoint. By Theorem 3.2.10, $(0, \infty) \subseteq \rho(A)$; hence $\sigma(A) \subseteq (-\infty, 0)$ and so $-A$ is positive. □

There is a class of Markov processes that will be important in Section 3.6, where we study Dirichlet forms. Let $X = (X(t), t \geq 0)$ be a Markov process with associated semigroup $(T(t), t \geq 0)$ and let μ be a Borel measure on $\mathbb{R}^d$. We say that X is a *μ-symmetric process* if

$$\int_{\mathbb{R}^d} f(x)(T_t g)(x) \mu(dx) = \int_{\mathbb{R}^d} (T_t f)(x) g(x) \mu(dx) \tag{3.17}$$

for all $t \geq 0$ and all $f, g \in \mathcal{B}_b(\mathbb{R}^d)$ with $f, g \geq 0$ (a.e. μ). Readers should be clear that the integrals in (3.17) may be (simultaneously) infinite.

In the case where μ is a Lebesgue measure, we simply say that X is *Lebesgue symmetric*.

Exercise 3.4.7 Let X be a normal Markov process with a transition density ρ for which $\rho(x, y) = \rho(y, x)$ for all $x, y \in \mathbb{R}^d$. Show that X is Lebesgue symmetric.

3.4 L^p-Markov semigroups

Theorem 3.4.8 *If X is a μ-symmetric Markov process with associated semigroup $(T_t, t \geq 0)$ and $\|T_t f\|_2 < \infty$ for all $f \in C_c(\mathbb{R}^d)$ with $f \geq 0$, then $(T_t, t \geq 0)$ is self-adjoint in $L^2(\mathbb{R}^d, \mu)$.*

Proof By linearity, if $f, g \in C_c(\mathbb{R}^d)$, with $f \geq 0$ and $g \leq 0$, we still have that (3.17) holds and both integrals are finite. Now let $f, g \in C_c(\mathbb{R}^d)$ be arbitrary; then, writing $f = f^+ - f^-$ and $g = g^+ - g^-$, we again deduce by linearity that (3.17) holds in this case. Finally let $f, g \in L^2(\mathbb{R}^d, \mu)$; then, by the density therein of $C_c(\mathbb{R}^d)$, we can find sequences $(f_n, n \in \mathbb{N})$ and $(g_n, n \in \mathbb{N})$ in $C_c(\mathbb{R}^d)$ that converge to f and g respectively. Using the continuity of T_t and of the inner product, we find for each $t \geq 0$ that

$$\langle f, T_t g \rangle = \lim_{n \to \infty} \lim_{m \to \infty} \langle f_n, T_t g_m \rangle = \lim_{n \to \infty} \lim_{m \to \infty} \langle T_t f_n, g_m \rangle = \langle T_t f, g \rangle.$$

$\square$

Now let $X = (X(t), t \geq 0)$ be a Lévy process taking values in $\mathbb{R}^d$. We have already seen in Theorem 3.4.2 that $(T_t, t \geq 0)$ is an L^p-Markov semigroup where $(T_t f)(x) = \mathbb{E}(f(X(t) + x))$ for each $f \in L^p(\mathbb{R}^d)$, $x \in \mathbb{R}^d$, $t \geq 0$. We recall that a Lévy process with laws $(q_t, t \geq 0)$ is symmetric if $q_t(A) = q_t(-A)$ for all $A \in \mathcal{B}(\mathbb{R}^d)$.

Exercise 3.4.9 Deduce that every symmetric Lévy process is a Lebesgue-symmetric Markov process.

Although we could use the result of Exercise 3.4.9 and Theorem 3.4.8 to establish the first part of Theorem 3.4.10 below, we will find it more instructive to give an independent proof.

Theorem 3.4.10 *If X is a Lévy process, then its associated semigroup $(T_t, t \geq 0)$ is self-adjoint in $L^2(\mathbb{R}^d)$ if and only if X is symmetric.*

Proof Suppose that X is symmetric; then $q_t(A) = q_t(-A)$ for each $A \in \mathcal{B}(\mathbb{R}^d)$, $t \geq 0$, where q_t is the law of $X(t)$. Then for each $f \in L^2(\mathbb{R}^d)$, $x \in \mathbb{R}^d$, $t \geq 0$,

$$(T_t f)(x) = \mathbb{E}(f(x + X(t))) = \int_{\mathbb{R}^d} f(x+y) q_t(dy)$$

$$= \int_{\mathbb{R}^d} f(x+y) q_t(-dy) = \int_{\mathbb{R}^d} f(x-y) q_t(dy)$$

$$= \mathbb{E}(f(x - X(t))).$$

So for each $f, g \in L^2(\mathbb{R}^d), t \geq 0$, using Fubini's theorem, we obtain

$$\langle T_t f, g \rangle = \int_{\mathbb{R}^d} (T_t f)(x)g(x)dx = \int_{\mathbb{R}^d} \mathbb{E}(f(x - X(t))\,g(x)dx$$

$$= \int_{\mathbb{R}^d} \left[\int_{\mathbb{R}^d} f(x-y)g(x)dx \right] q_t(dy)$$

$$= \int_{\mathbb{R}^d} \left[\int_{\mathbb{R}^d} f(x)g(x+y)dx \right] q_t(dy)$$

$$= \langle f, T_t g \rangle.$$

Conversely, suppose that $(T_t, t \geq 0)$ is self-adjoint. Then by a similar argument to the one above, we deduce that for all $f, g \in L^2(\mathbb{R}^d), t \geq 0$,

$$\int_{\mathbb{R}^d} \mathbb{E}(f(x + X(t))\,g(x)dx = \int_{\mathbb{R}^d} \mathbb{E}(f(x - X(t))\,g(x)dx.$$

Now define a sequence of functions $(g_n, n \in \mathbb{N})$ by

$$g_n(x) = n^{-d/2} \exp\left(-\frac{\pi x^2}{n}\right).$$

Then each $g_n \in S(\mathbb{R}^d) \subset L^2(\mathbb{R}^d)$, and

$$\lim_{n \to \infty} \int_{\mathbb{R}^d} \mathbb{E}(f(x \pm X(t))g_n(x)dx = \mathbb{E}(f(\pm X(t));$$

see e.g. Lieb and Loss [196], Theorem 2.16, p. 58 and the argument in the proof of Theorem 5.3, p. 118 therein.

We thus deduce that $\mathbb{E}(f(X(t)) = \mathbb{E}(f(-X(t)))$ and, if we take $f = \chi_A$ where $A \in \mathcal{B}(\mathbb{R}^d)$, we obtain

$$P(X(t) \in A) = P(X(t) \in -A),$$

i.e. X is symmetric. □

Corollary 3.4.11 *If A is the infinitesimal generator of a Lévy process with Lévy symbol η, then $-A$ is positive and self-adjoint if and only if*

$$\eta(u) = -\tfrac{1}{2}(u, au) + \int_{\mathbb{R}^d - \{0\}} \left[\cos(u, y) - 1\right]\nu(dy).$$

for each $u \in \mathbb{R}^d$, where a is a positive definite symmetric matrix and v is a symmetric Lévy measure.

Proof This follows immediately from Theorems 3.4.10 and 3.4.6 and Exercise 2.4.23. □

Equivalently, we see that A is self-adjoint if and only if $\Im \eta = 0$.

In particular, we find that the discussion of this section has yielded a probabilistic proof of the self-adjointness of some important operators in $L^2(\mathbb{R}^d)$, as is shown in the following examples.

Example 3.4.12 (The Laplacian) In fact, we consider multiples of the Laplacian and let $a = 2\gamma I$ where $\gamma > 0$; then, for all $u \in \mathbb{R}^d$,

$$\eta(u) = -\gamma |u|^2 \quad \text{and} \quad A = \gamma \Delta.$$

Example 3.4.13 (Fractional powers of the Laplacian) Let $0 < \alpha < 2$; then, for all $u \in \mathbb{R}^d$,

$$\eta(u) = |u|^\alpha \quad \text{and} \quad A = -(-\Delta)^{\alpha/2}.$$

Example 3.4.14 (Relativistic Schrödinger operators) Let $m, c > 0$; then, for all $u \in \mathbb{R}^d$,

$$E_{m,c}(u) = \sqrt{m^2 c^4 + c^2 |u|^2} - mc^2 \quad \text{and} \quad A = -(\sqrt{m^2 c^4 - c^2 \Delta} - mc^2);$$

recall Example 3.3.9.

Note that in all three of the above examples the domain of the operator is the appropriate non-isotropic Sobolev space of Theorem 3.4.4.

Examples 3.4.12 and 3.4.14 are important in quantum mechanics as the observables (modulo a minus sign) that describe the kinetic energy of a particle moving at non-relativistic speeds (for a suitable value of γ) and relativistic speeds, respectively. We emphasise that it is vital that we know that such operators really are self-adjoint (and not just symmetric, say) so that they legitimately satisfy the quantum-mechanical formalism.

Note that, in general, if A^X is the self-adjoint generator of a Lévy process and $(S(t), t \geq 0)$ is an independent subordinator then the generator A^Z of the subordinated process Z is also self-adjoint. This follows immediately from (3.14) in the proof of Theorem 3.3.15(2).

3.5 Lévy-type operators and the positive maximum principle

3.5.1 The positive maximum principle and Courrège's theorems

Let X be a Lévy process with characteristics (b, a, ν), Lévy symbol η and generator A. We remind the reader of key results from Theorem 3.3.3. For each $f \in S(\mathbb{R}^d), x \in \mathbb{R}^d$,

$$(Af)(x) = b^i \partial_i f(x) + \tfrac{1}{2} a^{ij} \partial_i \partial_j f(x)$$
$$+ \int_{\mathbb{R}^d - \{0\}} [f(x+y) - f(x) - y^i \partial_i f(x) \chi_{\hat{B}}(y)] \nu(dy), \tag{3.18}$$

and A is a pseudo-differential operator with symbol η, i.e.

$$(Af)(x) = (2\pi)^{-d/2} \int_{\mathbb{R}^d} e^{i(u,x)} \eta(u) \hat{f}(u) du. \tag{3.19}$$

In this section, we turn our attention to general Feller processes and ask the question, to what extent are the above representations typical of these? Clearly Lévy processes are a special case and, to go beyond these, we must abandon translation invariance (see Theorem 3.3.1), in which case we would expect variable coefficients $(b(x), a(x), \nu(x))$ in (3.18) and a variable symbol $\eta(x, \cdot)$ in (3.19). In this section, we will survey some of the theoretical structure underlying Feller processes having generators with such a form.

The key to this is the following analytical concept.

Let S be a linear operator in $C_0(\mathbb{R}^d)$ with domain D_S. We say that S satisfies the *positive maximum principle* if, whenever $f \in D_S$ and there exists $x_0 \in \mathbb{R}^d$ such that $f(x_0) = \sup_{x \in \mathbb{R}^d} f(x) \geq 0$, we have $(Sf)(x_0) \leq 0$.

Our first hint that the positive maximum principle may be of some use in probability comes from the following variant on the Hille–Yosida theorem (Theorem 3.2.10).

Theorem 3.5.1 (Hille–Yosida–Ray) *A densely defined closed linear operator A is the generator of a strongly continuous positivity-preserving contraction semigroup on $C_0(\mathbb{R}^d)$ if and only if*

(1) $(0, \infty) \subseteq \rho(A)$,
(2) *A satisfies the positive maximum principle.*

For a proof, see Ethier and Kurtz [99], pp. 165–6, or Jacob [149], Section 4.5. Just as in the case of Theorem 3.2.10, the above, theorem can be generalised in such a way as to weaken the condition on A and simply require it to be closable.

3.5 Lévy-type operators and the positive maximum principle

Now we return to probability theory and make a direct connection between the positive maximum principle and the theory of Markov processes.

Theorem 3.5.2 *If X is a Feller process, then its generator A satisfies the positive maximum principle.*

Proof We follow Revuz and Yor [260], Section 7.1. Let $f \in D_A$ and suppose there exists $x_0 \in \mathbb{R}^d$ such that $f(x_0) = \sup_{x \in \mathbb{R}^d} f(x) \geq 0$.

Let $(T_t, t \geq 0)$ be the associated Feller semigroup and $(p_t, t \geq 0)$ be the transition probabilities; then, by (3.3), we have

$$(T_t f)(x_0) = \int_{\mathbb{R}^d} f(y) p_t(x_0, dy).$$

Hence

$$(Af)(x_0) = \lim_{t \downarrow 0} \frac{1}{t} \int_{\mathbb{R}^d} [f(y) - f(x_0)] p_t(x_0, dy).$$

However, for each $y \in \mathbb{R}^d$,

$$f(y) - f(x_0) \leq \sup_{y \in \mathbb{R}^d} f(y) - f(x_0) = 0,$$

and so $(Af)(x_0) \leq 0$ as required. $\square$

We will now present some fundamental results due to Courrège [75], which classify linear operators that satisfy the positive maximum principle. First we need some preliminary concepts.

(1) A C^∞ mapping $\phi : \mathbb{R}^d \times \mathbb{R}^d \to [0, 1]$ will be called a *local unit* if:
   (i) $\phi(x, y) = 1$ for all (x, y) in a neighbourhood of the diagonal $D = \{(x, x); x \in \mathbb{R}^d\}$;
   (ii) for every compact set K in $\mathbb{R}^d$ the mappings $y \to \phi(x, y)$, where $x \in K$, have their support in a fixed compact set in $\mathbb{R}^d$.
(2) A mapping $f : \mathbb{R}^d \to \mathbb{R}$ is said to be *upper semicontinuous* if $f(x) \geq \limsup_{y \to x} f(y)$ for all $x \in \mathbb{R}^d$.
(3) A *Lévy kernel* is a family $\{\mu(x, \cdot), x \in \mathbb{R}^d\}$, where each $\mu(x, \cdot)$ is a Borel measure on $\mathbb{R}^d - \{x\}$, such that:
   (i) the mapping $x \to \int_{\mathbb{R}^d - \{x\}} |y - x|^2 f(y) \mu(x, dy)$ is Borel measurable and locally bounded for each $f \in C_c(\mathbb{R}^d)$;
   (ii) for each $x \in \mathbb{R}^d$, and for every neighbourhood V_x of x, $\mu(x, \mathbb{R}^d - V_x) < \infty$.

Now we can state the first of Courrège's remarkable theorems.

Theorem 3.5.3 (Courrège's first theorem) *If A is a linear operator in $C_0(\mathbb{R}^d)$ and $C_c^\infty(\mathbb{R}^d) \subseteq D_A$, then A satisfies the positive maximum principle if and only if there exist*

- *continuous functions c and b_j, $1 \leq j \leq d$, from $\mathbb{R}^d$ to $\mathbb{R}$ such that $c(x) \leq 0$ for all $x \in \mathbb{R}^d$,*
- *mappings $a^{ij} : \mathbb{R}^d \to \mathbb{R}$, $1 \leq i, j \leq d$, such that $(a^{ij}(x))$ is a positive definite symmetric matrix for each $x \in \mathbb{R}^d$ and the map $x \to (y, a(x)y)$ is upper semicontinuous for each $y \in \mathbb{R}^d$,*
- *a Lévy kernel μ,*
- *a local unit ϕ,*

such that for all $f \in C_c^\infty(\mathbb{R}^d)$, $x \in \mathbb{R}^d$,

$$(Af)(x) = c(x)f(x) + b^i(x)\partial_i f(x) + a^{ij}(x)\partial_i\partial_j f(x)$$
$$+ \int_{\mathbb{R}^d - \{x\}} [f(y) - f(x) - \phi(x, y)(y^i - x^i)\partial_i f(x)]\mu(x, dy). \quad (3.20)$$

A full proof of this result can be found in Courrège [75] or Section 4.5 of Jacob [149].

It is tempting to interpret (3.20) probabilistically, in terms of a killing rate c, a drift vector b, a diffusion matrix a and a jump term controlled by μ. We will return to this later.

Note that both Courrège and Jacob write the integral term in (3.20) as

$$\int_{\mathbb{R}^d - \{x\}} [f(y) - \phi(x, y)f(x) - \phi(x, y)(y^i - x^i)\partial_i f(x)]\mu(x, dy).$$

This is equivalent to the form we have given since, by definition of ϕ, we can find a neighbourhood N_x of each $x \in \mathbb{R}^d$ such that $\phi(x, y) = 1$ for all $y \in N_x$; then

$$\int_{N_x^c} [\phi(x, y) - 1]\mu(x, dy) < \infty,$$

and so this integral can be absorbed into the 'killing term'.

Suppose that we are give a linear operator A in $C_0(\mathbb{R}^d)$ for which $C_c^\infty(\mathbb{R}^d) \subseteq D_A$. We say that it is of *Lévy type* if it has the form (3.20).

Exercise 3.5.4 Show that the generator of a Lévy process can be written in the form (3.20).

Now we turn our attention to pseudo-differential operator representations.

3.5 Lévy-type operators and the positive maximum principle

Theorem 3.5.5 (Courrège's second theorem) *Let A be a linear operator in $C_0(\mathbb{R}^d)$; suppose that $C_c^\infty(\mathbb{R}^d) \subseteq D_A$ and that A satisfies the positive maximum principle. For each $x, u \in \mathbb{R}^d$, define*

$$\eta(x, u) = e^{-i(x,u)}(Ae^{i(\cdot,u)})(x). \tag{3.21}$$

Then:

- *for every $x \in \mathbb{R}^d$, the map $u \to \eta(x, u)$ is continuous, hermitian and conditionally positive definite;*
- *there exists a positive locally bounded function $h : \mathbb{R}^d \to \mathbb{R}$ such that, for each $x, u \in \mathbb{R}^d$,*

$$|\eta(x, u)| \leq h(x)|u|^2;$$

- *for every $f \in C_c^\infty(\mathbb{R}^d)$, $x \in \mathbb{R}^d$,*

$$(Af)(x) = (2\pi)^{-d/2} \int_{\mathbb{R}^d} e^{i(u,x)} \eta(x, u) \hat{f}(u) du. \tag{3.22}$$

Conversely, if η is a continuous map from $\mathbb{R}^d \times \mathbb{R}^d$ to $\mathbb{C}$ that is hermitian and conditionally positive definite in the second variable then the linear operator defined by (3.22) satisfies the positive maximum principle.

Note that it is implicit in the statement of the theorem that A is such that (3.21) makes sense.

Probabilistically, the importance of Courrège's theorems derives from Theorem 3.5.2. If A is the generator of a Feller process and satisfies the domain condition $C_c^\infty(\mathbb{R}^d) \subseteq D_A$ then it can be represented as a Lévy-type operator of the form (3.20), by Theorem 3.5.3, or a pseudo-differential operator of the form (3.22), by Theorem 3.5.5.

In recent years, there has been considerable interest in the converse to the last statement. Given a pseudo-differential operator A whose symbol η is continuous from $\mathbb{R}^d \times \mathbb{R}^d$ to $\mathbb{C}$ and hermitian and conditionally positive definite in the second variable, under what further conditions does A generate a (sub-) Feller process? One line of attack follows immediately from Theorem 3.5.5: since A must satisfy the positive maximum principle we can try to fulfil the other condition of the Hille–Yosida–Ray theorem (Theorem 3.5.1) and then use positivity of the semigroup to generate transition probabilities from which a process can be built using Kolmogorov's construction, as in Theorem 3.1.7. Other approaches to constructing a process include the use of Dirichlet forms (see Section 3.6) and martingale problems (see Subsection 6.7.3). The pioneers in investigating these questions have been Niels Jacob and his collaborators René Schilling and Walter Hoh. To go more deeply into their methods and results would be beyond the scope of the present volume but interested readers are referred to the monograph by Jacob [148], the review article by Jacob and Schilling in [23] and references therein.

Schilling has also used the analytical behaviour of the generator, in its pseudo-differential operator representation, to obtain sample-path properties of the associated Feller process. In [277] he studied the limiting behaviour, for both $t \downarrow 0$ and $t \to \infty$, while estimates on the Hausdorff dimension of the paths were obtained in [278].

3.5.2 Examples of Lévy-type operators

Here we will consider three interesting examples of Lévy-type operators.

Example 3.5.6 (Diffusion operators) Consider a second-order differential operator of the form

$$(Af)(x) = b^i(x)\partial_i f(x) + a^{ij}(x)\partial_i \partial_j f(x),$$

for each $f \in C_c^\infty(\mathbb{R}^d)$, $x \in \mathbb{R}^d$. In general, it is possible to construct a Markov process X in $\mathbb{R}^d$ that is naturally associated with A under quite general conditions on b and a. We call X a *diffusion process* and A the associated *diffusion operator*. Specifically, we require only that each b^i be bounded and measurable and that the a^{ij} are bounded and continuous, the matrix $(a^{ij}(x))$ being positive definite and symmetric for each $x \in \mathbb{R}^d$. We will discuss this in greater detail in Chapter 6. Conditions under which X is a Feller process will also be investigated there.

Example 3.5.7 (Feller's pseudo-Poisson process) Here we give an example of a genuine Feller process whose generator is a Lévy-type operator. It was called the *pseudo-Poisson process* by Feller [102], pp. 333–5.

Let $S = (S(n), n \in \mathbb{N})$ be a homogeneous Markov chain taking values in $\mathbb{R}^d$. For each $n \in \mathbb{N}$, we denote its n-step transition probabilities by $q^{(n)}$ so that for each $x \in \mathbb{R}^d$, $B \in \mathcal{B}(\mathbb{R}^d)$,

$$q^{(n)}(x, B) = P\big(S(n) \in B | S(0) = x\big).$$

We define the *transition operator* Q of the chain by the prescription

$$(Qf)(x) = \int_{\mathbb{R}^d} f(y) q(x, dy)$$

for each $f \in B_b(\mathbb{R}^d)$, $x \in \mathbb{R}^d$, where we have used the notation $q = q^{(1)}$.

3.5 Lévy-type operators and the positive maximum principle

Exercise 3.5.8 Deduce that for all $f \in B_b(\mathbb{R}^d)$, $x \in \mathbb{R}^d$,

$$(Q^n f)(x) = \int_{\mathbb{R}^d} f(y) q^{(n)}(x, dy).$$

Now let $N = (N(t), t \geq 0)$ be a Poisson process, of intensity $\lambda > 0$, that is independent of the Markov chain S and define a new process $X = (X(t), t \geq 0)$ by subordination,

$$X(t) = S(N(t)),$$

for all $t \geq 0$. Then X is a Feller process by Exercise 3.3.17. Clearly, if S is a random walk then X is nothing but a compound Poisson process. More generally, using independence and the results of Exercises 3.5.8 and 3.2.4, we obtain for each $t \geq 0$, $f \in B_b(\mathbb{R}^d)$, $x \in \mathbb{R}^d$,

$$(T_t f)(x) = \mathbb{E}\big(f(X(t)) \big| X(0) = x\big)$$

$$= \sum_{n=0}^{\infty} \mathbb{E}\big(f(S(n)) \big| S(0) = x\big) P(N(t) = n)$$

$$= e^{-\lambda t} \sum_{n=0}^{\infty} \mathbb{E}\big(f(S(n)) \big| S(0) = x\big) \frac{(\lambda t)^n}{n!}$$

$$= e^{-\lambda t} \sum_{n=0}^{\infty} (Q^n f)(x) \frac{(\lambda t)^n}{n!}$$

$$= e^{t[\lambda(Q-I)]} f(x).$$

Hence, if A is the infinitesimal generator of the restriction of $(T_t, t \geq 0)$ to $C_0(\mathbb{R}^d)$ then A is bounded and, for all $f \in C_0(\mathbb{R}^d)$, $x \in \mathbb{R}^d$,

$$(Af)(x) = \lambda((Q - I)f)(x) = \int_{\mathbb{R}^d} [f(y) - f(x)] \lambda q(x, dy).$$

Clearly A is of the form (3.20) with finite Lévy kernel $\mu = \lambda q$.

The above construction has a converse. Define a bounded operator B on $C_0(\mathbb{R}^d)$, by

$$(Bf)(x) = \int_{\mathbb{R}^d} [f(y) - f(x)] \lambda(x) q(x, dy),$$

where λ is a non-negative bounded measurable function on $\mathbb{R}^d$ and q is a *transition function*, i.e. $q(x, \cdot)$ is a probability measure on $\mathcal{B}(\mathbb{R}^d)$, for each $x \in \mathbb{R}^d$ and the map $x \to q(x, A)$ is Borel measurable for each $A \in \mathcal{B}(\mathbb{R}^d)$. It is shown in Ethier and Kurtz [99], pp. 162–4, that B is the infinitesimal generator of a Feller process that has the same finite-dimensional distributions as a certain pseudo-Poisson process.

Example 3.5.9 (Stable-like processes) Recall from Subsection 1.2.5 that if $X = (X(t), t \geq 0)$ is a rotationally invariant stable process of index $\alpha \in (0, 2)$ then it has Lévy symbol $\eta(u) = -|u|^\alpha$, for all $u \in \mathbb{R}^d$ (where we have taken $\sigma = 1$ for convenience) and Lévy–Khintchine representation

$$-|u|^\alpha = K(\alpha) \int_{\mathbb{R}^d - \{0\}} (e^{i(u,y)} - 1 - iuy\chi_{\hat{B}}) \frac{dy}{|y|^{d+\alpha}},$$

where $K(\alpha) > 0$.

Now let $\alpha : \mathbb{R}^d \to (0, 2)$ be continuous; then we can assert the existence of a positive function K on $\mathbb{R}^d$ such that

$$-|u|^{\alpha(x)} = \int_{\mathbb{R}^d - \{0\}} (e^{i(u,y)} - 1 - iuy\chi_{\hat{B}}) \frac{K(x)dy}{|y|^{d+\alpha(x)}}.$$

Define a mapping $\zeta : \mathbb{R}^d \times \mathbb{R}^d \to \mathbb{C}$ by $\zeta(x, u) = -|u|^{\alpha(x)}$. Then ζ clearly satisfies the conditions of Theorem 3.5.5 and so is the symbol of a linear operator A that satisfies the positive maximum principle.

Using the representation

$$(Af)(x) = -(2\pi)^{d/2} \int_{\mathbb{R}^d} |u|^{\alpha(x)} \hat{f}(u) e^{i(x,u)} du$$

for each $f \in S(\mathbb{R}^d)$, $x \in \mathbb{R}^d$, we see that $S(\mathbb{R}^d) \subseteq D_A$. An exercise in the use of the Fourier transform then yields

$$(Af)(x) = \int_{\mathbb{R}^d - \{0\}} [f(y+x) - f(x) - y^j \partial_j f(x)\chi_{\hat{B}}] \frac{K(x)dy}{|y|^{d+\alpha(x)}}.$$

It can now be easily verified that this operator is of Lévy type, with associated Lévy kernel

$$\mu(x, dy) = \frac{K(x)dy}{|y - x|^{d+\alpha(x)}}$$

for each $x \in \mathbb{R}^d$.

By the usual correspondence, we can also write

$$(Af)(x) = (-(-\Delta)^{\alpha(x)/2} f)(x).$$

Of course, we cannot claim at this stage that A is the generator of a Feller semigroup associated with a Feller process. Bass [29] associated a Markov process with X by solving the associated martingale problem under the additional constraint that $0 < \inf_{x \in \mathbb{R}^d} \alpha(x) < \sup_{x \in \mathbb{R}^d} \alpha(x) < 2$. Tsuchiya [296] then obtained the process as the solution of a stochastic differential equation under the constraint that α be Lipschitz continuous. For further studies of properties of these processes see Negoro [235], Kolokoltsov [174, 175] and Uemura [298].

3.5.3 The forward equation

For completeness, we include a brief non-rigorous account of the forward equation. Let $(T_t, t \geq 0)$ be the semigroup associated with a Lévy-type Feller process and, for each $f \in \text{Dom}(A)$, $x \in \mathbb{R}^d$, $t \geq 0$ write $u(t, x) = (T_t f)(x)$; then we have the initial-value problem

$$\frac{\partial u(t, x)}{\partial t} = Au(t, x), \tag{3.23}$$

with initial condition $u(0, x) = f(x)$. Let $(p_t, t \geq 0)$ be the transition probability measures associated with $(T_t, t \geq 0)$. We assume that each $p_t(x, \cdot)$ is absolutely continuous with respect to Lebesgue measure with density $\rho_t(x, \cdot)$. We also assume that, for each $y \in \mathbb{R}^d$, the mapping $t \to \rho_t(x, y)$ is differentiable and that its derivative is uniformly bounded with respect to y. By (3.3) and dominated convergence, for each $t \geq 0$, $x \in \mathbb{R}^d$, we have for all $f \in C_c^\infty(\mathbb{R}^d)$,

$$\frac{\partial u(t, x)}{\partial t} = \frac{\partial (T_t f)(x)}{\partial t}$$
$$= \frac{\partial}{\partial t} \int_{\mathbb{R}^d} f(y) p_t(x, y) dy = \int_{\mathbb{R}^d} f(y) \frac{\partial p_t(x, y)}{\partial t} dy.$$

On the other hand,

$$Au(t, x) = (T_t A f)(x) = \int_{\mathbb{R}^d} (Af)(y) p_t(x, y) \, dy$$
$$= \int_{\mathbb{R}^d} f(y) A^\dagger p_t(x, y) \, dy,$$

where $A^\dagger$ is the 'formal adjoint' of A, which acts on p_t through the y-variable. We thus conclude from (3.23) that

$$\int_{\mathbb{R}^d} f(y) \left[\frac{\partial p_t(x, y)}{\partial t} - A^\dagger p_t(x, y) \right] dy = 0.$$

In the general case, there appears to be no nice form for $A^\dagger$; however, if X is a killed diffusion (so that $\mu \equiv 0$ in (3.20)), integration by parts yields

$$A^\dagger p_t(x, y) = c(y) p_t(x, y) - \partial_i \left[b^i(y) p_t(x, y) \right] + \partial_i \partial_j \left[a^{ij}(y) p_t(x, y) \right].$$

In this case, the partial differential equation

$$\frac{\partial p_t(x, y)}{\partial t} = A^\dagger p_t(x, y) \tag{3.24}$$

is usually called the *Kolmogorov forward equation* by probabilists and the *Fokker–Planck equation* by physicists. In principle, we can try to solve it with the initial condition $p_0(x, y) = \delta(x - y)$ and then use the density to construct the process from its transition probabilities. Notice that all the action in (3.24) is with respect to the 'forward variables' t and y. An alternative equation, which can be more tractable analytically, is the *Kolmogorov backward equation*

$$\frac{\partial p_{t-s}(x, y)}{\partial s} = -A p_{t-s}(x, y). \tag{3.25}$$

Note that, on the right-hand side of (3.25), A operates with respect to the variable x, so this time all the action takes place in the 'backward variables' s and x.

A nice account of this partial differential equation approach to constructing Markov processes can be found in Chapter 3 of Stroock and Varadhan [289]. The discussion of forward and backward equations in their introductory chapter is also highly recommended.

Exercise 3.5.10 Find an explicit form for $A^\dagger$ in the case where A is the generator of Lévy process.

3.6 Dirichlet forms

In this section, we will attempt to give a gentle and somewhat sketchy introduction to the deep and impressive modern theory of Dirichlet forms. We will simplify matters by restricting ourselves to the symmetric case and also by continuing with our programme of studying processes whose state space is $\mathbb{R}^d$. However, readers should bear in mind that some of the most spectacular applications of the theory have been to the construction of processes in quite general contexts such as fractals and infinite-dimensional spaces.

For more detailed accounts, we recommend the reader to Fukushima *et al.* [112], Bouleau and Hirsch [55], Ma and Röckner [204], Chapter 3 of Jacob [148] and Chapter 4 of Jacob [149].

3.6.1 Dirichlet forms and sub-Markov semigroups

If you are unfamiliar with the notion of a closed symmetric form in a Hilbert space, then you should begin by reading Subsection 3.8.3. We fix the real Hilbert space $H = L^2(\mathbb{R}^d)$. By Theorem 3.8.9 there is a one-to-one correspondence between closed symmetric forms $\mathcal{E}$ in H and positive self-adjoint operators T in H, given by $\mathcal{E}(f) = ||T^{1/2} f||^2$ for each $f \in D_{T^{1/2}}$. When we combine this with Theorem 3.4.6, we deduce that there is a one-to-one correspondence between closed symmetric forms in H and self-adjoint semigroups $(T_t, t \geq 0)$ in H.

Now suppose that $(T_t, t \geq 0)$ is a self-adjoint sub-Markovian semigroup in H, so that $0 \leq f \leq 1$ (a.e.) $\Rightarrow 0 \leq T_t f \leq 1$ (a.e.). We can 'code' the self-adjoint semigroup property into a closed symmetric form $\mathcal{E}$. How can we also capture the sub-Markov property? The answer to this question is contained in the following definition.

Let $\mathcal{E}$ be a closed symmetric form in H with domain D. We say that it is a *Dirichlet form* if $f \in D \Rightarrow (f \vee 0) \wedge 1 \in D$ and

$$\mathcal{E}((f \vee 0) \wedge 1) \leq \mathcal{E}(f) \qquad (3.26)$$

for all $f \in D$.

A closed densely defined linear operator A in H with domain D_A is called a *Dirichlet operator* if

$$\langle Af, (f-1)^+ \rangle \leq 0$$

for all $f \in D_A$.

The following theorem describes the analytic importance of Dirichlet forms and operators. We will move on later to their probabilistic value.

Theorem 3.6.1 *The following are equivalent:*

(1) $(T_t, t \geq 0)$ *is a self-adjoint sub-Markovian semigroup in* $L^2(\mathbb{R}^d)$ *with infinitesimal generator* A;
(2) A *is a Dirichlet operator and* $-A$ *is positive self-adjoint;*
(3) $\mathcal{E}(f) = ||(-A)^{1/2} f||^2$ *is a Dirichlet form with domain* $D = D_{A^{1/2}}$.

A proof of this can be found in Bouleau and Hirsch [55], pp. 12–13.

Example 3.6.2 (Symmetric Lévy processes) Let $X = (X(t), t \geq 0)$ be a symmetric Lévy process. In Theorem 3.4.10 we showed that these give rise to self-adjoint L^2-Markov semigroups and so they will induce a Dirichlet form $\mathcal{E}$, by Theorem 3.6.1(3). Let A be the infinitesimal generator of the process. It follows from Corollary 3.4.11 that, for each $f \in C_c^\infty(\mathbb{R}^d)$,

$$(Af)(x) = \tfrac{1}{2} a^{ij} \partial_i \partial_j f(x) + \int_{\mathbb{R}^d - \{0\}} [f(x+y) - f(x)] \nu(dy).$$

The following formula is of course just a consequence of the Lévy–Khintchine formula. Intriguingly, as we will see in the next section, it is a paradigm for the structure of symmetric Dirichlet forms.

For all $f, g \in C_c^\infty(\mathbb{R}^d)$,

$$\mathcal{E}(f, g) = \tfrac{1}{2} a^{ij} \int_{\mathbb{R}^d} (\partial_i f)(x)(\partial_j g)(x) \, dx$$
$$+ \tfrac{1}{2} \int_{(\mathbb{R}^d \times \mathbb{R}^d) - D} [f(x) - f(x+y)]$$
$$\times [g(x) - g(x+y)] \nu(dy) dx, \quad (3.27)$$

where D is the diagonal $\{(x, x), x \in \mathbb{R}^d\}$.

To verify (3.27) we just use integration by parts, the symmetry of ν and a change of variable, to obtain

$$\mathcal{E}(f, g) = -\langle f, Ag \rangle$$
$$= \tfrac{1}{2} a^{ij} \int_{\mathbb{R}^d} (\partial_i f)(x) (\partial_j g)(x) \, dx$$
$$- \int_{\mathbb{R}^d} \int_{\mathbb{R}^d - \{0\}} f(x)[g(x+y) - g(x)] \nu(dy) \, dx$$
$$= \tfrac{1}{2} a^{ij} \int_{\mathbb{R}^d} (\partial_i f)(x)(\partial_j g)(x) \, dx$$
$$+ \tfrac{1}{2} \int_{\mathbb{R}^d} \int_{\mathbb{R}^d - \{0\}} [f(x+y)(g(x+y) - f(x+y)g(x)$$
$$- f(x)g(x+y) + f(x)g(x)] \nu(dy) dx,$$

and the result follows.

A special case of Example 3.6.2 merits particular attention:

Example 3.6.3 (The energy integral) Take X to be a standard Brownian motion; then, for all $f \in \mathcal{H}_2(\mathbb{R}^d)$,

$$\mathcal{E}(f) = \frac{1}{2} \sum_{i=1}^d \int_{\mathbb{R}^d} (\partial_i f)(x)^2 \, dx = \frac{1}{2} \int_{\mathbb{R}^d} |\nabla f(x)|^2 \, dx.$$

This form is often called the *energy integral* or the *Dirichlet integral*.

3.6.2 The Beurling–Deny formula

In this section we will see how the structure of symmetric Lévy processes generalises to a natural class of Dirichlet forms. First we need a definition.

A *core* of a symmetric closed form $\mathcal{E}$ with domain D in $L^2(\mathbb{R}^d)$ is a subset C of $D \cap C_c(\mathbb{R}^d)$ that is dense in D with respect to the norm $||\cdot||_\mathcal{E}$ (see Subsection 3.8.3) and dense in $C_c(\mathbb{R}^d)$ in the uniform norm. If $\mathcal{E}$ possesses such a core, it is said to be *regular*.

Example 3.6.4 Let T be a positive symmetric linear operator in $L^2(\mathbb{R}^d)$ and suppose that $C_c^\infty(\mathbb{R}^d)$ is a core for T in the usual operator sense; then it is also a core for the closed form given by $\mathcal{E}(f) = \langle f, Tf \rangle$, where $f \in D_T$.

We can now state the celebrated Beurling–Deny formula,

Theorem 3.6.5 (Beurling–Deny) *If $\mathcal{E}$ is a regular Dirichlet form in $L^2(\mathbb{R}^d)$ with domain D, then, for all $f, g \in D \cap C_c(\mathbb{R}^d)$,*

$$\mathcal{E}(f,g) = \int_{\mathbb{R}^d} \partial_i f(x) \partial_j g(x) \mu^{ij}(dx) + \int_{\mathbb{R}^d} f(x)g(x)k(dx)$$

$$+ \int_{\mathbb{R}^d \times \mathbb{R}^d - D} [f(x) - f(y)][g(x) - g(y)] J(dx, dy) \quad (3.28)$$

where k is a Borel measure on $\mathbb{R}^d$, J is a Borel measure on $\mathbb{R}^d \times \mathbb{R}^d - D$ ($D = \{(x, x), x \in \mathbb{R}^d\}$ being the diagonal); the $\{\mu^{ij}, 1 \leq i, j \leq d\}$ are Borel measures in $\mathbb{R}^d$ with each $\mu^{ij} = \mu^{ji}$ and $(u, \mu(K)u) \geq 0$ for all $u \in \mathbb{R}^d$ and all compact $K \in \mathcal{B}(\mathbb{R}^d)$, where $\mu(K)$ denotes the $d \times d$ matrix with (i, j)th entry $\mu^{ij}(K)$.

This important result clearly generalises (3.27), and its probabilistic interpretation is clear from the names given to the various measures that appear in (3.28): the μ^{ij} are called *diffusion measures*, J is called the *jump measure* and k is the *killing measure*.

In general, a Dirichlet form $\mathcal{E}$ with domain D is said to be *local* if, for all $f, g \in D$ for which $\text{supp}(f)$ and $\text{supp}(g)$ are disjoint compact sets, we have $\mathcal{E}(f, g) = 0$. A Dirichlet form that fails to be local is often called *non-local*. Since partial differentiation cannot increase supports, the non-local part of the Beurling–Deny form (3.28) is that controlled by the jump measure J.

3.6.3 Closable Markovian forms

In concrete applications it is quite rare to have a closed form. Fortunately, many forms that have to be dealt with have the pleasing property of being closable (see Subsection 3.8.3). In this case, we need an analogue of definition (3.26) for such forms, so that they can code probabilistic information.

Let $\mathcal{E}$ be a closable positive symmetric bilinear form with domain D. We say that it is *Markovian* if, for each $\varepsilon > 0$, there exists a family of infinitely differentiable functions $(\phi_\varepsilon(x), x \in \mathbb{R})$ such that:

(1) $\phi_\varepsilon(x) = x$ for all $x \in [0, 1]$;
(2) $-\varepsilon \leq \phi_\varepsilon(x) \leq 1 + \varepsilon$ for all $x \in \mathbb{R}$;
(3) $0 \leq \phi_\varepsilon(y) - \phi_\varepsilon(x) \leq y - x$ whenever $x, y \in \mathbb{R}$ with $x < y$.

Furthermore, for all $f \in D$, $\phi_\varepsilon(f) \in D$ and

$$\mathcal{E}(\phi_\varepsilon(f)) \leq \mathcal{E}(f).$$

Exercise 3.6.6 Given $(\phi_\varepsilon(x), x \in \mathbb{R})$ as above, show that, for each $x, y \in \mathbb{R}^d$, $|\phi_\varepsilon(x)| \leq |x|$, $|\phi_\varepsilon(y) - \phi_\varepsilon(x)| \leq |y - x|$ and $0 \leq \phi'_\varepsilon(x) \leq 1$.

Note that when $D = C_c^\infty(\mathbb{R}^d)$, a family $(\phi_\varepsilon(x), x \in \mathbb{R})$ satisfying the conditions (1) to (3) above can always be constructed using mollifiers (see e.g. Fukushima *et al.* [112], p. 8.) In this case we also have $\phi_\varepsilon(f) \in C_c^\infty(\mathbb{R}^d)$ whenever $f \in C_c^\infty(\mathbb{R}^d)$.

If we are given a closable Markovian form then we have the following result, which allows us to obtain a bona fide Dirichlet form.

Theorem 3.6.7 *If $\mathcal{E}$ is a closable Markovian symmetric form on $L^2(\mathbb{R}^d)$ then its closure $\overline{\mathcal{E}}$ is a Dirichlet form.*

Proof See Fukushima *et al.* [112], pp. 98–9. □

Example 3.6.8 (Symmetric diffusions) Let $a(x) = (a^{ij}(x))$ be a matrix-valued function from $\mathbb{R}^d$ to itself such that each $a(x)$ is a positive definite symmetric matrix and for each $1 \leq i, j \leq d$ the mapping $x \to a^{ij}(x)$ is Borel measurable. We consider the positive symmetric bilinear form on D given by

$$\mathcal{E}(f) = \int_{\mathbb{R}^d} a^{ij}(x) \partial_i f(x) \partial_j f(x) \, dx, \qquad (3.29)$$

on the domain $D = \{f \in C^1(\mathbb{R}^d) \cap L^2(\mathbb{R}^d), \mathcal{E}(f) < \infty\}$.

Forms of the type (3.29) appear in association with elliptic second-order differential operators in divergence form, i.e.

$$(Af)(x) = \sum_{i,j=1}^d \partial_i \big[a^{ij}(x) \partial_j f(x) \big],$$

3.6 Dirichlet forms

for each $f \in C_c^\infty(\mathbb{R}^d)$, $x \in \mathbb{R}^d$, where we assume for convenience that each a^{ij} is bounded and differentiable. In this case, it is easily verified that A is symmetric and hence that $\mathcal{E}$ is closable on the domain $C_c^\infty(\mathbb{R}^d)$ by Theorem 3.8.10.

More generally, it is shown in Fukushima et al. [112], pp. 100–1, that $\mathcal{E}$ as given by (3.29) is closable if either of the following two conditions is satisfied:

- for each $1 \leq i, j \leq d$, $a^{ij} \in L^2_{\text{loc}}(\mathbb{R}^d)$ and $\partial_i a^{ij} \in L^2_{\text{loc}}(\mathbb{R}^d)$.
- (uniform ellipticity) there exists $K > 0$ such that $(\xi, a(x)\xi) \geq K|\xi|^2$ for all $x, \xi \in \mathbb{R}^d$.

If $\mathcal{E}$ is closable on $C_c^\infty(\mathbb{R}^d)$ then it is Markovian. To see this, we use the result of Exercise 3.6.6 to obtain for all $\varepsilon > 0$, $f \in C_c^\infty(\mathbb{R}^d)$,

$$\mathcal{E}(\phi_\varepsilon(f)) = \int_{\mathbb{R}^d} a^{ij}(x) \partial_i \phi_\varepsilon(f)(x) \, \partial_j \phi_\varepsilon(f)(x) \, dx$$

$$= \int_{\mathbb{R}^d} a^{ij}(x) |\phi'_\varepsilon(f(x))|^2 \partial_i f(x) \partial_j f(x) \, dx$$

$$\leq \int_{\mathbb{R}^d} a^{ij}(x) \partial_i f(x) \partial_j f(x) \, dx = \mathcal{E}(f).$$

It then follows by Theorem 3.6.7 that $\overline{\mathcal{E}}$ is indeed a Dirichlet form.

Note that from a probabilistic point of view, a form of this type contains both diffusion and drift terms (unless a is constant). This is clear when $\mathcal{E}$ is determined by the differential operator A in divergence form.

Example 3.6.9 (Symmetric jump operators) Let ϱ be a Borel measurable mapping from $\mathbb{R}^d \times \mathbb{R}^d \to \mathbb{R}$ that satisfies the symmetry condition $\varrho(x, y) = \varrho(y, x)$ for all $x, y \in \mathbb{R}^d$. We introduce the form

$$\mathcal{E}(f) = \frac{1}{2} \int_{\mathbb{R}^d \times \mathbb{R}^d - D} [f(y) - f(x)]^2 \varrho(x, y) dx, \qquad (3.30)$$

with domain $C_c^\infty(\mathbb{R}^d)$.

We examine the case where $\mathcal{E}$ is induced by a linear operator A for which

$$(Af)(x) = \int_{\mathbb{R}^d - \{x\}} [f(y) - f(x)] \varrho(x, y) dy$$

for $f \in C_c^\infty(\mathbb{R}^d)$.

Let us suppose that ϱ is such that A is a bona fide operator in $L^2(\mathbb{R}^d)$; then, by the symmetry of ϱ, it follows easily that A is symmetric on $C_c^\infty(\mathbb{R}^d)$, with

$$\mathcal{E}(f) = \langle f, Af \rangle.$$

We can now proceed as in Example 3.6.8 and utilise Theorem 3.8.10 to deduce that $\mathcal{E}$ is closable, Exercise 3.6.6 to show that it is Markovian and Theorem 3.6.7 to infer that $\overline{\mathcal{E}}$ is indeed a Dirichlet form.

We will now look at some conditions under which A operates in $L^2(\mathbb{R}^d)$.

We impose a condition on ϱ that is related to the Lévy kernel concept considered in Section 3.5. For each $f \in C_c^\infty(\mathbb{R}^d)$ we require that the mapping

$$x \to \int_{\mathbb{R}^d - \{x\}} |y - x| \varrho(x, y) dy$$

is in $L^2(\mathbb{R}^d)$. Using the mean value theorem, there exists $0 < \theta < 1$ such that, for all $x \in \mathbb{R}^d$,

$$|(Af)(x)| \le \int_{\mathbb{R}^d - \{x\}} |y^i - x^i| |\partial_i f(x + \theta(y - x))| \varrho(x, y) dy$$

$$\le \int_{\mathbb{R}^d - \{x\}} |y - x| \left(\sum_{i=1}^d |\partial_i f(x + \theta(y - x))|^2 \right)^{1/2} \varrho(x, y) dy$$

$$\le d^{1/2} \max_{1 \le i \le n} \sup_{z \in \mathbb{R}^d} |\partial_i f(z)| \left| \int_{\mathbb{R}^d - \{x\}} |y - x| \varrho(x, y) dy \right|,$$

from which we easily deduce that $\|Af\|_2 < \infty$, as required.

Another condition is given in the following exercise.

Exercise 3.6.10 Suppose that

$$\int_{\mathbb{R}^d} \int_{\mathbb{R}^d} |\varrho(x, y)|^2 dy dx < \infty.$$

Deduce that $\|Af\|_2 < \infty$ for each $f \in C_c^\infty(\mathbb{R}^d)$.

A generalisation of this example was studied by René Schilling in [279]. He investigated operators, of a similar type to A above, that are symmetric and satisfy the positive maximum principle. He was able to show that the closure of A is a Dirichlet operator, which then gives rise to a Dirichlet form by Theorem 3.6.1 (see Schilling [279], pp. 89–90).

3.6 Dirichlet forms

We also direct readers to the paper by Albeverio and Song [3], where general conditions for the closability of positive symmetric forms of jump type are investigated.

3.6.4 Dirichlet forms and Hunt processes

Most of this subsection is based on Appendix A.2 of Fukushima *et al.* [112], pp. 310–31. We begin with the key definition. Let $X = (X(t), t \geq 0)$ be a homogeneous sub-Markov process defined on a probability space $(\Omega, \mathcal{F}, P)$ and adapted to a right-continuous filtration $(\mathcal{F}_t, t \geq 0)$. We will also require the augmented natural filtration $(\mathcal{G}_t^X, t \geq 0)$. We say that X is a *Hunt process* if:

(1) X is right-continuous;
(2) X has the *strong Markov property* with respect to $(\mathcal{G}_t^X, t \geq 0)$, i.e. given any $\mathcal{G}_t^X$-adapted stopping time T,

$$P\big(X(T+s) \in B \big| \mathcal{G}_t^X\big) = P\big(X(s) \in B \big| X(T)\big)$$

for all $s \geq 0$, $B \in \mathcal{B}(\mathbb{R}^d)$;
(3) X is *quasi-left-continuous*, i.e. if given any $\mathcal{G}_t^X$-adapted stopping time T and any sequence of $\mathcal{G}_t^X$-adapted stopping times $(T_n, n \in \mathbb{N})$ that are increasing to T we have

$$P\left(\lim_{n \to \infty} X(T_n) = X(T),\ T < \infty\right) = P(T < \infty).$$

If the notion of Hunt process seems a little unfamiliar and obscure, the good news is:

Theorem 3.6.11 *Every sub-Feller process is a Hunt process.*

In particular, then, every Lévy process is a Hunt process.

We now briefly summarise the connection between Hunt processes and Dirichlet forms.

First suppose that X is a Hunt process with associated semigroup $(T_t, t \geq 0)$ and transition probabilities $(p_t, t \geq 0)$. We further assume that X is symmetric. It can then be shown (see Fukushima *et al.* [112], pp. 28–9) that there exists $M \subseteq B_b(\mathbb{R}^d) \cap L^1(\mathbb{R}^d)$, with M dense in $L^2(\mathbb{R}^d)$, such that $\lim_{t \downarrow 0} \int_{\mathbb{R}^d} f(y) p_t(x, dy) = f(x)$ (a.e.) for all $f \in M, x \in \mathbb{R}^d$. From this it follows that the semigroup $(T_t, t \geq 0)$ is strongly continuous. Now since X is symmetric, $(T_t, t \geq 0)$ is self-adjoint and hence we can associate a Dirichlet form $\mathcal{E}$ with X by Theorem 3.6.1.

We note two interesting consequences of this construction.

- Every Lebesgue-symmetric Feller process induces a Dirichlet form in $L^2(\mathbb{R}^d)$.
- Every Lebesgue-symmetic sub-Feller semigroup in $C_0(\mathbb{R}^d)$ induces a self-adjoint sub-Markov semigroup in $L^2(\mathbb{R}^d)$.

The converse, whereby a symmetric Hunt process is associated with an arbitrary regular Dirichlet form, is much deeper and goes beyond the scope of the present volume. The full story can be found in Chapter 7 of Fukushima et al. [112] but there is also a nice introduction in Chapter 3 of Jacob [148].

We give here the briefest of outlines. Let $\mathcal{E}$ be a regular Dirichlet form in $L^2(\mathbb{R}^d)$. By Theorem 3.6.1, we can associate a sub-Markov semigroup $(T_t, t \geq 0)$ to $\mathcal{E}$. Formally, we can try to construct transition probabilities by the usual procedure $p_t(x, A) = (T_t \chi_A)(x)$ for all $t \geq 0$, $A \in \mathcal{B}(\mathbb{R}^d)$ and (Lebesgue) almost all $x \in \mathbb{R}^d$. The problem is that the Chapman–Kolmogorov equations are only valid on sets of 'capacity zero'. It is only by systematically avoiding sets of non-zero capacity that we can associate a Hunt process $X = (X(t), t \geq 0)$ with $\mathcal{E}$. Even when such a process is constructed, the mapping $x \to \mathbb{E}(f(X(t))|X(0) = x)$ for $f \in B_b(\mathbb{R}^d) \cap L^2(\mathbb{R}^d)$ is only defined up to a 'set of capacity zero', and this causes difficulties in giving a sense to the uniqueness of X. For further discussion and also an account of the important notion of capacity, see Fukushima et al. [112], Jacob [148] and Ma and Röckner [204].

3.6.5 Non-symmetric Dirichlet forms

The material in this subsection is mostly based on Ma and Röckner [204], but see also Chapter 3 of Jacob [148] and Section 4.7 of Jacob [149]).

Let D be a dense domain in $L^2(\mathbb{R}^d)$. We want to consider bilinear forms $\mathcal{E}$ with domain D that are positive, i.e. $\mathcal{E}(f, f) \geq 0$ for all $f \in D$, but not necessarily symmetric. We introduce the *symmetric* and *antisymmetric* parts of $\mathcal{E}$, which we denote as $\mathcal{E}_s$ and $\mathcal{E}_a$, respectively, for each $f, g \in D$, by

$$\mathcal{E}_s(f, g) = \tfrac{1}{2}\big[\mathcal{E}(f, g) + \mathcal{E}(g, f)\big] \quad \text{and} \quad \mathcal{E}_a(f, g) = \tfrac{1}{2}\big[\mathcal{E}(f, g) - \mathcal{E}(g, f)\big].$$

Note that $\mathcal{E} = \mathcal{E}_s + \mathcal{E}_a$ and that $\mathcal{E}_s$ is a positive symmetric bilinear form.

In order to obtain a good theory of non-symmetric Dirichlet forms, we need to impose more structure than in the symmetric case.

3.6 Dirichlet forms

We recall (see Subsection 3.8.3) the inner product $\langle \cdot, \cdot \rangle_{\mathcal{E}}$ induced by $\mathcal{E}$ and given by

$$\langle f, g \rangle_{\mathcal{E}} = \langle f, g \rangle + \mathcal{E}(f, g)$$

for each $f, g \in D$. We say that $\mathcal{E}$ satisfies the *weak-sector condition* if there exists $K > 0$ such that, for each $f, g \in D$,

$$|\langle f, g \rangle_{\mathcal{E}}| \leq K \|f\|_{\mathcal{E}} \|g\|_{\mathcal{E}}.$$

Exercise 3.6.12 Show that if $\mathcal{E}$ satisfies the weak-sector condition then there exists $K_1 \geq 0$ such that, for all $f \in D$,

$$|\mathcal{E}(f)| \leq K_1 \big[\|f\|_2 + \mathcal{E}_s(f) \big].$$

A positive bilinear form $\mathcal{E}$ with dense domain D is termed *coercive* if:

(1) $\mathcal{E}_s$ is a closed symmetric form;
(2) $\mathcal{E}$ satisfies the weak sector condition.

A coercive form $\mathcal{E}$ with domain D is said to be a *Dirichlet form* if, for all $f \in D$, we have $(f \vee 0) \wedge 1 \in D$ and:

(D1) $\mathcal{E}\big(f + (f \vee 0) \wedge 1, f - (f \vee 0) \wedge 1\big) \geq 0$;
(D2) $\mathcal{E}\big(f - (f \vee 0) \wedge 1, f + (f \vee 0) \wedge 1\big) \geq 0$.

We comment below on why there are two conditions of this type. Note however that if $\mathcal{E}$ is symmetric then both (D1) and (D2) coincide with the earlier definition (3.26).

Many results of the symmetric case carry over to the more general formalism, but the development is more complicated. For example, Theorem 3.6.1 generalises as follows:

Theorem 3.6.13

(1) $(T_t, t \geq 0)$ *is a sub-Markovian semigroup in* $L^2(\mathbb{R}^d)$ *with generator A if and only if A is a Dirichlet operator.*
(2) *If $\mathcal{E}$ is a coercive form with domain D, then there exists a Dirichlet operator A such that $D = D_A$ and $\mathcal{E}(f, g) = -\langle f, Ag \rangle$ for all $f, g \in D_A$ if and only if $\mathcal{E}$ satisfies* (D1).

See Ma and Röckner [204], pp. 31–2, for a proof.

You can clarify the relationship between (D1) and (D2) through the following exercise.

Exercise 3.6.14 Let A be a closed operator in $L^2(\mathbb{R}^d)$ for which there exists a dense linear manifold D such that $D \subseteq D_A \cap D_{A^*}$. Define two bilinear forms $\mathcal{E}_A$ and $\mathcal{E}_{A^*}$ with domain D by

$$\mathcal{E}_A(f, g) = -\langle f, Ag \rangle, \qquad \mathcal{E}_{A^*}(f, g) = -\langle f, A^*g \rangle.$$

Deduce that $\mathcal{E}_A$ satisfies (D1) if and only if $\mathcal{E}_{A^*}$ satisfies (D2).

We can also associate Hunt processes with non-symmetric Dirichlet forms. This is again more complex than in the symmetric case and, in fact, we need to identify a special class of forms called *quasi-regular* from which processes can be constructed. The details can be found in Chapter 3 of Ma and Röckner [204].

There are many interesting examples of non-symmetric Dirichlet forms on both $\mathbb{R}^d$ and in infinite-dimensional settings, and readers can consult Chapter 2 of Ma and Röckner [204] for some of these. We will consider the case of a Lévy process $X = (X(t), t \geq 0)$ with infinitesimal generator A and Lévy symbol η. We have already seen that if X is symmetric then it gives rise to the prototype symmetric Dirichlet form.

Define a bilinear form $\mathcal{E}$ on the domain $S(\mathbb{R}^d)$ by

$$\mathcal{E}(f, g) = -\langle f, Ag \rangle$$

for all $f, g \in S(\mathbb{R}^d)$. We will find that the relationship between Lévy processes and Dirichlet forms is not so clear cut as in the symmetric case. First, though, we need a preliminary result.

Lemma 3.6.15 *For all $f, g \in S(\mathbb{R}^d)$,*

$$\mathcal{E}_s(f, g) = -\int_{\mathbb{R}^d} \overline{\hat{f}(u)}\, \Re(\eta(u))\, \hat{g}(u)\, du.$$

Proof For all $f, g \in S(\mathbb{R}^d)$, by Theorem 3.3.3(2),

$$\mathcal{E}_s(f, g) = \frac{1}{2}[\mathcal{E}(f, g) + \mathcal{E}(g, f)] = -\frac{1}{2}[\langle f, Ag \rangle + \langle g, Af \rangle]$$

$$= -\frac{1}{2}\left[\int_{\mathbb{R}^d} \overline{\hat{f}(u)} \eta(u) \hat{g}(u) du + \int_{\mathbb{R}^d} \overline{\hat{g}(u)} \eta(u) \hat{f}(u) du\right].$$

In particular, we have

$$\mathcal{E}_s(f) = -\int_{\mathbb{R}^d} |\hat{f}(u)|^2 \eta(u) du$$

$$= -\int_{\mathbb{R}^d} |\hat{f}(u)|^2\, \Re(\eta(u))\, du - i\int_{\mathbb{R}^d} |\hat{f}(u)|^2\, \Im(\eta(u))\, du.$$

However, $A : S(\mathbb{R}^d) \to C_0(\mathbb{R}^d)$ and hence $\mathcal{E}_s(f) \in \mathbb{R}$, so we must have $\int_{\mathbb{R}^d} |\hat{f}(u)|^2 \Im(\eta(u))\, du = 0$. The result then follows by polarisation. □

Exercise 3.6.16 Deduce that $\mathcal{E}$ is positive, i.e. that $\mathcal{E}(f) \geq 0$ for all $f \in S(\mathbb{R}^d)$.

The following result is based on Jacob [149], Example 4.7.32.

Theorem 3.6.17 *If $\mathcal{E}$ satisfies the weak-sector condition, then, for all $u \in \mathbb{R}^d$, there exists $C > 0$ such that*

$$|\Im(\eta(u))| \leq C[1 + \Re(\eta(u))]. \tag{3.31}$$

Proof Suppose that $\mathcal{E}$ satisfies the weak-sector condition; then, by Exercise 3.6.12, there exists $K_1 > 0$ such that $|\mathcal{E}(f)| \leq K_1[\|f\|_2 + \mathcal{E}_s(f)]$ for all $f \in S(\mathbb{R}^d)$. Using Parseval's identity and the result of Lemma 3.6.15, we thus obtain

$$\left| \int_{\mathbb{R}^d} |\hat{f}(u)|^2 \eta(u) du \right| \leq K_1 \int_{\mathbb{R}^d} [1 + \Re(\eta(u))]|\hat{f}(u)|^2 du.$$

Hence

$$\left[\int_{\mathbb{R}^d} |\hat{f}(u)|^2 \Re(\eta(u))\, du \right]^2 + \left[\int_{\mathbb{R}^d} |\hat{f}(u)|^2 \Im(\eta(u))\, du \right]^2$$

$$\leq K_1^2 \left[\int_{\mathbb{R}^d} [1 + \Re(\eta(u))]|\hat{f}(u)|^2 du \right]^2.$$

We thus deduce that there exists $C > 0$ such that

$$\left| \int_{\mathbb{R}^d} |\hat{f}(u)|^2 \Im(\eta(u))\, du \right| \leq C \int_{\mathbb{R}^d} [1 + \Re(\eta(u))]|\hat{f}(u)|^2 du,$$

from which the required result follows. □

Theorem 3.6.17 indicates that the theory of non-symmetric Dirichlet forms is not powerful enough to cover all Lévy processes: indeed, if we take X to be a 'pure drift' with characteristics $(b, 0, 0)$ then it clearly fails to satisfy equation (3.31) and so cannot yield a Dirichlet form. In Jacob [149], Example 4.7.32, it is shown that the condition (3.31) is both necessary and sufficient for $\mathcal{E}$ to be a Dirichlet form. A result of similar type, but under slightly stronger hypotheses, was first established by Berg and Forst in [34].

3.7 Notes and further reading

The general theory of Markov processes is a deep and extensive subject and we have only touched on the basics here. The classic text by Dynkin [86] is a fundamental and groundbreaking study. Indeed, Dynkin is one of the giants of the subject, and I also recommend the collection of his papers [87] for insight into his contribution. Another classic text for Markov process theory is Blumenthal and Getoor [51]. A more modern approach, which is closer to the themes of this book, is the oft-cited Ethier and Kurtz [99].

A classic resource for the analytic theory of semigroups is Hille and Phillips [129]. In fact the idea of studying semigroups of linear mappings in Banach spaces seems to be due to Hille [130]. It is not clear which author first realised that semigroups could be used as a tool to investigate Markov processes; however, the idea certainly seems to have been known to Feller in the 1950s (see Chapters 9 and 10 of [102]).

The modern theory of Dirichlet forms originated with the work of Beurling and Deny [39, 40] and Deny [79] provides a very nice expository account from a potential-theoretic point of view. The application of these to construct Hunt processes was developed by Fukushima and is, as discussed above, described in Fukushima *et al.* [112]. The notion of a Hunt process is, of course, due to G.A. Hunt and can be found in [138].

3.8 Appendix: Unbounded operators in Banach spaces

In this section, we aim to give a primer on all the results about linear operators that are used in Chapter 3. In order to keep the book as self-contained as possible, we have included proofs of some key results; however, our account is, by its very nature, somewhat limited and those who require more sustenance should consult a dedicated book on functional analysis. Our major sources, at least for the first two subsections, are Chapter 8 of Reed and Simon [257] and Chapters 1 and 7 of Yosida [309]. The classic text by Kato [168] is also a wonderful resource. We assume a basic knowledge of Banach and Hilbert spaces.

3.8.1 Basic concepts: operators domains, closure, graphs, cores, resolvents

Let B_1 and B_2 be Banach spaces over either $\mathbb{R}$ or $\mathbb{C}$. An *operator* from B_1 to B_2 is a mapping T from a subset D_T of B_1 into B_2. We call D_T the *domain* of T. T is said to be *linear* if D_T is a linear space and

$$T(\alpha \psi_1 + \beta \psi_2) = \alpha T \psi_1 + \beta T \psi_2$$

3.8 Appendix: Unbounded operators in Banach spaces

for all $\psi_1, \psi_2 \in D_T$ and all scalars α and β. Operators that fail to be linear are usually called *non-linear*.

From now on, all our operators will be taken to be linear and we will take $B_1 = B_2 = B$ to be a real Banach space, as this is usually sufficient in probability theory (although not in quantum probability, see e.g. Meyer [226] or Parthasarathy [249]). Readers can check that almost all ideas extend naturally to the more general case (although one needs to be careful with complex conjugates when considering adjoint operators in complex spaces). When considering the Fourier transform, in the final section, we will in fact need some spaces of complex functions, but these should present no difficulty to the reader.

Linear operators from B to B are usually said to *operate in B*. The norm in B will always be denoted as $||\cdot||$.

Let T_1 and T_2 be linear operators in B with domains D_{T_1} and D_{T_2}, respectively. We say that T_2 is an *extension* of T_1 if

(1) $D_{T_1} \subseteq D_{T_2}$,
(2) $T_1\psi = T_2\psi$ for all $\psi \in D_{T_1}$.

We write $T_1 \subseteq T_2$ in this case. If T_2 is an extension of T_1, we often call T_1 the *restriction* of T_2 to D_{T_1} and write $T_1 = T_2|_{D_{T_1}}$.

Linear operators can be added and composed so long as we take care with domains. Let S and T be operators in B with domains D_S and D_T, respectively. Then $S + T$ is an operator with domain $D_S \cap D_T$ and

$$(S+T)\psi = S\psi + T\psi$$

for all $\psi \in D_S \cap D_T$.

The composition ST has domain $D_{ST} = D_T \cap T^{-1}(D_S)$ and

$$(ST)\psi = S(T\psi),$$

for all $\psi \in D_{ST}$.

Let T be a linear operator in B. It is said to be *densely defined* if its domain D_T is dense in B. Note that even if S and T are both densely defined, $S + T$ may not be.

A linear operator T in B with domain D_T is *bounded* if there exists $K \geq 0$ such that

$$||T\psi|| \leq K||\psi||$$

for all $\psi \in D_T$.

Operators that fail to be bounded are often referred to as *unbounded*.

Proposition 3.8.1 *A densely defined bounded linear operator T in B has a unique bounded extension whose domain is the whole of B.*

Proof (Sketch) Let $\psi \in B$; then since D_T is dense there exists $(\psi_n, n \in \mathbb{N})$ in D_T with $\lim_{n\to\infty} \psi_n = \psi$. Since T is bounded, we deduce easily that $(T\psi_n, n \in \mathbb{N})$ is a Cauchy sequence in B and so converges to a vector $\phi \in B$. Define an operator $\tilde{T}$ with domain B by the prescription

$$\tilde{T}\psi = \phi.$$

Then it is easy to see that $\tilde{T}$ is linear and extends T. Moreover, T is bounded since

$$||\tilde{T}\psi|| = ||\phi|| = \lim_{n\to\infty} ||T\psi_n|| \leq K \lim_{n\to\infty} ||\psi_n|| = K||\psi||,$$

where we have freely used the Banach-space inequality

$$|(||a|| - ||b||)| \leq ||a - b|| \text{ for all } a, b \in B.$$

It is clear that $\tilde{T}$ is unique. □

In the light of Proposition 3.8.1, whenever we speak of a bounded operator in B, we will implicitly assume that its domain is the whole of B.

Let T be a bounded linear operator in B. We define its *norm* $||T||$ by

$$||T|| = \sup\{||T\psi||; \psi \in B, ||\psi|| = 1\};$$

then the mapping $T \to ||T||$ really is a norm on the linear space $L(B)$ of all bounded linear operators in B. $L(B)$ is itself a Banach space (and in fact, a Banach algebra) with respect to this norm.

A bounded operator T is said to be a *contraction* if $||T|| \leq 1$ and an *isometry* if $||T|| = 1$ (the Itô stochastic integral as constructed in Chapter 4 is an example of an isometry between two Hilbert spaces). An operator T in B that is isometric and bijective is easily seen to have an isometric inverse. Such operators are called *isometric isomorphisms*.

Proposition 3.8.2 *A linear operator T in B is bounded if and only if it is continuous.*

Proof (Sketch) Suppose that T is bounded in B and let $\psi \in B$ and $(\psi_n, n \in \mathbb{N})$ be any sequence in B converging to ψ. Then by linearity

$$||T\psi - T\psi_n|| \leq ||T|| \, ||\psi - \psi_n||,$$

from which we deduce that $(T\psi_n, n \in \mathbb{N})$ converges to $T\psi$ and the result follows.

Conversely, suppose that T is continuous but not bounded; then for each $n \in \mathbb{N}$ we can find $\psi_n \in B$ with $||\psi_n|| = 1$ and $||T\psi_n|| \geq n$. Now let $\phi_n = \psi_n/n$; then $\lim_{n\to\infty} \phi_n = 0$ but $||T\phi_n|| > 1$ for each $n \in \mathbb{N}$. Hence T is not continuous at the origin and we have obtained the desired contradiction.
$\square$

For unbounded operators, the lack of continuity is somewhat alleviated if the operator is closed, which we may regard as a weak continuity property. Before defining this explicitly, we need another useful concept. Let T be an operator in B with domain D_T. Its *graph* is the set $G_T \subseteq B \times B$ defined by

$$G_T = \{(\psi, T\psi); \psi \in D_T\}.$$

We say that T is *closed* if G_T is closed in $B \times B$. Clearly this is equivalent to the requirement that, for every sequence $(\psi_n, n \in \mathbb{N})$ which converges to $\psi \in B$ and for which $(T\psi_n, n \in \mathbb{N})$ converges to $\phi \in B$, $\psi \in D_T$ and $\phi = T\psi$. If T is a closed linear operator then it is easy to check that its domain D_T is itself a Banach space with respect to the *graph norm* $|||\cdot|||$ where

$$|||\psi||| = ||\psi|| + ||T\psi||$$

for each $\psi \in D_T$.

In many situations, a linear operator only fails to be closed because its domain is too small. To accommodate this we say that a linear operator T in B is *closable* if it has a closed extension $\tilde{T}$. Clearly T is closable if and only if there exists a closed operator $\tilde{T}$ for which $\overline{G_T} \subseteq G_{\tilde{T}}$. Note that there is no reason why $\tilde{T}$ should be unique, and we define the *closure* $\overline{T}$ of a closable T to be its smallest closed extension, so that $\overline{T}$ is the closure of T if and only if the following hold:

(1) $\overline{T}$ is a closed extension of T;
(2) if $\tilde{T}$ is any other closed extension of T then $D_{\overline{T}} \subseteq D_{\tilde{T}}$.

The next theorem gives a useful practical criterion for establishing closability.

Theorem 3.8.3 *A linear operator T in B with domain D_T is closable if and only if for every sequence $(\psi_n, n \in \mathbb{N})$ in D_T which converges to 0 and for which $(T\psi_n, n \in \mathbb{N})$ converges to some $\phi \in B$, we always have $\phi = 0$.*

Proof If T is closable then the result is immediate from the definition. Conversely, let (x, y_1) and (x, y_2) be two points in $\overline{G_T}$. Our first task is to show that we always have $y_1 = y_2$. Let $(x_n^1, n \in \mathbb{N})$ and $(x_n^2, n \in \mathbb{N})$ be two sequences in D_T that converge to x; then $(x_n^1 - x_n^2, n \in \mathbb{N})$ converges to 0 and $(Tx_n^1 - Tx_n^2, n \in \mathbb{N})$ converges to $y_1 - y_2$. Hence $y_1 = y_2$ by the criterion.

From now on, we write $y = y_1 = y_2$ and define $T_1 x = y$. Then T_1 is a well-defined linear operator with $D_{T_1} = \{x \in B;$ there exists $y \in B$ such that $(x, y) \in \overline{G_T}\}$. Clearly T_1 extends T and by construction we have $G_{T_1} = \overline{G_T}$, so that T_1 is minimal, as required. $\square$

From the proof of Theorem 3.8.3, we see that a linear operator T is closable if and only if

$$G_{\overline{T}} = \overline{G_T}.$$

Having dealt with the case where the domain is too small, we should also consider the case where we know that an operator T is closed, but the domain is too large or complicated for us to work in it with ease. In that case it is very useful to have a core available.

Let T be a closed linear operator in B with domain D_T. A linear subspace C of D_T is a *core* for T if

$$\overline{T|_C} = T,$$

i.e. given any $\psi \in D_T$, there exists a sequence $(\psi_n, n \in \mathbb{N})$ in C such that $\lim_{n \to \infty} \psi_n = \psi$ and $\lim_{n \to \infty} T\psi_n = T\psi$.

Example 3.8.4 Let $B = C_0(\mathbb{R})$ and define

$$D_T = \{f \in C_0(\mathbb{R}); f \text{ is differentiable and } f' \in C_0(\mathbb{R})\}$$

and

$$Tf = f'$$

for all $f \in D_T$; then T is closed and $C_c^\infty(\mathbb{R})$ is a core for T.

The final concept we need in this subsection is that of a resolvent. Let T be a linear operator in B with domain D_T. Its *resolvent set* $\rho(T) = \{\lambda \in \mathbb{C}; \lambda I - T \text{ is invertible}\}$. The *spectrum* of T is the set $\sigma(T) = \rho(T)^c$. Note that every eigenvalue of T is an element of $\sigma(T)$. If $\lambda \in \rho(T)$, the linear operator $R_\lambda(T) = (\lambda I - T)^{-1}$ is called the *resolvent* of T.

3.8 Appendix: Unbounded operators in Banach spaces

Proposition 3.8.5 *If T is a closed linear operator in B with domain D_T and resolvent set $\rho(T)$, then, for all $\lambda \in \rho(T)$, $R_\lambda(T)$ is a bounded operator from B into D_T.*

Proof We will need the inverse mapping theorem, which states that a continuous bijection between two Banach spaces always has a continuous inverse (see e.g. Reed and Simon [257], p. 83). Now since T is closed, D_T is a Banach space under the graph norm and we find that for each $\lambda \in \rho(T)$, $\psi \in D_T$,

$$||(\lambda I - T)\psi|| \leq |\lambda|\, ||\psi|| + ||T\psi|| \leq \max\{1, |\lambda|\}\, |||\psi|||.$$

So $\lambda I - T$ is bounded and hence continuous (by Proposition 3.8.2) from D_T to B. The result then follows by the inverse mapping theorem. □

3.8.2 Dual and adjoint operators – self-adjointness

Let B be a real Banach space and recall that its *dual space* B^* is the linear space comprising all continuous linear functionals from B to $\mathbb{R}$. The space B^* is itself a Banach space with respect to the norm

$$||l|| = \sup\{|l(x)|;\, ||x|| = 1\}.$$

Now let T be a densely defined linear operator in B with domain D_T. We define the *dual operator* T^c of T to be the linear operator in B^* with $D_{T^c} = \{l \in B^*;\, l \circ T \in B^*\}$ and for which

$$T^c l = l \circ T$$

for each $l \in D_{T^c}$, so that $T^c l(\psi) = l(T(\psi))$ for each $l \in D_{T^c}$, $\psi \in D_T$.

One of the most important classes of dual operators occurs when B is a Hilbert space with inner product $\langle \cdot, \cdot \rangle$. In this case the Riesz representation theorem ensures that B and B^* are isometrically isomorphic and that every $l \in B^*$ is of the form l_ψ for some $\psi \in B$, where $l_\psi(\phi) = \langle \psi, \phi \rangle$ for each $\psi \in B$. We may then define the *adjoint operator* T^* of T with domain $D_{T^*} = \{\psi \in B;\, l_\psi \in D_{T^c}\}$ by the prescription $T^*\psi = T^c(l_\psi)$ for each $\psi \in D_{T^*}$. If S and T are both linear operators in a Hilbert space B, and $\alpha \in \mathbb{R}$, we have

$$(S + \alpha T)^* \subseteq S^* + \alpha T^*, \qquad (ST)^* \subseteq T^* S^*, \quad \text{if } S \subseteq T \text{ then } T^* \subseteq S^*.$$

Note that $T^{**} = (T^*)^*$ is an extension of T.

The following result is very useful.

Theorem 3.8.6 *Let T be a linear operator in a Hilbert space B. Then*

(1) *T^* is closed,*
(2) *T is closable if and only if T^* is densely defined, in which case we have $\overline{T} = T^{**}$,*
(3) *if T is closable then $(\overline{T})^* = T^*$.*

Proof See Reed and Simon [257], pp. 252–3, or Yosida [309], p. 196. □

In applications, we frequently encounter linear operators T that satisfy the condition

$$\langle \psi_1, T\psi_2 \rangle = \langle T\psi_1, \psi_2 \rangle$$

for all $\psi_1, \psi_2 \in D_T$. Such operators are said to be *symmetric* and the above condition is clearly equivalent to the requirement that $T \subseteq T^*$. Note that if T is densely defined and symmetric then it is closable by Theorem 3.8.6(2) and we can further deduce that $T \subseteq T^{**} \subseteq T^*$.

We often require more than this, and a linear operator is said to be *self-adjoint* if $T = T^*$. We emphasise that for T to be self-adjoint we must have $D_T = D_{T^*}$.

The problem of extending a given symmetric operator to be self-adjoint is sometimes fraught with difficulty. In particular, a given symmetric operator may have many distinct self-adjoint extensions; see e.g. Reed and Simon [257], pp. 257–9. We say that a symmetric operator is *essentially self-adjoint* if it has a unique self-adjoint extension.

Proposition 3.8.7 *A symmetric operator T in a Hilbert space B is essentially self-adjoint if and only if $\overline{T} = T^*$.*

Proof We prove only sufficiency here. To establish this, observe that if S is another self-adjoint extension of T then $\overline{T} \subseteq S$ and so, on taking adjoints,

$$S = S^* \subseteq \overline{T}^* = T^{**} = \overline{T}.$$

Hence $S = \overline{T}$. □

Readers should be warned that for a linear operator T to be closed and symmetric does not imply that it is essentially self-adjoint. Of course, if T is bounded then it is self-adjoint if and only if it is symmetric. The simplest example of a bounded self-adjoint operator is a *projection*. This is a linear self-adjoint operator P that is also idempotent, in that $P^2 = P$. In fact any

3.8 Appendix: Unbounded operators in Banach spaces

self-adjoint operator can be built in a natural way from projections. To see this, we need the idea of a *projection-valued measure*. This is a family of projections $\{P(A), A \in \mathcal{B}(\mathbb{R})\}$ that satisfies the following:

(1) $P(\emptyset) = 0$, $P(\mathbb{R}) = 1$;
(2) $P(A_1 \cap A_2) = P(A_1)P(A_2)$ for all $A_1, A_2 \in \mathcal{B}(\mathbb{R})$;
(3) if $(A_n, n \in \mathbb{N})$ is a Borel partition of $A \in \mathcal{B}(\mathbb{R})$ then

$$P(A)\psi = \sum_{n=1}^{\infty} P(A_n)\psi$$

for all $\psi \in B$.

For each $\phi, \psi \in B$, a projection-valued measure gives rise to a finite measure $\mu_{\phi,\psi}$ on $\mathcal{B}(\mathbb{R})$ via the prescription $\mu_{\phi,\psi}(A) = \langle \phi, P(A)\psi \rangle$ for each $A \in \mathcal{B}(\mathbb{R})$.

Theorem 3.8.8 (The spectral theorem) *If T is a self-adjoint operator in a Hilbert space B, then there exists a projection-valued measure $\{P(A), A \in \mathcal{B}(\mathbb{R})\}$ in B such that for all $f \in B, g \in D_T$,*

$$\langle \phi, T\psi \rangle = \int_{\mathbb{R}} \lambda \mu_{\phi,\psi}(d\lambda).$$

We write this symbolically as

$$T = \int_{\mathbb{R}} \lambda P(d\lambda).$$

Note that the support of the measure $\mu_{\phi,\phi}$ for each $\phi \in D_T$ is $\sigma(T)$, the spectrum of T.

Spectral theory allows us to develop a functional calculus for self-adjoint operators; specifically, if f is a Borel function from $\mathbb{R}$ to $\mathbb{R}$ then $f(T)$ is again self-adjoint, where

$$f(T) = \int_{\mathbb{R}} f(\lambda) P(d\lambda).$$

Note that $||f(T)\psi||^2 = \int_{\mathbb{R}} |f(\lambda)|^2 \, ||P(d\lambda)\psi||^2$ for all $\psi \in D_{f(T)}$.

A self-adjoint operator T is said to be *positive* if $\langle f, Tf \rangle \geq 0$ for all $f \in D_T$. We have that T is positive if and only if $\sigma(T) \subseteq [0, \infty)$.

It is easily verified that a bounded linear operator T in a Hilbert space B is isometric if and only if $T^*T = I$. We say that T is a *co-isometry* if $TT^* = I$ and *unitary* if it is isometric and co-isometric. T is an isometric isomorphism of B if and only if it is unitary, and in this case we have $T^{-1} = T^*$.

3.8.3 Closed symmetric forms

A useful reference for this subsection is Chapter 1 of Bouleau and Hirsch [55].

Let B be a real Hilbert space and suppose that D is a dense linear subspace of B. A *closed symmetric form* in B is a bilinear map $\mathcal{E} : D \times D \to \mathbb{R}$ such that:

(1) $\mathcal{E}$ is symmetric, i.e. $\mathcal{E}(f, g) = \mathcal{E}(g, f)$ for all $f, g \in D$;
(2) $\mathcal{E}$ is positive, i.e. $\mathcal{E}(f, f) \geq 0$ for all $f \in D$;
(3) D is a real Hilbert space with respect to the inner product $\langle \cdot, \cdot \rangle_\mathcal{E}$, where, for each $f, g \in D$,

$$\langle f, g \rangle_\mathcal{E} = \langle f, g \rangle + \mathcal{E}(f, g).$$

With respect to the inner product in (3), we have the associated norm $\|\cdot\|_\mathcal{E} = \langle \cdot, \cdot \rangle_\mathcal{E}^{1/2}$.

For each $f \in D$, we write $\mathcal{E}(f) = \mathcal{E}(f, f)$ and note that $\mathcal{E}(\cdot)$ determines $\mathcal{E}(\cdot, \cdot)$ by polarisation.

An important class of closed symmetric forms is generated as follows. Let T be a positive self-adjoint operator in B; then, by the spectral theorem, we can obtain its square root $T^{1/2}$, which is also a positive self-adjoint operator in B. We have $D_T \subseteq D_{T^{1/2}}$ since, for all $f \in D_T$,

$$\|T^{1/2} f\|^2 = \langle f, Tf \rangle \leq \|f\| \|Tf\|.$$

Now take $D = D_{T^{1/2}}$; then $\mathcal{E}(f) = \|T^{1/2} f\|^2$ is a closed symmetric form. Indeed (3) above is just the statement that $D_{T^{1/2}}$ is complete with respect to the graph norm. Suppose that $\mathcal{E}$ is a closed symmetric form in B with domain D. Then we can define a positive self-adjoint operator A in B by the prescription

$$D_A = \{f \in D, \exists g \in B \text{ such that } \mathcal{E}(f, h) = (g, h), \forall h \in D\},$$
$$Af = g \quad \text{for all} \quad f \in D_A.$$

Our conclusion from the above discussion is

Theorem 3.8.9 *There is a one-to-one correspondence between closed symmetric forms in B and positive self-adjoint operators in B.*

Sometimes we need a weaker concept than the closed symmetric form. Let $\mathcal{E}$ be a positive symmetric bilinear form on B with domain D. We say that it is *closable* if there exists a closed form $\mathcal{E}_1$ with domain D_1 that extends $\mathcal{E}$, in the sense that $D \subseteq D_1$ and $\mathcal{E}(f) = \mathcal{E}_1(f)$ whenever $f \in D$. Just as in the case of closable operators, we can show that a closable $\mathcal{E}$ has a smallest closed extension, which we call the *closure* of $\mathcal{E}$ and denote as $\overline{\mathcal{E}}$. We always write its domain as $\overline{D}$. Here are some useful practical techniques for proving that a form is closable.

- A necessary and sufficient condition for a positive symmetric bilinear form to be closable is that for every sequence $(f_n, n \in \mathbb{N})$ in D for which $\lim_{n\to\infty} f_n = 0$ and $\lim_{m,n\to\infty} \mathcal{E}(f_n - f_m) = 0$ we have $\lim_{n\to\infty} \mathcal{E}(f_n) = 0$.
- A sufficient condition for a positive symmetric bilinear form to be closable is that for every sequence $(f_n, n \in \mathbb{N})$ in D for which $\lim_{n\to\infty} f_n = 0$ we have $\lim_{n\to\infty} \mathcal{E}(f_n, g) = 0$ for every $g \in D$.

The following result is useful in applications.

Theorem 3.8.10 *Let T be a densely defined symmetric positive operator in B, so that $\langle f, Tf \rangle \geq 0$ for all $f \in D_T$. Define a form $\mathcal{E}$ by $D = D_T$ and $\mathcal{E}(f) = \langle f, Tf \rangle$ for all $f \in D$. Then:*

(1) *$\mathcal{E}$ is closable;*
(2) *there exists a positive self-adjoint operator T_F that extends T such that $\overline{\mathcal{E}}(f) = \langle f, T_F f \rangle$ for all $f \in \overline{D}$.*

The operator T_F of Theorem 3.8.10 is called the *Friedrichs extension* of T.

3.8.4 The Fourier transform and pseudo-differential operators

The material in the first part of this section is largely based on Rudin [267].

Let $f \in L^1(\mathbb{R}^d, \mathbb{C})$; then its *Fourier transform* is the mapping $\hat{f}$, defined by

$$\hat{f}(u) = (2\pi)^{-d/2} \int_{\mathbb{R}^d} e^{-i(u,x)} f(x) dx \qquad (3.32)$$

for all $u \in \mathbb{R}^d$. If we define $\mathcal{F}(f) = \hat{f}$ then $\mathcal{F}$ is a linear mapping from $L^1(\mathbb{R}^d, \mathbb{C})$ to the space of all continuous complex-valued functions on $\mathbb{R}^d$ called the *Fourier transformation*.

We introduce two important families of linear operators in $L^1(\mathbb{R}^d, \mathbb{C})$, *translations* $(\tau_x, x \in \mathbb{R}^d)$ and *phase multiplications* $(e_x, x \in \mathbb{R}^d)$, by

$$(\tau_x f)(y) = f(y - x), \qquad (e_x f)(y) = e^{i(x,y)} f(y)$$

for each $f \in L^1(\mathbb{R}^d, \mathbb{C})$ and $x, y \in \mathbb{R}^d$.

It is easy to show that each of τ_x and e_x are isometric isomorphisms of $L^1(\mathbb{R}^d, \mathbb{C})$. Two key, easily verified, properties of the Fourier transform are

$$\widehat{\tau_x f} = e_{-x}\hat{f} \quad \text{and} \quad \widehat{e_x f} = \tau_x \hat{f} \tag{3.33}$$

for each $x \in \mathbb{R}^d$.

Furthermore, if we define the *convolution* $f * g$ of $f, g \in L^1(\mathbb{R}^d, \mathbb{C})$ by

$$(f * g)(x) = (2\pi)^{-d/2} \int_{\mathbb{R}^d} f(x-y) g(y) dy$$

for each $x \in \mathbb{R}^d$, then we have $\widehat{(f * g)} = \hat{f}\hat{g}$.

If μ is a finite measure on $\mathbb{R}^d$, we can define its Fourier transform by

$$\hat{\mu}(u) = (2\pi)^{-d/2} \int_{\mathbb{R}^d} e^{-i(u,x)} \mu(dx)$$

for each $u \in \mathbb{R}^d$, and $\hat{\mu}$ is then a continuous positive definite mapping from $\mathbb{R}^d$ to $\mathbb{C}$.

The *convolution* $f * \mu$, for $f \in L^1(\mathbb{R}^d, \mathbb{C}) \cap C_0(\mathbb{R}^d, \mathbb{C})$, is defined by

$$(f * \mu)(x) = (2\pi)^{-d/2} \int_{\mathbb{R}^d} f(x-y) \mu(dy),$$

and we have again that $\widehat{(f * \mu)} = \hat{f}\hat{\mu}$.

Valuable information about the range of $\mathcal{F}$ is given by the following key theorem.

Theorem 3.8.11 (Riemann–Lebesgue lemma) *If $f \in L^1(\mathbb{R}^d, \mathbb{C})$ then $\hat{f} \in C_0(\mathbb{R}^d, \mathbb{C})$ and $\|\hat{f}\|_0 \leq \|f\|_1$.*

If $f \in L^2(\mathbb{R}^d, \mathbb{C})$, we can also define its Fourier transform as in (3.32). It then transpires that $\mathcal{F} : L^2(\mathbb{R}^d, \mathbb{C}) \to L^2(\mathbb{R}^d, \mathbb{C})$ is a unitary operator. The fact that $\mathcal{F}$ is isometric is sometimes expressed by

Theorem 3.8.12 (Plancherel) *If $f \in L^2(\mathbb{R}^d, \mathbb{C})$ then*

$$\int_{\mathbb{R}^d} |f(x)|^2 dx = \int_{\mathbb{R}^d} |\hat{f}(u)|^2 du;$$

or

Theorem 3.8.13 (Parseval) *If $f, g \in L^2(\mathbb{R}^d, \mathbb{C})$ then*

$$\int_{\mathbb{R}^d} \overline{f(x)} g(x) dx = \int_{\mathbb{R}^d} \overline{\hat{f}(u)} \hat{g}(u) du.$$

3.8 Appendix: Unbounded operators in Banach spaces

Although, as we have seen, $\mathcal{F}$ has nice properties in both L^1 and L^2, perhaps the most natural context in which to discuss it is the Schwartz space of rapidly decreasing functions. These are smooth functions such that they, and all their derivatives, decay to zero at infinity faster than any negative power of $|x|$. To make this precise, we first need some standard notation for partial differential operators. Let $\alpha = (\alpha_1, \ldots, \alpha_d)$ be a *multi-index*, so that $\alpha \in (\mathbb{N} \cup \{0\})^d$. We define $|\alpha| = \alpha_1 + \cdots + \alpha_d$ and

$$D^\alpha = \frac{1}{i^{|\alpha|}} \frac{\partial^{\alpha_1}}{\partial x_1^{\alpha_1}} \cdots \frac{\partial^{\alpha_d}}{\partial x_d^{\alpha_d}}.$$

Similarly, if $x = (x_1, \ldots, x_d) \in \mathbb{R}^d$ then $x^\alpha = x_1^{\alpha_1} \cdots x_d^{\alpha_d}$.

Now we define *Schwartz space* $S(\mathbb{R}^d, \mathbb{C})$ to be the linear space of all $f \in C^\infty(\mathbb{R}^d, \mathbb{C})$ for which

$$\sup_{x \in \mathbb{R}^d} |x^\beta D^\alpha f(x)| < \infty$$

for all multi-indices α and β. Note that $C_c^\infty(\mathbb{R}^d, \mathbb{C}) \subset S(\mathbb{R}^d, \mathbb{C})$ and that the 'Gaussian function' $x \to \exp(-x^2)$ is in $S(\mathbb{R}^d, \mathbb{C})$. The space $S(\mathbb{R}^d, \mathbb{C})$ is dense in $C_0(\mathbb{R}^d, \mathbb{C})$ and in $L^p(\mathbb{R}^d, \mathbb{C})$ for all $1 \leq p < \infty$. These statements remain true when $\mathbb{C}$ is replaced by $\mathbb{R}$.

The space $S(\mathbb{R}^d, \mathbb{C})$ is a Fréchet space with respect to the family of norms $\{\|\cdot\|_N, N \in \mathbb{N} \cup \{0\}\}$, where for each $f \in S(\mathbb{R}^d, \mathbb{C})$

$$\|f\|_N = \max_{|\alpha| \leq N} \sup_{x \in \mathbb{R}^d} (1 + |x|^2)^N |D^\alpha f(x)|.$$

The dual of $S(\mathbb{R}^d, \mathbb{C})$ with this topology is the space $S'(\mathbb{R}^d, \mathbb{C})$ of *tempered distributions*.

The operator $\mathcal{F}$ is a continuous bijection of $S(\mathbb{R}^d, \mathbb{C})$ into itself with a continuous inverse, and we have the following important result.

Theorem 3.8.14 (Fourier inversion) *If $f \in S(\mathbb{R}^d, \mathbb{C})$ then*

$$f(x) = (2\pi)^{-d/2} \int_{\mathbb{R}^d} \hat{f}(u) e^{i(u,x)} du.$$

In the final part of this subsection, we show how the Fourier transform allows us to build pseudo-differential operators. We begin by examining the Fourier transform of differential operators. More or less everything flows from the following simple fact:

$$D^\alpha e^{i(u,x)} = u^\alpha e^{i(u,x)},$$

for each $x, u \in \mathbb{R}^d$ and each multi-index α.

Using Fourier inversion and dominated convergence, we then find that

$$(D^\alpha f)(x) = (2\pi)^{-d/2} \int_{\mathbb{R}^d} u^\alpha \hat{f}(u) e^{i(u,x)} du$$

for all $f \in S(\mathbb{R}^d, \mathbb{C}), x \in \mathbb{R}^d$.

If p is a polynomial in u of the form $p(u) = \sum_{|\alpha| \leq k} c_\alpha u^\alpha$, where $k \in \mathbb{N}$ and each $c_\alpha \in \mathbb{C}$, we can form the associated differential operator $P(D) = \sum_{|\alpha| \leq k} c_\alpha D^\alpha$ and, by linearity,

$$(P(D)f)(x) = (2\pi)^{-d/2} \int_{\mathbb{R}^d} p(u) \hat{f}(u) e^{i(u,x)} du.$$

The next step is to employ variable coefficients. If each $c_\alpha \in C^\infty(\mathbb{R}^d)$, for example, we may define $p(x, u) = \sum_{|\alpha| \leq k} c_\alpha(x) u^\alpha$ and $P(x, D) = \sum_{|\alpha| \leq k} c_\alpha(x) D^\alpha$. We then find that

$$(P(x, D)f)(x) = (2\pi)^{-d/2} \int_{\mathbb{R}^d} p(x, u) \hat{f}(u) e^{i(u,x)} du.$$

The passage from D to $P(x, D)$ has been rather straightforward, but now we will take a leap into the unknown and abandon formal notions of differentiation. So we replace p by a more general function $\sigma : \mathbb{R}^d \times \mathbb{R}^d \to \mathbb{C}$. Informally, we may then define a *pseudo-differential operator* $\sigma(x, D)$ by the prescription:

$$(\sigma(x, D)f)(x) = (2\pi)^{-d/2} \int_{\mathbb{R}^d} \sigma(x, u) \hat{f}(u) e^{i(u,x)} du,$$

and σ is then called the *symbol* of this operator. Of course we have been somewhat cavalier here, and we should make some further assumptions on the symbol σ to ensure that $\sigma(x, D)$ really is a bona fide operator. There are various classes of symbols that may be defined to achieve this. One of the most useful is the *Hörmander class* $S^m_{\rho,\delta}$. This is defined to be the set of all $\sigma \in C^\infty(\mathbb{R}^d)$ such that, for each multi-index α and β,

$$|D^\alpha_x D^\beta_u \sigma(x, u)| \leq C_{\alpha,\beta} (1 + |u|^2)^{(m - \rho|\alpha| + \delta|\beta|)/2}$$

for each $x, u \in \mathbb{R}^d$, where $C_{\alpha,\beta} > 0$, $m \in \mathbb{R}$ and $\rho, \delta \in [0, 1]$. In this case $\sigma(x, D) : S(\mathbb{R}^d, \mathbb{C}) \to S(\mathbb{R}^d, \mathbb{C})$ and extends to an operator $S'(\mathbb{R}^d, \mathbb{C}) \to S'(\mathbb{R}^d, \mathbb{C})$.

3.8 Appendix: Unbounded operators in Banach spaces

For those who hanker after operators in Banach spaces, note the following:

- if $\rho > 0$ and $m < -d + \rho(d-1)$ then $\sigma(x, D) : L^p(\mathbb{R}^d, \mathbb{C}) \to L^p(\mathbb{R}^d, \mathbb{C})$ for $1 \leq p \leq \infty$;
- if $m = 0$ and $0 \leq \delta < \rho \leq 1$ then $\sigma(x, D) : L^2(\mathbb{R}^d, \mathbb{C}) \to L^2(\mathbb{R}^d, \mathbb{C})$.

Proofs of these and more general results can be found in Taylor [295]. However, note that this book, like most on the subject, is written from the point of view of partial differential equations, where it is natural for the symbol to be smooth in both variables. For applications to Markov processes this is too restrictive, and we usually impose much weaker requirements on the dependence of σ in the x-variable. A systematic treatment of these can be found in Section 2.3 of Jacob [150].

4
Stochastic integration

Summary We will now study the stochastic integration of predictable processes against martingale-valued measures. Important examples are the Brownian, Poisson and Lévy-type cases. In the case where the integrand is a sure function, we investigate the associated Wiener–Lévy integrals, particularly the important example of the Ornstein–Uhlenbeck process and its relationship with self-decomposable random variables. In Section 4.4, we establish Itô's formula, which is one of the most important results in this book. Immediate spin-offs from this are Lévy's characterisation of Brownian motion and Burkholder's inequality. We also introduce the Stratonovitch, Marcus and backwards stochastic integrals and indicate the role of local time in extending Itô's formula beyond the class of twice-differentiable functions.

4.1 Integrators and integrands

In Section 2.6 we identified the need to develop a theory of integration against martingales that is not based on the usual Stieltjes integral. Given that our aim is to study stochastic differential equations driven by Lévy processes, our experience with Poisson integrals suggests that it might be profitable to integrate against a class of real-valued martingale-valued measures M defined on $\mathbb{R}^+ \times E$, where $E \in \mathcal{B}(\mathbb{R}^d)$. At this stage, readers should recall the definition of martingale-valued measure from Subsection 2.3.1. We will frequently employ the notation

$$M((s, t], A) = M(t, A) - M(s, A)$$

for all $0 \leq s < t < \infty$, $A \in \mathcal{B}(E)$.

In order to get a workable stochastic integration theory, we will need to impose some conditions on M. These are as follows:

(M1) $M(\{0\}, A) = 0$ (a.s.);
(M2) $M((s, t], A)$ is independent of $\mathcal{F}_s$;
(M3) there exists a σ-finite measure ρ on $\mathbb{R}^+ \times E$ for which
$$\mathbb{E}(M(t, A)^2) = \rho(t, A)$$
for all $0 \leq s < t < \infty$, $A \in \mathcal{B}(E)$. Here we have introduced the abbreviated notation $\rho(t, A) = \rho((0, t] \times A)$.

Martingale-valued measures satisfying (M1) to (M3) are said to be of type $(2, \rho)$, as the second moment always exists and can be expressed in terms of the measure ρ.

In all the examples that we will consider, ρ will be a product measure taking the form
$$\rho((0, t] \times A) = t\mu(A)$$
for each $t \geq 0$, $A \in \mathcal{B}(E)$, where μ is a σ-finite measure on E, and we will assume that this is the case from now on.

By Theorem 2.2.3, we see that $\mathbb{E}(\langle M(t, A), M(t, A)\rangle) = \rho(t, A)$ provided $\rho(t, A) < \infty$.

A martingale-valued measure is said to be *continuous* if the sample paths $t \to M(t, A)(\omega)$ are continuous for almost all $\omega \in \Omega$ and all $A \in \mathcal{B}(E)$.

Example 4.1.1 (Lévy martingale-valued measures) Let X be a Lévy process with Lévy–Itô decomposition given by (2.25) and take $E = \hat{B} - \{0\}$, where we recall that $\hat{B} = \{x \in \mathbb{R}^d, |x| < 1\}$. For each $1 \leq i \leq d$, $A \in \mathcal{B}(E)$, define
$$M_i(t, A) = \alpha \tilde{N}(t, A - \{0\}) + \beta \sigma_i^j B_j(t) \delta_0(A),$$
where $\alpha, \beta \in \mathbb{R}$ are fixed and $\sigma\sigma^T = a$. Then each M_i is a real-valued martingale-valued measure and we have
$$\rho_i(t, A) = t[\alpha^2 \nu(A - \{0\}) + \beta^2 a_i^i \delta_0(A)].$$
Note that $\rho_i(t, A) < \infty$ whenever $A - \{0\}$ is bounded below.

In most of the applications that we consider henceforth, we will have $(\alpha, \beta) = (1, 0)$ or $(0, 1)$.

Example 4.1.2 (Gaussian space–time white noise) Although we will not use them directly in this book, we will briefly give an example of a class of martingale-valued measures that do not arise from Lévy processes.

Let (S, Σ, μ) be a measure space; then a *Gaussian space–time white noise* is a random measure W on $(S \times \mathbb{R}^+ \times \Omega, \Sigma \otimes \mathcal{B}(\mathbb{R}^+) \otimes \mathcal{F})$ for which:

(1) $W(A)$ and $W(B)$ are independent whenever A and B are disjoint sets in $\mathcal{B}(\mathbb{R}^+) \otimes \mathcal{F}$;
(2) each $W(A)$ is a centred Gaussian random variable.

For each $t \geq 0$, $A \in \Sigma$, we can consider the process $(W_A(t), t \geq 0)$ where $W_A(t) = W(A \times [0, t])$ and this is clearly a martingale-valued measure. In concrete applications, we may want to impose the requirements (M1)–(M3) above.

The simplest non-trivial example of a space–time white noise is a *Brownian sheet*, for which $S = (\mathbb{R}^+)^d$ and Σ is its Borel σ-algebra. Writing $W_\mathbf{t} = W([0, t_1] \times [0, t_2] \times \cdots \times [0, t_{d+1}])$, for each $\mathbf{t} = (t_1, t_2, \ldots, t_{d+1}) \in (\mathbb{R}^+)^{d+1}$, the Brownian sheet is defined to be a Gaussian white noise with covariance structure

$$\mathbb{E}(W_\mathbf{t} W_\mathbf{s}) = (s_1 \wedge t_1)(s_2 \wedge t_2) \cdots (s_{d+1} \wedge t_{d+1}).$$

For further details and properties see Walsh [299], pp. 269–71. Some examples of non-Gaussian white noises, that are generalisations of Lévy processes, can be found in Applebaum and Wu [9].

Now we will consider the appropriate space of integrands. First we need to consider a generalisation of the notion of predictability, which was introduced earlier in Subsection 2.2.1.

Fix $E \in \mathcal{B}(\mathbb{R}^d)$ and $0 < T < \infty$ and let $\mathcal{P}$ denote the smallest σ-algebra with respect to which all mappings $F : [0, T] \times E \times \Omega \to \mathbb{R}$ satisfying (1) and (2) below are measurable:

(1) for each $0 \leq t \leq T$ the mapping $(x, \omega) \to F(t, x, \omega)$ is $\mathcal{B}(E) \otimes \mathcal{F}_t$-measurable;
(2) For each $x \in E$, $\omega \in \Omega$, the mapping $t \to F(t, x, \omega)$ is left-continuous.

We call $\mathcal{P}$ the *predictable σ-algebra*. A $\mathcal{P}$-measurable mapping $G : [0, T] \times E \times \Omega \to \mathbb{R}$ is then said to be *predictable*. Clearly the definition extends naturally to the case where $[0, T]$ is replaced by $\mathbb{R}^+$.

Note that, by (1), if G is predictable then the process $t \to G(t, x, \cdot)$ is adapted, for each $x \in E$. If G satisfies (1) and is left-continuous then it is clearly predictable. As the theory of stochastic integration unfolds below, we will see more clearly why the notion of predictability is essential. Some interesting observations about predictability are collected in Klebaner [170], pp. 214–15.

Now let M be a $(2, \rho)$-type martingale-valued measure. We also fix $T > 0$ and define $\mathcal{H}_2(T, E)$ to be the linear space of all equivalence classes of mappings $F : [0, T] \times E \times \Omega \to \mathbb{R}$ which coincide almost everywhere with respect to $\rho \times P$ and which satisfy the following conditions:

- F is predictable;
- $\int_0^T \int_E \mathbb{E}(|F(t,x)|^2)\, \rho(dt, dx) < \infty$.

We may now define the inner product $\langle \cdot, \cdot \rangle_{T,\rho}$ on $\mathcal{H}_2(T, E)$ by

$$\langle F, G \rangle_{T,\rho} = \int_0^T \int_E \mathbb{E}((F(t,x), G(t,x)))\, \rho(dt, dx)$$

for each $F, G \in \mathcal{H}_2(T, E)$, and we obtain a norm $||\cdot||_{T,\rho}$ in the usual way. Note that by Fubini's theorem we may also write

$$||F||_{T,\rho}^2 = \mathbb{E}\left(\int_0^T \int_E |F(t,x)|^2 \rho(dt, dx)\right).$$

Lemma 4.1.3 $\mathcal{H}_2(T, E)$ *is a real Hilbert space.*

Proof Clearly $\mathcal{H}_2(T, E)$ is a subspace of $L^2([0, T) \times E \times \Omega,\ \rho \times P))$. We need only prove that it is closed and the result follows. Let $(F_n, n \in \mathbb{N})$ be a sequence in $\mathcal{H}_2(T, E)$ converging to $F \in L^2$. It follows by the Chebyshev–Markov inequality that $(F_n, n \in \mathbb{N})$ converges to F in measure, with respect to $\rho \times P$, and hence (see e.g. Cohn [73], p. 86) there is a subsequence that converges to F almost everywhere. Since the subsequence comprises predictable mappings it follows that F is also predictable, hence $F \in \mathcal{H}_2(T, E)$ and we are done. □

Recall that ρ is always of the form $\rho(dx, dt) = \mu(dx)dt$. In the case where $E = \{0\}$ and $\mu(\{0\}) = 1$, we write $\mathcal{H}_2(T, E) = \mathcal{H}_2(T)$. The norm in $\mathcal{H}_2(T)$ is given by

$$||F||_T^2 = \mathbb{E}\left(\int_0^T |F(t)|^2 dt\right).$$

For the general case, we have the natural (i.e. basis-independent) Hilbert-space isomorphisms

$$\mathcal{H}_2(T, E) \cong L^2(E, \mu; \mathcal{H}_2(T)) \cong L^2(E, \mu) \otimes \mathcal{H}_2(T),$$

where $\otimes$ denotes the Hilbert space tensor product (see e.g. Reed and Simon [257], pp. 49–53), and $\cong$ means 'is isomorphic to'.

Define $S(T, E)$ to be the linear space of all simple processes in $\mathcal{H}_2(T, E)$ where F is *simple* if, for some $m, n \in \mathbb{N}$, there exists $0 \leq t_1 \leq t_2 \leq \cdots \leq t_{m+1} = T$ and disjoint Borel subsets $A_1, A_2, \ldots, A_n$ of E with each $\mu(A_i) <$

∞ such that

$$F = \sum_{j=1}^{m}\sum_{k=1}^{n} c_k F(t_j) \chi_{(t_j, t_{j+1}]} \chi_{A_k},$$

where each $c_k \in \mathbb{R}$ and each $F(t_j)$ is a bounded $\mathcal{F}_{t_j}$-measurable random variable. Note that F is left-continuous and $\mathcal{B}(E) \otimes \mathcal{F}_t$-measurable, hence it is predictable.

In the case of $\mathcal{H}_2(T)$, the space of simple processes is denoted $S(T)$ and comprises mappings of the form

$$F = \sum_{j=1}^{m} F(t_j) \chi_{(t_j, t_{j+1}]}.$$

Lemma 4.1.4 $S(T, E)$ *is dense in* $\mathcal{H}_2(T, E)$.

Proof We carry this out in four stages. In the first two of these, our aim is to prove a special case of the main result, namely that $S(T)$ is dense in $\mathcal{H}_2(T)$.

Step 1 (Approximation by bounded maps) We will denote Lebesgue measure in $[0, T)$ by l. Let $F \in \mathcal{H}_2(T)$ and, for each $1 \leq i \leq d, n \in \mathbb{N}$, define

$$F_n^i(s, \omega) = \begin{cases} F^i(s, \omega) & \text{if } |F(s, \omega)| < n, \\ 0 & \text{if } |F(s, \omega)| \geq n. \end{cases}$$

The sequence $(F_n, n \in \mathbb{N})$ converges to F pointwise almost everywhere (with respect to $l \times P$), since, given any $\delta > 0$, there exists $n_0 \in \mathbb{N}$ such that

$$(l \times P)\left(\bigcup_{\epsilon \in \mathbb{Q} \cap \mathbb{R}^+} \bigcap_{n_0 \in \mathbb{N}} \bigcup_{n \geq n_0} \{(t, \omega); |F_n(t, \omega) - F(t, \omega)| > \epsilon\}\right)$$

$$= (l \times P)\left(\bigcap_{n_0 \in \mathbb{N}} \bigcup_{n \geq n_0} \{(t, \omega); |F(t, \omega)| \geq n\}\right)$$

$$\leq \sum_{n=n_0}^{\infty} (\rho \times P)(|F(t, \omega)| \geq n) \leq \sum_{n=n_0}^{\infty} \frac{||F||_T^2}{n^2} < \delta,$$

where we have used the Chebyshev–Markov inequality.
By dominated convergence, we obtain

$$\lim_{n \to \infty} ||F_n - F||_T = 0.$$

4.1 Integrators and integrands

Step 2 (Approximation by step functions on $[0, T]$) Let $F \in \mathcal{H}_2(T)$ be bounded as above. Define

$$F_n(t, \omega) = 2^n T^{-1} \sum_{k=1}^{2^n-1} \left(\int_{2^{-n}(k-1)T}^{2^{-n}kT} F(s, \omega) ds \right) \chi_{\{2^{-n}kT < t \leq 2^{-n}(k+1)T\}}$$

for all $n \in \mathbb{N}$, $0 \leq t < T$, $\omega \in \Omega$; then it is easy to see that each $F_n \in S(T)$. For each $n \in \mathbb{N}$, define $A_n F = F_n$; then A_n is a linear operator in $\mathcal{H}_2(T)$ with range in $S(T)$. It is not difficult to verify that each A_n is a contraction, i.e. that $||A_n F||_T \leq ||F||_T$, for each $F \in \mathcal{H}_2(T)$. The result we seek follows immediately once it is established that, for each $F \in \mathcal{H}_2(T)$, $\lim_{n \to \infty} ||A_n F - F||_T = 0$. This is comparatively hard to prove and we direct the reader to Steele [288], pp. 90–3, for the full details. An outline of the argument is as follows. For each $n \in \mathbb{N}$, define a linear operator $B_n : \mathcal{H}_2(T) \to L^2([0, T) \times \Omega, l \times P)$ by

$$(B_n F)(t, \omega)$$
$$= 2^n T^{-1} \sum_{k=1}^{2^n} \left(\int_{2^{-n}(k-1)T}^{2^{-n}kT} F(s, \omega) ds \right) \chi_{\{2^{-n}(k-1)T < t \leq 2^{-n}kT\}}$$

for each $\omega \in \Omega$, $0 \leq t < T$. Note that the range of each B_n is not in $S(T)$. However, if we fix $\omega \in \Omega$ then each $((B_n F)(\cdot, \omega), n \in \mathbb{N})$ can be realised as a discrete-parameter martingale on the filtered probability space $(S_\omega, \mathcal{G}_\omega, (\mathcal{G}_\omega^{(n)}, n \in \mathbb{N}), Q_\omega)$, which is constructed as follows:

$$S_\omega = \{\omega\} \times [0, T], \quad \mathcal{G}_\omega = \{(\omega, A), A \in \mathcal{B}([0, T])\}, \quad Q_\omega(A) = \frac{l(A)}{T}$$

for each $A \in \mathcal{B}([0, T])$. For each $n \in \mathbb{N}$, $\mathcal{G}_\omega^{(n)}$ is the smallest σ-algebra with respect to which all mappings of the form

$$\sum_{k=1}^{2^n} c_k \chi_{\{2^{-n}(k-1)T < t \leq 2^{-n}kT\}},$$

where each $c_k \in \mathbb{R}$, are $\mathcal{G}_\omega^{(n)}$-measurable. Using the fact that conditional expectations are orthogonal projections, we deduce the martingale property from the observation that, for each $n \in \mathbb{N}$, $(B_n F)(t, \omega) = \mathbb{E}_Q(F(t, \omega)|\mathcal{G}^{(n)})$. By the martingale convergence theorem (see, for example, Steele [288] pp. 22–3), we can conclude that $(B_n F)(t, \omega), n \in \mathbb{N}$ converges to $F(t, \omega)$ for all $t \in [0, T)$ except a set of Lebesgue measure zero. The dominated convergence theorem then yields $\lim_{n\to\infty} ||B_n F - F||_T = 0$. Further manipulations lead to $\lim_{n\to\infty} ||A_n(B_m F) - F||_T = 0$ for each fixed $m \in \mathbb{N}$. Finally, using these two limiting results and the fact that each A_n is a contraction, we conclude that for each $n, m \in \mathbb{N}$

$$||A_n F - F||_T \leq ||A_n(F - B_m(F))||_T + ||A_n(B_m F) - F||_T$$
$$\leq ||F - B_m(F)||_T + ||A_n(B_m F) - F||_T$$

and the required result follows.

By steps 1 and 2 together, we see that $S(T)$ is dense in $\mathcal{H}_2(T)$.

Step 3 (Approximation by mappings with support having finite measure) Let $f \in L^2(E, \mu)$. Since μ is σ-finite, we can find a sequence $(A_n, n \in \mathbb{N})$ in $\mathcal{B}(E)$ such that each $\mu(A_n) < \infty$ and $A_n \uparrow E$ as $n \to \infty$. Define $(f_n, n \in \mathbb{N})$ by $f_n = f \chi_{A_n}$. Then we can use dominated convergence to deduce that

$$\lim_{n\to\infty} ||f_n - f||_2 = ||f||^2 - \lim_{n\to\infty} ||f_n||^2 = 0.$$

Step 4 ($S(T, E)$ is dense in $\mathcal{H}(T, E)$) Vectors of the form

$$\sum_{j=1}^{m} F(t_j) \chi_{(t_j, t_{j+1}]}$$

are dense in $\mathcal{H}_2(T)$ by steps 1 and 2, and vectors of the form $\sum_{k=1}^{n} c_k \chi_{A_k}$ are dense in $L^2(E, \mu)$, with each $\mu(A_k) < \infty$, by step 3. Hence vectors of the form

$$\left(\sum_{k=1}^{n} c_k \chi_{A_k} \right) \otimes \left(\sum_{j=1}^{m} F(t_j) \chi_{(t_j, t_{j+1}]} \right)$$

are total in $L^2(E, \mu) \otimes \mathcal{H}_2(T)$, and the result follows. $\square$

Henceforth, we will simplify the notation for vectors in $S(T, E)$ by writing each $c_k F(t_j) = F_k(t_j)$ and

$$\sum_{j=1}^{m} \sum_{k=1}^{n} c_k F(t_j) \chi_{(t_j, t_{j+1}]} \chi_{A_k} = \sum_{j,k=1}^{m,n} F_k(t_j) \chi_{(t_j, t_{j+1}]} \chi_{A_k}.$$

4.2 Stochastic integration

4.2.1 The L^2-theory

In this section our aim is to define, for fixed $T \geq 0$, the stochastic integral $I_T(F) = \int_0^T \int_E F(t, x) M(dt, dx)$ as a real-valued random variable, where $F \in \mathcal{H}_2(T, E)$ and M is a martingale-valued measure of type $(2, \rho)$.

We begin by considering the case where $F \in S(T, E)$, for which we can write

$$F = \sum_{j,k=1}^{m,n} F_k(t_j) \chi_{(t_j, t_{j+1}]} \chi_{A_k}$$

as above. We then define

$$I_T(F) = \sum_{j,k=1}^{m,n} F_k(t_j) M((t_j, t_{j+1}], A_k). \qquad (4.1)$$

Before we analyse this object, we should sit back and gasp at the breathtaking audacity of this prescription, due originally to K. Itô. The key point in the definition (4.1) is that, for each time interval $[t_j, t_{j+1}]$, $F_k(t_j)$ is adapted to the past filtration $\mathcal{F}_{t_j}$ while $M((t_j, t_{j+1}], A_k)$ 'sticks into the future' and is independent of $\mathcal{F}_{t_j}$. For a Stieltjes integral, we would have taken instead $F_k(u_j)$, where $t_j \leq u_j \leq t_{j+1}$ is arbitrary. It is impossible to exaggerate the importance for what follows of Itô's simple but highly effective idea.

Equation (4.1) also gives us an intuitive understanding of why we need the notion of predictability. The present t_j and the future $(t_j, t_{j+1}]$ should not overlap, forcing us to make our step functions left-continuous.

Exercise 4.2.1 Deduce that if $F, G \in S(T, E)$ and $\alpha, \beta \in \mathbb{R}$ then $\alpha F + \beta G \in S(T, E)$ and

$$I_T(\alpha F + \beta G) = \alpha I_T(F) + \beta I_T(G).$$

Lemma 4.2.2 For each $T \geq 0$, $F \in S(T, E)$,

$$\mathbb{E}(I_T(F)) = 0, \qquad \mathbb{E}(I_T(F)^2) = \int_0^T \int_E \mathbb{E}(|F(t, x)|^2) \rho(dt, dx).$$

Proof By the martingale property, for each $1 \leq j \leq m$, $1 \leq k \leq n$, we have

$$\mathbb{E}(M((t_j, t_{j+1}], A_k)) = 0.$$

Hence by linearity and (M2),

$$\mathbb{E}(I_T(F)) = \sum_{j,k=1}^{m,n} \mathbb{E}(F_k(t_j)) \, \mathbb{E}(M((t_j, t_{j+1}], A_k)) = 0.$$

By linearity again, we find that

$$\mathbb{E}(I_T(F)^2)$$

$$= \sum_{j,k=1}^{m,n} \sum_{l,p=1}^{m,n} \mathbb{E}(F_k(t_j) M((t_j, t_{j+1}], A_k) F(t_l) M((t_l, t_{l+1}], A_p))$$

$$= \sum_{j,k=1}^{m,n} \sum_{p=1}^{n} \sum_{l<j} \mathbb{E}(F_k(t_j) M((t_j, t_{j+1}], A_k) F(t_l) M((t_l, t_{l+1}], A_k))$$

$$+ \sum_{j,k=1}^{m,n} \sum_{p=1}^{n} \mathbb{E}(F_k(t_j) F_p(t_j) M((t_j, t_{j+1}], A_j) M((t_j, t_{j+1}], A_p))$$

$$+ \sum_{j,k=1}^{m,n} \sum_{p=1}^{n} \sum_{l>j} \mathbb{E}(F_k(t_j) M((t_j, t_{j+1}], A_k) F_p(t_l) M((t_l, t_{l+1}], A_p)).$$

Dealing with each of these three terms in turn, we find that by (M2) again

$$\sum_{j,k=1}^{m,n} \sum_{p=1}^{n} \sum_{l<j} \mathbb{E}(F_k(t_j) M((t_j, t_{j+1}], A_k) F_p(t_l) M((t_l, t_{l+1}], A_k))$$

$$= \sum_{j=1}^{m,n} \sum_{p=1}^{n} \sum_{l<j} \mathbb{E}(F_p(t_l) M((t_l, t_{l+1}], A_p) F_k(t_j)) \mathbb{E}(M((t_j, t_{j+1}], A_k)) = 0,$$

and a similar argument shows that

$$\sum_{j,k=1}^{m,n} \sum_{p=1}^{n} \sum_{l>j} \mathbb{E}(F_k(t_j) M((t_j, t_{j+1}], A_k) F(t_p) M((t_l, t_{l+1}], A_p)) = 0.$$

By (M2) and the independently scattered property of random measures,

$$\sum_{j,k=1}^{m,n} \sum_{p=1}^{n} \mathbb{E}(F_k(t_j) F_p(t_j) M((t_j, t_{j+1}], A_k) M((t_j, t_{j+1}], A_p))$$

$$= \sum_{j,k=1}^{m,n} \sum_{p=1}^{n} \mathbb{E}(F_k(t_j) F_p(t_j)) \, \mathbb{E}(M((t_j, t_{j+1}], A_k) M((t_j, t_{j+1}], A_p))$$

$$= \sum_{j,k=1}^{m,n} \mathbb{E}(F_k(t_j)^2)) \, \mathbb{E}(M((t_j, t_{j+1}], A_k)^2).$$

4.2 Stochastic integration

Finally, we use the martingale property and (M3) to obtain

$$\sum_{j,k=1}^{m,n} \sum_{p=1}^{n} \mathbb{E}\big(F_k(t_j)F_p(t_j)M((t_j, t_{j+1}], A_k)M((t_j, t_{j+1}], A_p)\big)$$

$$= \sum_{j,k=1}^{n} \mathbb{E}(F_k(t_j)^2)\big[\mathbb{E}(M(t_{j+1}, A_k)^2) - \mathbb{E}(M(t_j, A_k)^2)\big]$$

$$= \sum_{j=1}^{n} \mathbb{E}(F_k(t_j)^2)\rho((t_j, t_{j+1}], A_k),$$

and this is the required result. □

We deduce from Lemma 4.2.2 and Exercise 4.2.1 that I_T is a linear isometry from $S(T, E)$ into $L^2(\Omega, \mathcal{F}, P)$, and hence by Lemma 4.1.4 it extends to an isometric embedding of the whole of $\mathcal{H}_2(T, E)$ into $L^2(\Omega, \mathcal{F}, P)$. We will continue to denote this extension as I_T and will call $I_T(F)$ the *(Itô) stochastic integral* of $F \in \mathcal{H}_2(T, E)$. When convenient, we will use the Leibniz notation $I_T(F) = \int_0^T \int_E F(t, x)M(dt, dx)$. We have

$$\mathbb{E}(|I_T(F)|^2) = ||F||_{T,\rho}^2$$

for all $F \in \mathcal{H}_2(T, E)$, and this identity is sometimes called *Itô's isometry*. It follows from Lemma 4.1.4 that for any $F \in \mathcal{H}_2(T, E)$ we can find a sequence $(F_n, n \in \mathbb{N}) \in S(T, E)$ such that $\lim_{n\to\infty} ||F - F_n||_{T,\rho} = 0$ and

$$\int_0^T \int_E F(t, x)M(dt, dx) = L^2 - \lim_{n \to \infty} \int_0^T \int_E F_n(t, x)M(dt, dx).$$

If $0 \leq a \leq b \leq T$, $A \in \mathcal{B}(E)$ and $F \in \mathcal{H}_2(T, E)$, it is easily verified that $\chi_{(a,b)}\chi_A F \in \mathcal{H}_2(T, E)$ and we may then define

$$I_{a,b;A}(F) = \int_a^b \int_A F(t, x)M(dt, dx) = I_T(\chi_{(a,b)}\chi_A F).$$

We will also write $I_{a,b} = I_{a,b;E}$.

If $||F||_{t,\rho} < \infty$ for all $t \geq 0$ it makes sense to consider $(I_t(F), t \geq 0)$ as a stochastic process, and we will implicitly assume that this condition is satisfied whenever we do this.

The following theorem summarises some useful properties of the stochastic integral.

Theorem 4.2.3 *If $F, G \in \mathcal{H}_2(T, E)$ and $\alpha, \beta \in \mathbb{R}$ then:*

(1) $I_T(\alpha F + \beta G) = \alpha I_T(F) + \beta I_T(G)$;
(2) $\mathbb{E}(I_T(F)) = 0$, $\quad \mathbb{E}(I_T(F)^2) = \int_0^T \int_E \mathbb{E}(|F(t, x)|^2) \rho(dt, dx)$;
(3) $(I_t(F), t \geq 0)$ is $\mathcal{F}_t$-adapted;
(4) $(I_t(F), t \geq 0)$ is a square-integrable martingale.

Proof (1) and (2) follow by continuity from Exercise 4.2.1 and Lemma 4.2.2.

For (3), let $(F_n, n \in \mathbb{N})$ be a sequence in $S(T, E)$ converging to F; then each process $(I_t(F_n), t \geq 0)$ is clearly adapted. Since each $I_t(F_n) \to I_t(F)$ in L^2 as $n \to \infty$, we can find a subsequence $(F_{n_k}, n_k \in \mathbb{N})$ such that $I_t(F_{n_k}) \to I_t(F)$ (a.s.) as $n_k \to \infty$, and the required result follows.

(4). Let $F \in S(T, E)$ and (without loss of generality) choose $0 < s = t_l < t_{l+1} < t$. Then it is easy to see that $I_t(F) = I_s(F) + I_{s,t}(F)$ and hence $\mathbb{E}_s(I_t(F)) = I_s(F) + \mathbb{E}_s(I_{s,t}(F))$ by (3). However, by (M2),

$$\mathbb{E}_s(I_{s,t}(F)) = \mathbb{E}_s \left(\sum_{j=l+1}^{m} \sum_{k=1}^{n} F_k(t_j) M((t_j, t_{j+1}], A_k) \right)$$

$$= \sum_{j=l+1}^{n} \sum_{k=1}^{n} \mathbb{E}_s(F_k(t_j)) \, \mathbb{E}(M((t_j, t_{j+1}], A_k)) = 0.$$

The result now follows by the contractivity of $\mathbb{E}_s$ in L^2. Indeed, let $(F_n, n \in \mathbb{N})$ be a sequence in $S(T, E)$ converging to F; then we have

$$\left\| \mathbb{E}_s(I_t(F)) - \mathbb{E}_s(I_t(F_n)) \right\|_2 \leq \| I_t(F) - I_t(F_n) \|_2$$
$$= \| F - F_n \|_{T, \rho} \to 0 \quad \text{as } n \to \infty.$$

$\square$

Exercise 4.2.4 Deduce that if $F, G \in \mathcal{H}_2(T, E)$ then

$$\mathbb{E}(I_T(F) I_T(G)) = \langle F, G \rangle_{T, \rho}.$$

Exercise 4.2.5 Let M be a martingale-valued measure that satisfies (M1) and (M2) but not (M3). Define the stochastic integral in this case as an isometric embedding of the space of all predictable mappings F for which $\int_0^T \int_E \mathbb{E}(|F(t, x)|^2) \langle M, M \rangle(dt, dx) < \infty$, where for each $A \in \mathcal{B}(E), t \geq 0$, we define

$$\langle M, M \rangle(t, E) = \langle M(\cdot, E), M(\cdot, E) \rangle(t).$$

4.2.2 The extended theory

We define $\mathcal{P}_2(T, E)$ to be the set of all equivalence classes of mappings $F : [0, T] \times E \times \Omega \to \mathbb{R}$ which coincide almost everywhere with respect to $\rho \times P$ and which satisfy the following conditions:

- F is predictable;
- $P\left(\int_0^T \int_E |F(t, x)|^2 \rho(dt, dx) < \infty\right) = 1.$

Exercise 4.2.6 Deduce that $\mathcal{P}_2(T, E)$ is a linear space and show that

$$\mathcal{H}_2(T, E) \subseteq \mathcal{P}_2(T, E).$$

Show also that $\mathcal{P}_2(T, E)$ is a topological space with topology generated by the basis $\{O_{a,F}; F \in \mathcal{P}_2(T, E), a > 0\}$, where $O_{a,F}$ equals

$$\left\{G \in \mathcal{P}_2(T, E); P\left(\int_0^T \int_E |G(t, x) - F(t, x)|^2 \rho(dt, dx) < a\right) = 1\right\}.$$

We have a good notion of convergence for sequences in $\mathcal{P}_2(T, E)$, i.e. $(F_n, n \in \mathbb{N})$ converges to F if

$$P\left(\lim_{n \to \infty} \int_0^T \int_E |F_n(t, x) - F(t, x)|^2 \rho(dt, dx) = 0\right) = 1.$$

Exercise 4.2.7 Imitate the argument in the proof of Lemma 4.1.4 to show that $S(T, E)$ is dense in $\mathcal{P}_2(T, E)$.

Lemma 4.2.8 (cf. Gihman and Skorohod [118], p. 20) *If $F \in S(T, E)$ then for all $C, K \geq 0$*

$$P\left(\left|\int_0^T \int_E F(t, x) M(dt, dx)\right| > C\right)$$
$$\leq \frac{K}{C^2} + P\left(\int_0^T \int_E |F(t, x)|^2 \rho(dt, dx) > K\right).$$

Proof Fix $K > 0$ and define $\tilde{F}^K$ by

$$\tilde{F}_p^K(t_j) = \begin{cases} F_p(t_j) & \text{if } \sum_{i,l=1}^{j,p} F_l(t_i)^2 \rho((t_i, t_{i+1}], A_l) \leq K, \\ 0 & \text{otherwise.} \end{cases}$$

Then $\tilde{F}^K \in S(T, E)$ and
$$\int_0^T \int_E |\tilde{F}^K(t,x)|^2 \rho(dt, dx) = \sum_{i=1}^{m_K} \sum_{l=1}^{n_K} F_l(t_i)^2 \rho((t_i, t_{i+1}], A_l),$$
where m_K and n_K are the largest integers for which
$$\sum_{i=1}^{m_K} \sum_{l=1}^{n_K} F_l(t_i)^2 \rho((t_i, t_{i+1}], A_i) \leq K.$$
By definition, we have
$$F = \tilde{F}_K \quad \text{if and only if} \quad \int_0^T \int_E |F(t,x)|^2 \rho(dt, dx) \leq K;$$
then, by the Chebychev–Markov inequality,
$$P\left(\left|\int_0^T \int_E F(t,x) M(dt, dx)\right| > C\right)$$
$$= P\left(\left|\int_0^T \int_E F(t,x) M(dt, dx)\right| > C, \ F = \tilde{F}_K\right)$$
$$+ P\left(\left|\int_{0,T} \int_E F(t,x) M(dt, dx)\right| > C, \ F \neq \tilde{F}_K\right)$$
$$\leq P\left(\left|\int_0^T \int_E \tilde{F}_K(t,x) M(dt, dx)\right| > C\right) + P(F \neq \tilde{F}_K)$$
$$\leq \frac{\mathbb{E}(I_T(\tilde{F}_K)^2)}{C^2} + P\left(\int_0^T \int_E |F(t,x)|^2 \rho(dt, dx) > K\right)$$
$$\leq \frac{K}{C^2} + P\left(\int_0^T \int_E |F(t,x)|^2 \rho(dt, dx) > K\right),$$
as required. □

Now let $F \in \mathcal{P}_2(T, E)$; then by Exercise 4.2.7 we can find $(F_n, n \in \mathbb{N})$ in $S(T, E)$ such that $\lim_{n \to \infty} \alpha(F)_n = 0$ (a.s.), where for each $n \in \mathbb{N}$ $\alpha(F)_n = \int_0^T \int_E |F(t,x) - F_n(t,x)|^2 \rho(dt, dx)$. Hence $\lim_{n \to \infty} \alpha(F)_n = 0$ in probability and so $(\alpha(F)_n, n \in \mathbb{N})$ is a Cauchy sequence in probability.

By Lemma 4.2.8, for any $m, n \in \mathbb{N}, K, \beta > 0$,
$$P\left(\left|\int_0^T \int_E [F_n(t,x) - F_m(t,x)] M(dt, dx)\right| > \beta\right)$$
$$\leq \frac{K}{\beta^2} + P\left(\int_0^T \int_E |F_n(t,x) - F_m(t,x)|^2 \rho(dt, dx) > K\right). \quad (4.2)$$

Hence, for any $\gamma > 0$, given $\epsilon > 0$ we can find $m_0 \in \mathbb{N}$ such that whenever $n, m > m_0$

$$P\left(\int_0^T \int_E |F_n(t,x) - F_m(t,x)|^2 \rho(dt, dx) > \gamma \beta^2\right) < \epsilon.$$

Now choose $K = \gamma \beta^2$ in (4.2) to deduce that the sequence

$$\left(\int_{0,T} \int_E F_n(t,x) M(dt, dx), \ n \in \mathbb{N}\right)$$

is Cauchy in probability and thus has a unique limit in probability (up to almost-sure agreement). We denote this limit by

$$\hat{I}_T(F) = \int_0^T \int_E F(t,x) M(dt, dx),$$

so that

$$\int_0^T \int_E F(t,x) M(dt, dx) = \lim_{n \to \infty} \int_0^T \int_E F_n(t,x) M(dt, dx) \quad \text{in probability.}$$

We call $\hat{I}_T(F)$ an *(extended) stochastic integral* and drop the qualifier 'extended' when the context is clear.

We can again consider (extended) stochastic integrals as stochastic processes $(\hat{I}_t(F), t \geq 0)$, provided that we impose the condition

$$P\left(\int_0^t \int_E |F(t,x)|^2 \rho(dt, dx) < \infty\right) = 1$$

for all $t \geq 0$.

Exercise 4.2.9 Show that (1) and (3) of Theorem 4.2.3 continue to hold for extended stochastic integrals.

Exercise 4.2.10 Extend the result of Lemma 4.2.8 to arbitrary $F \in \mathcal{P}_2(T, E)$.

Exercise 4.2.11 Let Y be an adapted càdlàg process on $[0, T]$ and let $F \in \mathcal{P}_2(T, E)$. Confirm that the mapping $(s, x, \cdot) \to Y(s-)(\cdot) F(s, x)(\cdot)$ is in $\mathcal{P}_2(T, E)$.

Of course we cannot expect the processes $(\hat{I}_t(F), t \geq 0)$ to be martingales in general, but we have the following:

Theorem 4.2.12

(1) $(\hat{I}_t(F), t \geq 0)$ is a local martingale.
(2) $(\hat{I}_t(F), t \geq 0)$ has a càdlàg modification.

Proof (1) Define a sequence of stopping times $(T_n, n \in \mathbb{N})$ by:

$$T_n(\omega) = \inf\left\{t \geq 0; \int_0^t \int_E |F(t,x)(\omega)|^2 \rho(dt, dx) > n\right\}$$

for all $\omega \in \Omega, n \in \mathbb{N}$. Then $\lim_{n \to \infty} T_n = \infty$ (a.s.). Define $F_n(t, x) = F(t, x) \chi_{\{T_n \geq t\}}$ for all $x \in E, t \geq 0, n \in \mathbb{N}$; then

$$\int_0^t \int_E |F_n(t,x)(\omega)|^2 \rho(dt, dx) \leq n,$$

hence $F_n \in \mathcal{H}_2(t, E)$ for all $t \geq 0$. By Theorem 4.2.3(4), each $(\hat{I}_t(F_n), t \geq 0)$ is an L^2-martingale, but $\hat{I}_t(F_n) = \hat{I}_{t \wedge T_n}(F)$ and so we have our required result.

(2) Since $(T_n, n \in \mathbb{N})$ is increasing, for each $\omega \in \Omega$ we can find $n_0(\omega) \in \mathbb{N}$ such that $t_0 = t_0 \wedge T_{n_0}(\omega)$. But by Theorem 4.2.3 each $(\hat{I}_{t \wedge T_n}(F), t \geq 0)$ is a martingale and so has a càdlàg modification by Theorem 2.1.6, and the required result follows. $\square$

We finish this section by looking at the special case when our martingale-valued measure is continuous.

Exercise 4.2.13 Show that if M is continuous then $\hat{I}_t(F)$ is continuous at each $0 \leq t \leq T$, when $F \in S(T, E)$.

Theorem 4.2.14 *If M is continuous and $F \in \mathcal{P}_2(T, E)$, then $\hat{I}_t(F)$ is continuous on $[0, T]$.*

Proof First we consider the case where $F \in \mathcal{H}_2(T, E)$. Let $(F_n, n \in \mathbb{N})$ be a sequence in $S(T, E)$ converging to F; then by the Chebyshev–Markov inequality and Doob's martingale inequality we have, for each $\epsilon > 0$,

$$P\left(\sup_{0 \leq t \leq T} |I_t(F_n) - I_t(F)| > \epsilon\right) \leq \frac{1}{\epsilon^2} \mathbb{E}\left(\sup_{0 \leq t \leq T} |I_t(F_n) - I_t(F)|^2\right)$$

$$\leq \frac{4}{\epsilon^2} \mathbb{E}(|I_T(F_n) - I_T(F)|^2)$$

$$\to 0 \quad \text{as} \quad n \to \infty.$$

Hence we can find a subsequence $(F_{n_k}, n_k \in \mathbb{N})$ such that

$$\lim_{n_k \to \infty} \sup_{0 \leq t \leq T} |I_t(F_{n_k}) - I_t(F)| = 0 \quad \text{a.s.,}$$

and the required continuity follows from the result of Exercise 4.2.13 by an $\epsilon/3$ argument. The extension to $\mathcal{P}_2(T, E)$ follows by the stopping argument given in the proof of Theorem 4.2.12(2). □

4.3 Stochastic integrals based on Lévy processes

In this section our aim is to examine various stochastic integrals for which the integrator is a Lévy process.

4.3.1 Brownian stochastic integrals

In this case we take $E = \{0\}$ and we write $\mathcal{P}_2(T, \{0\}) = \mathcal{P}_2(T)$, so that this space comprises all predictable mappings $F : [0, T] \times \Omega \to \mathbb{R}$ for which $P\left(\int_0^T |F(t)|^2 dt < \infty\right) = 1$.

For our martingale-valued measure M we take any of the components $(B^1, B^2, \ldots, B^m)$ of an m-dimensional standard Brownian motion $B = (B(t), t \geq 0)$. For most of the applications in which we will be interested, we will want to consider integrals of the type

$$Y^i(t) = \int_0^t F^i_j(s) dB^j(s)$$

for $1 \leq i \leq d, 0 \leq t \leq T$, where $F = (F^i_j)$ is a $d \times m$ matrix with entries in $\mathcal{P}_2(T)$. This stochastic integral generates an $\mathbb{R}^d$-valued process $Y = (Y(t), 0 \leq t \leq T)$ with components $(Y^1, Y^2, \ldots, Y^d)$, and Y is clearly continuous at each $0 \leq t \leq T$ by Theorem 4.2.14. Furthermore, if $G = (G(t), t \geq 0)$ is an $\mathbb{R}^d$-valued predictable process with each $G(t) \in L^1[0, T]$ then $Z = (Z(t), t \geq 0)$ is adapted and has continuous sample paths, where for each $1 \leq i \leq d$

$$Z^i(t) = \int_0^t F^i_j(s) dB^j(s) + \int_0^t G^i(s) ds \tag{4.3}$$

(see e.g. Royden [266], p. 105, for a proof of the almost-sure continuity of $t \to \int_0^t G^i(s)(\omega) ds$, where $\omega \in \Omega$).

In the next section we will meet the following situation. Let $(\mathcal{P}_n, n \in \mathbb{N})$ be a sequence of partitions of $[0, T]$ of the form

$$\mathcal{P}_n = \left\{0 = t_0^{(n)} < t_1^{(n)} < \cdots < t_{m(n)}^{(n)} < t_{m(n)+1}^{(n)} = T\right\}$$

and suppose that $\lim_{n\to\infty} \delta(\mathcal{P}_n) = 0$, where the mesh

$$\delta(\mathcal{P}_n) = \max_{0 \leq j \leq m(n)} \left| t_{j+1}^{(n)} - t_j^{(n)} \right|.$$

Let $(F(t), t \geq 0)$ be a left-continuous adapted process and define a sequence of simple processes $(F_n, n \in \mathbb{N})$ by writing

$$F_n(t) = \sum_{j=0}^{m(n)} F(t_j^{(n)}) \chi_{(t_j^{(n)}, t_{j+1}^{(n)}]}(t)$$

for each $n \in \mathbb{N}, 0 \leq t \leq T$.

Lemma 4.3.1 $F_n \to F$ in $\mathcal{P}_2(T)$ as $n \to \infty$.

Proof This is left as an exercise for the reader. □

It follows by Lemma 4.3.1 via Exercise 4.2.10 that $\hat{I}_t(F_n) \to \hat{I}_t(F)$ in probability as $n \to \infty$, for each $0 \leq t \leq T$.

4.3.2 Poisson stochastic integrals

In this section we will take $E = \hat{B} - \{0\}$. Let N be a Poisson random measure on $\mathbb{R}^+ \times (\mathbb{R}^d - \{0\})$ with intensity measure ν. We will find it convenient to assume that ν is a Lévy measure. Now take M to be the associated compensated Poisson random measure $\tilde{N}$. In this case, $\mathcal{P}_2(T, E)$ is the space of all predictable mappings $F : [0, T] \times E \times \Omega \to \mathbb{R}$ for which $P\left(\int_0^T \int_E |F(t, x)|^2 \nu(dx) dt < \infty\right) = 1$. Let H be a vector with components $(H^1, H^2, \ldots, H^d)$ taking values in $\mathcal{P}_2(T, E)$; then we may construct an $\mathbb{R}^d$-valued process $Z = (Z(t), t \geq 0)$ with components $(Z^1, Z^2, \ldots, Z^d)$ where each

$$Z^i(T) = \int_0^T \int_{|x|<1} H^i(t, x) \tilde{N}(dt, dx).$$

We can gain greater insight into the structure of Z by using our knowledge of the jumps of N.

Let A be a Borel set in $\mathbb{R}^d - \{0\}$ that is bounded below, and introduce the compound Poisson process $P = (P(t), t \geq 0)$, where each $P(t) = \int_A x N(t, dx)$. Let K be a predictable mapping; then, generalising equation (2.5), we define

$$\int_0^T \int_A K(t, x) N(dt, dx) = \sum_{0 \leq u \leq T} K(u, \Delta P(u)) \chi_A(\Delta P(u)) \quad (4.4)$$

as a random finite sum.

4.3 Stochastic integrals based on Lévy processes

In particular, if H satisfies the square-integrability condition given above we may then define, for each $1 \leq i \leq d$,

$$\int_0^T \int_A H^i(t,x)\tilde{N}(dt,dx)$$
$$= \int_0^T \int_A H^i(t,x)N(dt,dx) - \int_0^T \int_A H^i(t,x)\nu(dx)dt.$$

Exercise 4.3.2 Confirm that the above integral is finite (a.s.) and verify that this is consistent with the earlier definition (2.5) based on martingale-valued measures. (Hint: Begin with the case where H is simple.)

The definition (4.4) can, in principle, be used to define stochastic integrals for a more general class of integrands than we have been considering. For simplicity, let $N = (N(t), t \geq 0)$ be a Poisson process of intensity 1 and let $f : \mathbb{R} \to \mathbb{R}$; then we may define

$$\int_0^t f(N(s))dN(s) = \sum_{0 \leq s \leq t} f\big(N(s-) + \Delta N(s)\big)\Delta N(s).$$

Exercise 4.3.3 Show that, for each $t \geq 0$,

$$\int_0^t N(s)d\tilde{N}(s) - \int_0^t N(s-)d\tilde{N}(s) = N(t).$$

Hence deduce that the process whose value at time t is $\int_0^t N(s)d\tilde{N}(s)$ cannot be a local martingale.

Within any theory of stochastic integration, it is highly desirable that the stochastic integral of a process against a martingale as integrator should at least be a local martingale. The last example illustrates the perils of abandoning the requirement of predictability of our integrands, which, as we have seen in Theorem 4.2.12, always ensures that this is the case. The following result allows us to extend the interlacing technique to stochastic integrals.

Theorem 4.3.4

(1) If $F \in \mathcal{P}_2(T, E)$ then for every sequence $(A_n, n \in \mathbb{N})$ in $\mathcal{B}(E)$ with $A_n \uparrow E$ as $n \to \infty$ we have

$$\lim_{n \to \infty} \int_0^T \int_{A_n} F(t,x)\tilde{N}(dt,dx) = \int_0^T \int_E F(t,x)\tilde{N}(dt,dx)$$

in probability.

(2) *If* $F \in \mathcal{H}_2(T, E)$ *then there exists a sequence* $(A_n, n \in \mathbb{N})$ *in* $\mathcal{B}(E)$ *with each* $\nu(A_n) < \infty$ *and* $A_n \uparrow E$ *as* $n \to \infty$ *for which*

$$\lim_{n \to \infty} \int_0^T \int_{A_n} F(t,x) \tilde{N}(dt, dx) = \int_0^T \int_E F(t,x) \tilde{N}(dt, dx) \quad \text{a.s.}$$

and the convergence is uniform on compact intervals of $[0, T]$.

Proof (1) Using the result of Exercise 4.2.10, we find that for any $\delta, \epsilon > 0$, $n \in \mathbb{N}$,

$$P\left(\left|\int_0^T \int_E F(t,x) \tilde{N}(dt, dx) - \int_0^T \int_{A_n} F(t,x) \tilde{N}(dt, dx)\right| > \epsilon\right)$$
$$\leq \frac{\delta}{\epsilon^2} + P\left(\int_0^T \int_{E-A_n} |F(t,x)|^2 \nu(dx) dt > \delta\right),$$

from which the required result follows immediately.

(2) Define a sequence $(\epsilon_n, n \in \mathbb{N})$ that decreases monotonically to zero, with $\epsilon_1 = 1$ and, for $n \geq 2$,

$$\epsilon_n = \sup\left(y \geq 0, \int_0^T \int_{0 < |x| < y} \mathbb{E}(|F(t,x)|^2) \, \nu(dx) dt \leq 8^{-n}\right).$$

Define $A_n = \{x \in E; \epsilon_n < |x| < 1\}$ for each $n \in \mathbb{N}$. By Doob's martingale inequality, for each $n \in \mathbb{N}$,

$$\mathbb{E}\left(\sup_{0 \leq s \leq t} \left|\int_0^s \int_{A_{n+1}} F(u,x) \tilde{N}(du, dx) - \int_0^s \int_{A_n} F(u,x) \tilde{N}(du, dx)\right|^2\right)$$
$$\leq \mathbb{E}\left(\left|\int_0^t \int_{A_{n+1} - A_n} F(u,x) \tilde{N}(du, dx)\right|^2\right)$$
$$= \int_0^t \int_{A_{n+1} - A_n} \mathbb{E}(|F(u,x)|^2) \, \nu(dx) du.$$

The result then follows by the argument in the proof of Theorem 2.5.2. $\square$

4.3.3 *Lévy-type stochastic integrals*

We continue to take $E = \hat{B} - \{0\}$ throughout this section. We say that an $\mathbb{R}^d$-valued stochastic process $Y = (Y(t), t \geq 0)$ is a *Lévy-type stochastic integral*

4.3 Stochastic integrals based on Lévy processes

if it can be written in the following form, for each $1 \leq i \leq d, t \geq 0$:

$$Y^i(t) = Y^i(0) + \int_0^t G^i(s)ds + \int_0^t F^i_j(s)dB^j(s)$$
$$+ \int_0^t \int_{|x|<1} H^i(s,x)\tilde{N}(ds,dx)$$
$$+ \int_0^t \int_{|x|\geq 1} K^i(s,x)N(ds,dx), \quad (4.5)$$

where, for each $1 \leq i \leq d, 1 \leq j \leq m, t \geq 0$, we have $|G^i|^{1/2}, F^i_j \in \mathcal{P}_2(T)$, $H^i \in \mathcal{P}_2(T,E)$ and K is predictable. Here B is an m-dimensional standard Brownian motion and N is an independent Poisson random measure on $\mathbb{R}^+ \times (\mathbb{R}^d - \{0\})$ with compensator $\tilde{N}$ and intensity measure ν, which we will assume is a Lévy measure.

Let $(\tau_n, \mathbb{N} \cup \{\infty\})$ be the arrival times of the Poisson process $(N(t, E^c), t \geq 0)$. Then the process with value $\int_0^t \int_{|x|\geq 1} K^i(s,x)N(ds,dx)$ at time t is a fixed random variable on each interval $[\tau_n, \tau_{n+1})$ and hence it has càdlàg paths. It then follows from Theorems 4.2.12 and 4.2.14 that Y has a càdlàg modification, and from now on we will identify Y with this modification. We will assume that the random variable $Y(0)$ is $\mathcal{F}_0$-measurable, and then it is clear that Y is an adapted process.

We can often simplify complicated expressions by employing the notation of *stochastic differentials* (sometimes called *Itô differentials*) to represent Lévy-type stochastic integrals. We then write (4.5) as

$$dY(t) = G(t)dt + F(t)dB(t) + H(t,x)\tilde{N}(dt,dx) + K(t,x)N(dt,dx).$$

When we want particularly to emphasise the domains of integration with respect to x, we will use the equivalent notation

$$dY(t) = G(t)dt + F(t)dB(t)$$
$$+ \int_{|x|<1} H(t,x)\tilde{N}(dt,dx) + \int_{|x|\geq 1} K(t,x)N(dt,dx).$$

Clearly Y is a semimartingale.

Exercise 4.3.5 Deduce that Y is a local martingale if and only if G is identically zero and K vanishes on the support of the restriction of N to $\{|x| \geq 1\}$. Under what further conditions is Y a martingale?

Let $\mathcal{L}(\Omega)$ denote the set of all Lévy-type stochastic integrals on $(\Omega, \mathcal{F}, P)$.

Exercise 4.3.6 Show that $\mathcal{L}(\Omega)$ is a linear space.

Exercise 4.3.7 Let $(\mathcal{P}_n, n \in \mathbb{N})$ be a sequence of partitions of $[0, T]$ as above. Show that if Y is a Lévy-type stochastic integral then

$$\lim_{n \to \infty} \sum_{n=0}^{m(n)} Y(t_j^{(n)}) \left[Y(t_{j+1}^{(n)}) - Y(t_j^{(n)}) \right] = \int_0^T Y(s-)dY(s),$$

where the limit is taken in probability.

Let $M = (M(t), t \geq 0)$ be an adapted process that is such that $MJ \in \mathcal{P}_2(t, A)$ whenever $J \in \mathcal{P}_2(t, A)$, where $A \in \mathcal{B}(\mathbb{R}^d)$ is arbitrary. For example, it is sufficient to take M to be adapted and left-continuous.

For these processes we can define an adapted process $Z = (Z(t), t \geq 0)$ by the prescription that it has the stochastic differential

$$dZ(t) = M(t)G(t)dt + M(t)F(t)dB(t) + M(t)H(t,x)\tilde{N}(dt,dx) + M(t)K(t,x)N(dt,dx),$$

and we will adopt the natural notation

$$dZ(t) = M(t)dY(t).$$

Example 4.3.8 (Lévy stochastic integrals) Let X be a Lévy process with characteristics (b, a, ν) and Lévy–Itô decomposition given by equation (2.25):

$$X(t) = bt + B_a(t) + \int_{|x|<1} x\tilde{N}(t,dx) + \int_{|x|\geq 1} xN(t,dx),$$

for each $t \geq 0$. Let $L \in \mathcal{P}_2(t)$ for all $t \geq 0$ and in (4.5) choose each $F_j^i = \sigma_j^i L$, $H^i = K^i = x^i L$, where $\sigma \sigma^T = a$. Then we can construct processes with the stochastic differential

$$dY(t) = L(t)dX(t). \tag{4.6}$$

We call Y a *Lévy stochastic integral*.

In the case where X has finite variation (necessary and sufficient conditions for this are given at the end of Section 2.3), the Lévy stochastic integral Y can also be constructed as a Lebesgue–Stieltjes integral and this coincides (up to set of measure zero) with the prescription (4.6); see Millar [229], p. 314.

Exercise 4.3.9 Check that each $Y(t), t \geq 0$, is almost surely finite.

We can construct Lévy-type stochastic integrals by interlacing. Indeed if we let $(A_n, n \in \mathbb{N})$ be defined as in the hypothesis of Theorem 4.3.4, we may consider the sequence of processes $(Y_n, n \in \mathbb{N})$ defined by

$$Y_n^i(t) = \int_0^t G^i(s)ds + \int_0^t F_j^i(s)dB^j(s) + \int_0^t \int_{A_n} H^i(s,x)\tilde{N}(ds,dx)$$
$$+ \int_0^t \int_{|x|\geq 1} K^i(s,x)N(ds,dx)$$

for each $1 \leq i \leq d, t \geq 0$. We then obtain from Theorem 4.3.4 the following:

Corollary 4.3.10

(1) If $H \in \mathcal{P}_2(t, E)$, then, for every sequence $(A_n, n \in \mathbb{N})$ in $\mathcal{B}(E)$ with $A_n \uparrow E$ as $n \to \infty$, we have

$$\lim_{n \to \infty} Y_n(t) = Y(t) \quad \text{in probability.}$$

(2) If $F \in \mathcal{H}_2(T, E)$ then there exists a sequence $(A_n, n \in \mathbb{N})$ in $\mathcal{B}(E)$ with each $\nu(A_n) < \infty$ and $A_n \uparrow E$ as $n \to \infty$ for which

$$\lim_{n \to \infty} Y_n(t) = Y(t) \quad \text{a.s.}$$

and for which the convergence is uniform on compact intervals of $[0, T]$.

We can gain greater insight into the above result by directly constructing the path of the interlacing sequence but we will need to impose an additional restriction on the sequence $(A_n, n \in \mathbb{N})$ appearing in part (1), that each $\nu(A_n) < \infty$.

Let $C = (C(t), t \geq 0)$ be the process with stochastic differential $dC(t) = G(t)dt + F(t)dB(t)$; let $dW(t) = dC(t) + K(t,x)N(dt,dx)$. We can construct W from C by interlacing with the jumps of the compound Poisson process $P = (P(t), t \geq 0)$ for which $P(t) = \int_{|x|>1} xN(t,dx)$, as follows. Let $(S^n, n \in \mathbb{N})$ be the jump times of P; then we have

$$W(t) = \begin{cases} C(t) & \text{for } 0 \leq t < S^1, \\ C(S^1) + K(S^1, \Delta P(S^1)) & \text{for } t = S^1, \\ W(S^1) + C(t) - C(S^1) & \text{for } S^1 < t < S^2, \\ W(S^2-) + K(S^2, \Delta P(S^2)) & \text{for } t = S^2, \end{cases}$$

and so on recursively.

To construct the sequence $(Y_n, n \in \mathbb{N})$ we need a sequence of compound Poisson processes $Q_n = (Q_n(t), t \geq 0)$ for which each $Q_n(t) = \int_{A_n} x N(t, dx)$, and we will denote by $(T_n^m, m \in \mathbb{N})$ the corresponding sequence of jump times. We will also need the sequence of Lévy-type stochastic integrals $(Z_n, n \in \mathbb{N})$ wherein each

$$Z_n^i(t) = W^i(t) - \int_0^t \int_{A_n} H^i(s, x) \nu(dx) ds.$$

We construct the sequence $(Y_n, n \in \mathbb{N})$ appearing in Corollary 4.3.10 as follows:

$$Y_n(t) = \begin{cases} Z_n(t) & \text{for } 0 \leq t < T_n^1, \\ Z_n(T_n^1) + H(T_n^1, \Delta Q_n(T_n^1)) & \text{for } t = T_n^1, \\ Y_n(T_n^1) + Z_n(t) - Z_n(T_n^1) & \text{for } T_n^1 < t < T_n^2, \\ Y_n(T_n^2-) + H(T_n^2, \Delta Q_n(T_n^2)) & \text{for } t = T_n^2, \end{cases}$$

and so on recursively.

4.3.4 Stable stochastic integrals

The techniques we have developed above allow us to define stochastic integrals with respect to an α-stable Lévy process $(X_\alpha(t), t \geq 0)$ for $0 < \alpha < 2$; indeed, we can define such an integral as a Lévy stochastic integral of the form

$$Y(t) = \int_0^t L(s) dX_\alpha(s)$$

where $L \in \mathcal{P}_2(t)$ and, in the Lévy–Itô decomposition, we take $a = 0$ and the Poisson random measure N to have Lévy measure

$$\nu(dx) = C \frac{1}{|x|^{d+\alpha}} dx$$

where $C > 0$. There are alternative approaches to the problem of defining such stochastic integrals, (at least in the case $d = 1$) that start off as we do, by defining the integral on step functions as in equation (4.1). However, the corresponding limit is taken in a more subtle way, that exploits the form of the characteristic function given in Theorem 1.2.21 rather than the Lévy–Khintchine formula and which allows more subtle properties of the stable process X to pass through to the integral. In Samorodnitsky and Taqqu [271], pp. 121–6, this is carried out for sure measurable functions f, which, instead of being L^2, satisfy the requirement $\int_0^t |f(s)|^\alpha ds < \infty$ and in the case $\alpha = 1$ the additional constraint $\int_0^t |f(s) \log |f(s)|| ds < \infty$. It is shown that each $\int_0^t f(s) dX_\alpha(s)$ is itself α-stable.

The extension to predictable processes $(L(s), s \geq 0)$ satisfying the integrability property $(||L||_\alpha)^\alpha = \int_0^t \mathbb{E}(|L(s)|^\alpha) \, ds < \infty$ was carried out by Giné and Marcus [119]; see also Rosiński and Woyczyński [265] for further developments. The extent to which the structure of the stable integrator is carried over to the integral is reflected in the inequalities

$$c_1(||L||_\alpha)^\alpha \leq \sup_{\lambda > 0} \lambda^\alpha P \left(\sup_{0 \leq t \leq T} \left| \int_0^t L(s) dX_\alpha(s) \right| > \lambda \right) \leq c_2(||L||_\alpha)^\alpha,$$

where $c_1, c_2 > 0$. The left-hand side of this inequality is established in Rosiński and Woyczyński [265] and the right-hand side in Giné and Marcus [119].

4.3.5 Wiener–Lévy integrals, moving averages and the Ornstein–Uhlenbeck process

In this section we study stochastic integrals with sure integrands. These have a number of important applications, as we shall see. Let $X = (X(t), t \geq 0)$ be a Lévy process taking values in $\mathbb{R}^d$ and let $f \in L^2(\mathbb{R}^+)$; then we can consider the *Wiener–Lévy integral* $Y = (Y(t), t \geq 0)$ where each

$$Y(t) = \int_0^t f(s) dX(s). \tag{4.7}$$

These integrals are defined by the same procedure as that used above for random integrands. The terminology 'Wiener–Lévy integral' recognises that we are generalising *Wiener integrals*, which are obtained in (4.7) when X is a standard Brownian motion $B = (B(t), t \geq 0)$. In this latter case, we have the following useful result.

Lemma 4.3.11 *For each $t \geq 0$, we have $Y(t) \sim N\left(0, \int_0^t |f(s)|^2 ds\right)$.*

Proof Employing our usual sequence of partitions, we have

$$\int_0^t f(s) dB(s) = L^2 - \lim_{n \to \infty} \sum_{j=0}^{m(n)} f\left(t_j^{(n)}\right) \left[B\left(t_{j+1}^{(n)}\right) - B\left(t_j^{(n)}\right) \right],$$

so that each $Y(t)$ is the L^2-limit of a sequence of Gaussians and thus is itself Gaussian. The expressions for the mean and variance then follow immediately, from arguments similar to those that established Theorem 4.2.3(2). □

We now return to the general case (4.7). We write each $X(t) = M(t) + A(t)$, where M is a martingale and A has finite variation, and recall the precise form of these from the Lévy–Itô decomposition. Our first observation is that the process Y has independent increments.

Lemma 4.3.12 *For each $0 \leq s < t < \infty$, $Y(t) - Y(s)$ is independent of $\mathcal{F}_s$.*

Proof Utilising the partitions of $[s, t]$ from Lemma 4.3.11 we obtain

$$Y(t) - Y(s) = \int_s^t f(u) dX(u)$$

$$= \lim_{n \to \infty} \sum_{j=0}^{m(n)} f\left(t_j^{(n)}\right)\left[M\left(t_{j+1}^{(n)}\right) - M\left(t_j^{(n)}\right)\right]$$

$$+ \lim_{n \to \infty} \sum_{j=0}^{m(n)} f\left(t_j^{(n)}\right)\left[A\left(t_{j+1}^{(n)}\right) - A\left(t_j^{(n)}\right)\right],$$

where the first limit is taken in the L^2-sense and the second in the pathwise sense. In both cases, each term in the summand is adapted to the σ-algebra $\sigma\{X(v) - X(u); s \leq u < v \leq t\}$, which is independent of $\mathcal{F}_s$, and the result follows. □

From now on we assume that $f \in L^2(\mathbb{R}) \cap L^1(\mathbb{R})$ so that, for each $t \geq 0$, the shifted function $s \to f(s-t)$ is also in $L^2(\mathbb{R}) \cap L^1(\mathbb{R})$. We want to make sense of the *moving-average process* $Z = (Z(t), t \geq 0)$ given by

$$Z(t) = \int_{-\infty}^{\infty} f(s-t) dX(s)$$

for all $t \geq 0$, where the integral is defined by taking $(X(t), t < 0)$ to be an independent copy of $(-X(t), t > 0)$.[1]

Assumption 4.3.13 For the remainder of this subsection, we will impose the condition $\int_{|x|>1} |x| \nu(dx) < \infty$ on the Lévy measure ν of X.

Exercise 4.3.14 Show that for each $t \geq 0$, the following exist:

$$\int_{-\infty}^{\infty} f(s-t) dM(s) = L^2 - \lim_{T \to \infty} \int_{-T}^{T} f(s-t) dM(s),$$

$$\int_{-\infty}^{\infty} f(s-t) dA(s) = L^1 - \lim_{T \to \infty} \int_{-T}^{T} f(s-t) dA(s).$$

Exercise 4.3.15 Let $f \in L^2(\mathbb{R}) \cap L^1(\mathbb{R})$ and consider the Wiener–Lévy integral defined by $Y(t) = \int_0^t f(s) dX(s)$ for each $t \geq 0$. Show that

---

[1] If you want $(X(t), t \in \mathbb{R})$ to be càdlàg, when $t < 0$ take $X(t)$ to be an independent copy of $-X(-t-)$.

$Y = (Y(t), t \geq 0)$ is stochastically continuous. (Hint: Use

$$P\left(\left|\int_0^t f(s)dX(s)\right| > c\right) \leq P\left(\left|\int_0^t f(s)dM(s)\right| > \frac{c}{2}\right)$$
$$+ P\left(\left|\int_0^t f(s)dA(s)\right| > \frac{c}{2}\right)$$

for each $c \geq 0$, and then apply the appropriate Chebyshev–Markov inequality to each term.)

Recall that a stochastic process $C = (C(t), t \geq 0)$ is *strictly stationary* if, for each $n \in \mathbb{N}, 0 \leq t_1 < t_2 < \cdots < t_n < \infty, h > 0$, we have

$$(C(t_1 + h), C(t_2 + h), \ldots, C(t_n + h)) \stackrel{d}{=} (C(t_1), C(t_2), \ldots, C(t_n)).$$

Theorem 4.3.16 *The moving-average process $Z = (Z(t), t \geq 0)$ is strictly stationary.*

Proof Let $t \geq 0$ and fix $h > 0$, then

$$Z(t+h) = \int_{-\infty}^{\infty} f(s-t-h)dX(s) = \int_{-\infty}^{\infty} f(s-t)dX(s+h)$$
$$= \lim_{T \to \infty} \lim_{n \to \infty} \sum_{j=0}^{m(n)} f\left(s_j^{(n)} - t\right)\left[X\left(s_{j+1}^{(n)} + h\right) - X\left(s_j^{(n)} + h\right)\right],$$

where $\{-T = s_0^{(n)} < s_1^{(n)} < \cdots < s_{m(n)+1}^{(n)} = T\}$ is a sequence of partitions of each $[-T, T]$ and limits are taken in the L^2 (respectively, L^1) sense for the martingale (respectively, finite-variation) parts of X.

Since convergence in L^p (for $p \geq 1$) implies convergence in distribution and X has stationary increments, we find that for each $u \in \mathbb{R}^d$

$$\mathbb{E}(e^{i(u, Z(t+h))}) = \lim_{T \to \infty} \lim_{n \to \infty} \mathbb{E}\left(\exp\left[i\left(u, \sum_{j=0}^{m(n)} f\left(s_j^{(n)} - t\right)\right.\right.\right.$$
$$\left.\left.\left.\times \left[X\left(s_{j+1}^{(n)} + h\right) - X\left(s_j^{(n)} + h\right)\right]\right)\right]\right)$$
$$= \lim_{T \to \infty} \lim_{n \to \infty} \mathbb{E}\left(\exp\left[i\left(u, \sum_{j=0}^{m(n)} f\left(s_j^{(n)} - t\right)\right.\right.\right.$$
$$\left.\left.\left.\times \left[X\left(s_{j+1}^{(n)}\right) - X\left(s_j^{(n)}\right)\right]\right)\right]\right)$$
$$= \mathbb{E}(e^{i(u, Z(t))}),$$

so that $Z(t+h) \stackrel{d}{=} Z(t)$.

In the general case, let $0 \leq t_1 < t_2 < \cdots < t_n$ and $u_j \in \mathbb{R}^d$, $1 \leq j \leq n$. Arguing as above, we then find that

$$\mathbb{E}\left(\exp\left[\sum_{j=1}^n (u_j, Z(t_j + h))\right]\right)$$

$$= \mathbb{E}\left(\exp\left[\sum_{j=1}^n \left(u_j, \int_{-\infty}^\infty f(s - t_j - h) dX(s)\right)\right]\right)$$

$$= \mathbb{E}\left(\exp\left[\sum_{j=1}^n \left(u_j, \int_{-\infty}^\infty f(s - t_j) dX(s)\right)\right]\right)$$

$$= \mathbb{E}\left(\exp\left[\sum_{j=1}^n (u_j, Z(t_j))\right]\right),$$

from which the required result follows. □

Note In the case where X is α-stable ($0 < \alpha \leq 2$) and $\int_{-\infty}^\infty |f(s)|^\alpha ds < \infty$, then Z is itself α-stable; see Samorodnitsky and Taqqu [271], p. 138, for details.

The *Ornstein–Uhlenbeck process* is an important special case of the moving-average process. To obtain this, we fix $\lambda > 0$ and take $f(s) = e^{\lambda s} \chi_{(-\infty, 0]}(s)$ for each $s \leq 0$. Then we have, for each $t \geq 0$,

$$Z(t) = \int_{-\infty}^t e^{-\lambda(t-s)} dX(s) = \int_{-\infty}^0 e^{-\lambda(t-s)} dX(s) + \int_0^t e^{-\lambda(t-s)} dX(s)$$

$$= e^{-\lambda t} Z(0) + \int_0^t e^{-\lambda(t-s)} dX(s). \tag{4.8}$$

The Ornstein–Uhlenbeck process has interesting applications to finance and to the physics of Brownian motion, and we will return to these in later chapters. We now examine a remarkable connection with self-decomposable random variables. We write $Z = Z(0)$ so that, for each $t > 0$,

$$Z = \int_{-\infty}^0 e^{\lambda s} dX(s) = \int_{-\infty}^{-t} e^{\lambda s} dX(s) + \int_{-t}^0 e^{\lambda s} dX(s),$$

and we observe that these two stochastic integrals are independent by Lemma 4.3.12. Now since X has stationary increments, we can argue as in the proof of

Theorem 4.3.16 to show that

$$\int_{-\infty}^{-t} e^{\lambda s} dX(s) = e^{-\lambda t} \int_{-\infty}^{0} e^{\lambda s} dX(s-t)$$

$$\stackrel{d}{=} e^{-\lambda t} \int_{-\infty}^{0} e^{\lambda s} dX(s)$$

$$= e^{-\lambda t} Z.$$

Hence we have that $Z = Z_1 + Z_2$, where Z_1 and Z_2 are independent and $Z_2 \stackrel{d}{=} e^{-\lambda t} Z$. It follows that Y is self-decomposable (see Subsection 1.2.5). This result has a remarkable converse, namely that given any self-decomposable random variable Z there exists a Lévy process $X = (X(t), t \geq 0)$ such that

$$Z = \int_{-\infty}^{0} e^{-s} dX(s).$$

This result is due to Wolfe [306] in one dimension and to Jurek and Vervaat [162] in the many- (including infinite-) dimensional case (see also Jurek and Mason [161], pp. 116–44). When it is used to generate a self-decomposable random variable in this way, the Lévy process X is often called a *background-driving Lévy process*, or *BDLP* for short.

Note Our study of the Ornstein–Uhlenbeck process has been somewhat crude as, through our assumption on the Lévy measure v, we have imposed convergence in L^1 on $\lim_{t \to \infty} \int_{-t}^{0} e^{-s} dX(s)$. The following more subtle theorem can be found in Wolfe [306], Jurek and Vervaat [162] and Jacod [155]; see also Barndorff-Nielsen *et al.* [21] and Jeanblanc *et al.* [158].

Theorem 4.3.17 *The following are equivalent:*

(1) Z is a self-decomposable random variable;
(2) $Z = \lim_{t \to \infty} \int_{-t}^{0} e^{-s} dX(s)$ *in distribution, for some càdlàg Lévy process $X = (X(t), t \geq 0)$;*
(3) $\int_{|x|>1} \log(1+|x|) v(dx) < \infty$, *where v is the Lévy measure of X;*
(4) Z *can be represented as $Z(0)$ in a stationary Ornstein–Uhlenbeck process $(Z(t), t \geq 0)$.*

The term 'Ornstein–Uhlenbeck process' is also used to describe processes of the form $Y = (Y(t), t \geq 0)$ where, for each $t \geq 0$,

$$Y(t) = e^{-\lambda t} y_0 + \int_0^t e^{-\lambda(t-s)} dX(s), \qquad (4.9)$$

where $y_0 \in \mathbb{R}^d$ is fixed. Indeed, these were the first such processes to be studied historically, in the case where X is a standard Brownian motion (see Chapter 6 for more details). Note that such processes cannot be stationary, as illustrated by the following exercise.

Exercise 4.3.18 If X is a standard Brownian motion show that each $Y(t)$ is Gaussian with mean $e^{-\lambda t} y_0$ and variance $(1/2\lambda)(1 - e^{-2\lambda t} I)$.

The final topic in this section is the *integrated Ornstein–Uhlenbeck process* $I_Z = (I_Z(t), t \geq 0)$, defined as

$$I_Z(t) = \int_0^t Z(u) du.$$

Clearly I_Z has continuous sample paths. We derive an interesting relation due to Barndorff-Nielsen [28]. Note that the use of Fubini's theorem below to interchange integrals is certainly justified when X is of finite variation.

Integrating (4.8) yields, for each $t \geq 0$,

$$\begin{aligned}
I_Z(t) &= \frac{1}{\lambda}(1 - e^{-\lambda t})Z(0) + \int_0^t \int_0^u e^{-\lambda(u-s)} dX(s) du \\
&= \frac{1}{\lambda}(1 - e^{-\lambda t})Z(0) + \int_0^t \int_s^t e^{-\lambda(u-s)} du\, dX(s) \\
&= \frac{1}{\lambda}(1 - e^{-\lambda t})Z(0) + \frac{1}{\lambda}\int_0^t (1 - e^{-\lambda(t-s)}) dX(s) \\
&= \frac{1}{\lambda}\big[Z(0) - Z(t) + X(t)\big].
\end{aligned}$$

This result expresses the precise mechanism for the cancellation of jumps in the sample paths of Z and X to yield sample-path continuity for I_Z.

4.4 Itô's formula

In this section we will establish the rightly celebrated Itô formulae for sufficiently smooth functions of stochastic integrals. Some writers refer to this acclaimed result as *Itô's lemma*, but this author takes the point of view that the result is far more important than many others in mathematics that bear the title 'theorem'. As in drinking a fine wine, we will proceed slowly in gradual stages to bring out the full beauty of the result.

4.4.1 Itô's formula for Brownian integrals

Let $M = (M(t), t \geq 0)$ be a Brownian integral of the form

$$M^i(t) = \int_0^t F^i_j(s) dB^j(s)$$

4.4 Itô's formula

for $1 \le i \le d$, where $F = (F^i_j)$ is a $d \times m$ matrix taking values in $\mathcal{P}_2(T)$ and $B = (B^1, \ldots, B^m)$ is a standard Brownian motion in $\mathbb{R}^m$. Our goal is to analyse the structure of $(f(M(t)), t \ge 0)$, where $f \in C^2(\mathbb{R}^d)$.

We begin with a result of fundamental importance. Here we meet in disguise the notion of 'quadratic variation', which controls many of the algebraic properties of stochastic integrals.

Let $(\mathcal{P}_n, n \in \mathbb{N})$ be a sequence of partitions of the form

$$\mathcal{P}_n = \left\{ 0 = t_0^{(n)} < t_1^{(n)} < \cdots < t_{m(n)}^{(n)} < t_{m(n)+1}^{(n)} = T \right\},$$

and suppose that $\lim_{n \to \infty} \delta(\mathcal{P}_n) = 0$, where the mesh $\delta(\mathcal{P}_n)$ is given by $\max_{0 \le j \le m(n)} |t_{j+1}^{(n)} - t_j^{(n)}|$.

Lemma 4.4.1 *If $W_{kl} \in \mathcal{H}_2(T)$ for each $1 \le k, l \le m$ then*

$$L^2 - \lim_{n \to \infty} \sum_{j=0}^n W_{kl}\left(t_j^{(n)}\right) \left[B^k\left(t_{j+1}^{(n)}\right) - B^k\left(t_j^{(n)}\right) \right] \left[B^l\left(t_{j+1}^{(n)}\right) - B^l\left(t_j^{(n)}\right) \right]$$

$$= \sum_{k=1}^m \int_0^T W_{kk}(s) ds.$$

Proof To simplify the notation we will suppress n, write each $W_{kl}\left(t_j^{(n)}\right)$ as W_{kl}^j and introduce $\Delta B_j^k = B^k\left(t_{j+1}^{(n)}\right) - B^k\left(t_j^{(n)}\right)$ and $\Delta t_j = t_{j+1}^{(n)} - t_j^{(n)}$.

Now since B^k and B^l are independent Brownian motions, we find that

$$\mathbb{E}\left(\left[\sum_j W_{kl}^j(\Delta B_j^k)(\Delta B_j^l) - \sum_j \sum_k W_{kk}^j \Delta t_j\right]^2\right)$$

$$= \mathbb{E}\left(\left[\sum_j W_{kk}^j(\Delta B_j^k)^2 - \sum_j \sum_k W_{kk}^j \Delta t_j\right]^2\right)$$

$$= \sum_{i,j,k} \mathbb{E}\left(W_{kk}^i W_{kk}^j \left[(\Delta B_i^k)^2 - \Delta t_i\right]\left[(\Delta B_j^k)^2 - \Delta t_j\right]\right).$$

As in the proof of Lemma 4.2.2, we can split the sum in the last term into three cases: $i < j$; $j > i$; $i = j$. By use of (M2) and independent increments we see that the first two vanish. We then use the fact that each $\Delta B_j^k \sim N(0, \Delta t_j)$, which implies $\mathbb{E}((\Delta B_j^k)^4) = 3(\Delta t_j)^2$, to obtain

$$\sum_{i,j,k} \mathbb{E}\left(W_{kk}^i W_{kk}^j [(\Delta B_i^k)^2 - \Delta t_i][(\Delta B_j^k)^2 - \Delta t_j]\right)$$

$$= \sum_{j,k} \mathbb{E}\left((W_{kk}^j)^2 [(\Delta B_j^k)^2 - \Delta t_j]^2\right)$$

$$= \sum_{j,k} \mathbb{E}((W_{kk}^j)^2) \, \mathbb{E}\left([(\Delta B_j^k)^2 - \Delta t_j]^2\right) \quad \text{by (M2)}$$

$$= \sum_{j,k} \mathbb{E}((W_{kk}^j)^2) \, \mathbb{E}\left((\Delta B_j^k)^4 - 2(\Delta B_j^k)^2 \Delta t_j + (\Delta t_j)^2\right)$$

$$= 2 \sum_{j,k} \mathbb{E}((W_{kk}^j)^2) \, (\Delta t_j)^2$$

$$\to 0 \quad \text{as} \quad n \to \infty,$$

and the required result follows. □

Corollary 4.4.2 *Let B be a one-dimensional standard Brownian motion; then*

$$L^2 - \lim_{n \to \infty} \sum_{j=0}^{n} \left[B(t_{j+1}^{(n)}) - B(t_j^{(n)})\right]^2 = T.$$

Proof Immediate from the above. □

Now let M be a Brownian integral with drift of the form

$$M^i(t) = \int_0^t F_j^i(s) dB^j(s) + \int_0^t G^i(s) ds, \qquad (4.10)$$

where each $F_j^i, (G^i)^{1/2} \in \mathcal{P}_2(t)$ for all $t \geq 0$, $1 \leq i \leq d$, $1 \leq j \leq m$.

For each $1 \leq i \leq j$, we introduce the *quadratic variation process*, denoted as $([M^i, M^j](t), t \geq 0)$, by

$$[M^i, M^j](t) = \sum_{k=1}^{m} \int_0^t F_k^i(s) F_k^j(s) ds.$$

We will explore quadratic variation in greater depth as this chapter unfolds.

Now let $f \in C^2(\mathbb{R}^d)$ and consider the process $(f(M(t)), t \geq 0)$. The chain rule from elementary calculus leads us to expect that $f(M(t))$ will again have a stochastic differential of the form

$$df(M(t)) = \partial_i f(M(t)) \, dM^i(t).$$

In fact, Itô showed that $df(M(t))$ really is a stochastic differential but that in this case the chain rule takes a modified form. Additional second-order terms appear and these are described by the quadratic variation. More precisely we have the following.

Theorem 4.4.3 (Itô's theorem 1) *If $M = (M(t), t \geq 0)$ is a Brownian integral with drift of the form (4.10), then for all $f \in C^2(\mathbb{R}^d)$, $t \geq 0$, with probability 1 we have*

$$f(M(t)) - f(M(0))$$
$$= \int_0^t \partial_i f(M(s))\, dM^i(s) + \frac{1}{2} \int_0^t \partial_i \partial_j f(M(s))\, d[M^i, M^j](s).$$

Proof We follow closely the argument given in Kunita [182], pp. 64–5.

We begin by assuming that $M(t)$, $F_j^i(t)$ and $G^i(t)$ are bounded random variables for all $t \geq 0$. Let $(\mathcal{P}_n, n \in \mathbb{N})$ be a sequence of partitions of $[0, t]$ as above. By Taylor's theorem we have, for each such partition (where we again suppress the index n),

$$f(M(t)) - f(M(0)) = \sum_{k=0}^m f(M(t_{k+1})) - f(M(t_k)) = J_1(t) + \tfrac{1}{2} J_2(t)$$

where

$$J_1(t) = \sum_{k=0}^m \partial_i f(M(t_k))\big[M^i(t_{k+1}) - M^i(t_k)\big],$$

$$J_2(t) = \sum_{k=0}^m \partial_i \partial_j f(N_{ij}^k)\big[M^i(t_{k+1}) - M^i(t_k)\big]\big[M^j(t_{k+1}) - M^j(t_k)\big]$$

and where the N_{ij}^k are each $\mathcal{F}(t_{k+1})$-adapted $\mathbb{R}^d$-valued random variables satisfying $|N_{ij}^k - M(t_k)| \leq |M(t_{k+1}) - M(t_k)|$.

Now by Lemma 4.3.1 we find that, as $n \to \infty$,

$$J_1(t) \to \int_0^t \partial_i f(M(s))\, dM^i(s)$$

in probability.

We write $J_2(t) = K_1(t) + K_2(t)$, where

$$K_1(t) = \sum_{k=0}^m \partial_i \partial_j f(M(t_k))\big[M^i(t_{k+1}) - M^i(t_k)\big]\big[M^j(t_{k+1}) - M^j(t_k)\big],$$

$$K_2(t) = \sum_{k=0}^m \big[\partial_i \partial_j f(N_{ij}^k) - \partial_i \partial_j f(M(t_k))\big]$$
$$\times \big[M^i(t_{k+1}) - M^i(t_k)\big]\big[M^j(t_{k+1}) - M^j(t_k)\big].$$

Then by a slight extension of Lemma 4.4.1, we find that as $n \to \infty$,

$$K_1(t) \to \int_0^t \partial_i \partial_j f(M(s)) \, d[M^i, M^j](s).$$

By the Cauchy–Schwarz inequality, we have

$$|K_2(t)| \leq \max_{0 \leq k \leq m} \left| \partial_i \partial_j f(N_{ij}^k) - \partial_i \partial_j f(M(t_k)) \right|$$

$$\times \left\{ \sum_{k=0}^m [M^i(t_{k+1}) - M^i(t_k)]^2 \right\}^{1/2} \left\{ \sum_{k=0}^m [(M^j(t_{k+1}) - M^j(t_k)]^2 \right\}^{1/2}.$$

Now as $n \to \infty$, by continuity,

$$\max_{0 \leq k \leq m} \left| \partial_i \partial_j f(N_{ij}^k) - \partial_i \partial_j f(M(t_k)) \right| \to 0$$

while

$$\sum_{k=0}^m \left(M^i(t_{k+1}) - M^i(t_k) \right)^2 \to [M^i, M^i](t)$$

in L^2. Hence, by Proposition 1.1.9 we have $K_2(t) \to 0$ in probability.

To establish the general result, we introduce the sequence of stopping times $(T(n), n \in \mathbb{N})$ defined by

$$T(n) = \inf \{ t \geq 0; \max\{M^i(t), G^i(t), F_j^i(t); \\ 1 \leq i \leq d, 1 \leq j \leq m \} > n \} \wedge n$$

so that $\lim_{n \to \infty} T(n) = \infty$ (a.s.). Then the argument given above applies to each $(M(t \wedge T(n)), t \geq 0)$ and the required result follows by the continuity of f, when we take limits. □

Now let $C^{1,2}(\mathbb{R}^+, \mathbb{R}^d)$ denote the class of mappings from $\mathbb{R}^+ \times \mathbb{R}^d$ to $\mathbb{R}$ that are continuously differentiable with respect to the first variable and twice continuously differentiable with respect to the second.

Corollary 4.4.4 *If $M = (M(t), t \geq 0)$ is a Brownian integral with drift of the form (4.10), then for all $f \in C^{1,2}(\mathbb{R}^+, \mathbb{R}^d), t \geq 0$, with probability 1 we have*

$$f(t, M(t)) - f(0, M(0)) = \int_0^t \frac{\partial f}{\partial s}(s, M(s)) \, ds + \int_0^t \frac{\partial f}{\partial x_i}(s, M(s)) \, dM^i(s)$$

$$+ \frac{1}{2} \int_0^t \frac{\partial^2 f}{\partial x_i \partial x_j}(s, M(s)) \, d[M^i, M^j](s).$$

Proof Using the same notation as in the previous theorem, we write

$$f(t, M(t)) - f(0, M(0)) = \sum_{k=0}^{m}[f(t_{k+1}, M(t_{k+1})) - f(t_k, M(t_{k+1}))]$$

$$+ \sum_{k=0}^{m}[f(t_k, M(t_{k+1})) - f(t_k, M(t_k))].$$

By the mean value theorem, we can find $t_k < s_k < t_{k+1}$, for each $0 \leq k \leq m - 1$, such that

$$\sum_{k=0}^{m}[f(t_{k+1}, M(t_{k+1})) - f(t_k, M(t_{k+1}))]$$

$$= \sum_{k=0}^{m} \frac{\partial f}{\partial s}(s_k, M(t_{k+1}))(t_{k+1} - t_k)$$

$$\to \int_0^t \frac{\partial f}{\partial s}(s, M(s))\, ds \quad \text{as } n \to \infty.$$

The remaining terms are treated as in Theorem 4.4.3. $\square$

4.4.2 Itô's formula for Lévy-type stochastic integrals

We begin by considering Poisson stochastic integrals of the form

$$W^i(t) = W^i(0) + \int_0^t \int_A K^i(s, x) N(ds, dx) \tag{4.11}$$

for $1 \leq i \leq d$, where $t \geq 0$, A is bounded below and each K^i is predictable. Itô's formula for such processes takes a particularly simple form.

Lemma 4.4.5 *If W is a Poisson stochastic integral of the form (4.11), then for each $f \in C(\mathbb{R}^d)$, and for each $t \geq 0$, with probability 1 we have*

$$f(W(t)) - f(W(0))$$
$$= \int_0^t \int_A [f(W(s-) + K(s, x)) - f(W(s-))] N(ds, dx).$$

Proof Let $Y(t) = \int_A x N(dt, dx)$ and recall that the jump times for Y are defined recursively as $T_0^A = 0$ and, for each $n \in \mathbb{N}$, $T_n^A = \inf\{t > T_{n-1}^A;$

$\Delta Y(t) \in A\}$. We then find that

$$f((W(t)) - f(W(0))$$
$$= \sum_{0 \leq s \leq t} f(W(s)) - f(W(s-))$$
$$= \sum_{n=1}^{\infty} f(W(t \wedge T_n^A)) - f(W(t \wedge T_{n-1}^A))$$
$$= \sum_{n=1}^{\infty} \left[f(W(t \wedge T_n^A -)) + K(t \wedge T_n^A, \Delta Y(t \wedge T_n^A)) - f(t \wedge W(T_n^A -)) \right]$$
$$= \int_0^t \int_A [f(W(s-) + K(s,x)) - f(W(s-))] N(ds, dx).$$

$\square$

Now consider a Lévy-type stochastic integral of the form

$$Y^i(t) = Y^i(0) + Y_c^i(t) + \int_0^t \int_A K^i(s,x) N(ds, dx) \qquad (4.12)$$

where

$$Y_c^i(t) = \int_0^t G^i(s) ds + \int_0^t F_j^i(s) dB^j(s)$$

for each $t \geq 0$, $1 \leq i \leq d$.

In view of Theorem 4.2.14, the notation Y_c used to denote the *continuous part* of Y is not unreasonable.

We then obtain the following Itô formula.

Lemma 4.4.6 *If Y is a Lévy-type stochastic integral of the form (4.12), then, for each $f \in C^2(\mathbb{R}^d)$, $t \geq 0$, with probability 1 we have*

$$f(Y(t)) - f(Y(0))$$
$$= \int_0^t \partial_i f(Y(s-)) \, dY_c^i(s) + \frac{1}{2} \int_0^t \partial_i \partial_j f(Y(s-)) \, d[Y_c^i, Y_c^j](s)$$
$$+ \int_0^t \int_A [f(Y(s-) + K(s,x)) - f(Y(s-))] N(ds, dx).$$

Proof Using the stopping times from the previous lemma, we find that

$$f(Y(t)) - f(Y(0)) = \sum_{j=0}^{\infty} \left[f\left(Y\left(t \wedge T_{j+1}^A\right)\right) - f\left(Y\left(t \wedge T_j^A\right)\right) \right]$$

$$= \sum_{j=0}^{\infty} \left[f\left(Y\left(t \wedge T_{j+1}^A -\right)\right) - f\left(\left(Y\left(t \wedge T_j^A\right)\right)\right) \right]$$

$$+ \sum_{j=0}^{\infty} \left[f\left(Y\left(t \wedge T_{j+1}^A\right)\right) - f\left(Y\left(t \wedge T_{j+1}^A -\right)\right) \right].$$

Now using the interlacing structure, we observe that for each $T_j^A < t < T_{j+1}^A$ we have

$$Y(t) = Y(T_j^A -) + Y_c(t) - Y_c(T_j^A),$$

and the result then follows by applying Theorem 4.4.3 within the first sum and Lemma 4.4.5 within the second. $\square$

Now we are ready to prove Itô's formula for general Lévy-type stochastic integrals, so let Y be such a process with stochastic differential

$$dY(t) = G(t)\,dt + F(t)\,dB(t) + \int_{|x|<1} H(t,x)\tilde{N}(dt,dx)$$

$$+ \int_{|x|\geq 1} K(t,x)N(dt,dx), \qquad (4.13)$$

where, for each $1 \leq i \leq d$, $1 \leq j \leq m$, $t \geq 0$, $|G^i|^{1/2}, F_j^i \in \mathcal{P}_2(T)$ and $H^i \in \mathcal{P}_2(T, E)$. Furthermore, we take K to be predictable and $E = \hat{B} - \{0\}$. We will also continue to use the notation introduced above,

$$dY_c(t) = G(t)dt + F(t)dB(t),$$

and we will, later on, have need of the *discontinuous part* of Y, which we denote as Y_d and which is given by

$$dY_d(t) = \int_{|x|<1} H(t,x)\tilde{N}(dt,dx) + \int_{|x|\geq 1} K(t,x)N(dt,dx)$$

so that for each $t \geq 0$

$$Y(t) = Y(0) + Y_c(t) + Y_d(t).$$

Stochastic integration

Theorem 4.4.7 (Itô's theorem 2) *If Y is a Lévy-type stochastic integral of the form (4.13), then, for each $f \in C^2(\mathbb{R}^d)$, $t \geq 0$, with probability 1 we have*

$$f(Y(t)) - f(Y(0))$$
$$= \int_0^t \partial_i f(Y(s-)) \, dY_c^i(s) + \frac{1}{2} \int_0^t \partial_i \partial_j f(Y(s-)) \, d[Y_c^i, Y_c^j](s)$$
$$+ \int_0^t \int_{|x| \geq 1} [f(Y(s-) + K(s, x)) - f(Y(s-))] N(ds, dx)$$
$$+ \int_0^t \int_{|x| < 1} [f(Y(s-) + H(s, x)) - f(Y(s-))] \tilde{N}(ds, dx)$$
$$+ \int_0^t \int_{|x| < 1} [f(Y(s-) + H(s, x)) - f(Y(s-))$$
$$- H^i(s, x) \partial_i f(Y(s-))] \nu(dx) ds.$$

Proof We first assume that $Y(s)$ and $H(s, x)$ are bounded random variables for each $0 \leq s \leq t$ and $|x| \leq 1$. We can then assert that all terms in the formula are well defined. Indeed, for the penultimate term in the formula we can use the mean value theorem to show that the mapping $(t, x) \to [f(Y(s-)+H(s, x))-f(Y(s-))]$ is in $\mathcal{P}_2(t, E)$, and the finiteness of the final term in the formula follows by similar considerations, using Taylor's theorem.

Now, to establish the formula itself we recall the sets $(A_n, n \in \mathbb{N})$ defined as in the hypothesis of Theorem 4.3.4 and the sequence of interlacing processes $(Y_n, n \in \mathbb{N})$ defined by

$$Y_n^i(t) = \int_0^t G^i(s) ds + \int_0^t F_j^i(s) dB^j(s) + \int_0^t \int_{A_n} H^i(s, x) \tilde{N}(ds, dx)$$
$$+ \int_0^t \int_{|x| \geq 1} K^i(s, x) N(ds, dx)$$

for each $1 \leq i \leq d, t \geq 0$.

By Lemma 4.4.6, for each $n \in \mathbb{N}$,

$$f(Y_n(t)) - f(Y_n(0))$$
$$= \int_0^t \partial_i f(Y(s-)) \, dY_c^i(s) + \frac{1}{2} \int_0^t \partial_i \partial_j f(Y(s-)) \, d[Y_c^i, Y_c^j](s)$$
$$+ \int_0^t \int_{|x| \geq 1} [f(Y(s-) + K(s, x)) - f(Y(s-))] N(ds, dx)$$

$$+ \int_0^t \int_{A_n} [f(Y(s-) + H(s,x)) - f(Y(s-))] N(ds, dx)$$

$$- \int_0^t \int_{A_n} H^i(s,x) \partial_i f(Y(s-)) \nu(dx) ds$$

$$= \int_0^t \partial_i f(Y(s-)) \, dY_c^i(s) + \frac{1}{2} \int_0^t \partial_i \partial_j f(Y(s-)) \, d[Y_c^i, Y_c^j](s)$$

$$+ \int_0^t \int_{|x| \geq 1} [f(Y(s-) + K(s,x)) - f(Y(s-))] N(ds, dx)$$

$$+ \int_0^t \int_{A_n} [f(Y(s-) + H(s,x)) - f(Y(s-))] \tilde{N}(ds, dx)$$

$$+ \int_0^t \int_{A_n} [f(Y(s-) + H(s,x)) - f(Y(s-))$$

$$- H^i(s,x) \partial_i f(Y(s-))] \nu(dx) ds.$$

Now by Corollary 4.3.10 we have that $Y_n(t) \to Y(t)$ in probability as $n \to \infty$, and hence there is a subsequence that converges to $Y(t)$ almost surely. The required result follows by passage to the limit in the above along that subsequence. The case of general Y and H now follows by a straightforward stopping-time argument. □

Note Theorem 4.4.7 is extended to a more general class of semimartingales in Ikeda and Watanabe [140], Chapter II, Section 4.

We have not yet finished with Itô's formula, but before we probe it further we need some subsidiary results.

Proposition 4.4.8 *If $H^i \in \mathcal{P}_2(t, E)$ for each $1 \leq i \leq d$ then*

$$\int_0^t \int_{|x|<1} |H^i(s,x) H^j(s,x)| N(ds, dx) < \infty \quad \text{a.s.}$$

for each $1 \leq i, j \leq d, t \geq 0$.

Proof The required result follows by the inequality $2|xy| \leq |x|^2 + |y|^2$, for $x, y \in \mathbb{R}$, if we can first show that $\int_0^t \int_{|x|<1} |H^i(s,x)|^2 N(ds, dx) < \infty$ (a.s.) for each $1 \leq i \leq d$, and so we will aim to establish this result.

Suppose that each $H^i \in \mathcal{H}_2(t, E)$; then

$$\mathbb{E}\left(\int_0^t \int_{|x|<1} |H^i(s,x)|^2 N(ds, dx)\right)$$

$$= \int_0^t \int_{|x|<1} \mathbb{E}(|H^i(s,x)|^2)\, ds\nu(dx) < \infty.$$

Hence $\int_0^t \int_{|x|<1} |H^i(s,x)|^2 N(ds, dx) < \infty$ (a.s.). Now let each $H^i \in \mathcal{P}_2(t, E)$. For each $n \in \mathbb{N}$, define

$$T_n = \inf\left\{t > 0; \int_0^t \int_{|x|<1} |H^i(s,x)|^2 ds\nu(dx) > n\right\},$$

so that $\lim_{n\to\infty} T_n = \infty$. Then each

$$\int_0^{t\wedge T_n} \int_{|x|<1} |H^i(s,x)|^2 N(ds, dx) < \infty \qquad \text{a.s.}$$

and the required result follows on taking limits. $\square$

Corollary 4.4.9 *If Y is a Lévy-type stochastic integral then for $1 \leq i \leq d$, $t \geq 0$,*

$$\sum_{0 \leq s \leq t} \Delta Y^i(s)^2 < \infty \qquad \text{a.s.}$$

Proof By Proposition 4.4.8,

$$\sum_{0 \leq s \leq t} \Delta Y^i(s)^2$$

$$= \sum_{0 \leq s \leq t} \left[H^i(s, \Delta Y(s))\chi_{\{\Delta Y(s) \in E\}} + \sum_{0 \leq s \leq t} K^i(s, \Delta Y(s))\chi_{\{\Delta Y(s) \in E^c\}} \right]^2$$

$$= \sum_{0 \leq s \leq t} H^i(s, \Delta Y(s))^2 \chi_{\{\Delta Y(s) \in E\}} + \sum_{0 \leq s \leq t} K^i(s, \Delta Y(s))^2 \chi_{\{\Delta Y(s) \in E^c\}}$$

$$= \int_0^t \int_E H^i(s,x)^2 N(ds, dx) + \int_0^t \int_{E^c} K^i(s,x)^2 N(ds, dx) < \infty \quad \text{a.s.}$$

$\square$

We will use Proposition 4.4.8 to transform Itô's formula in Theorem 4.4.7 to a more general form.

4.4 Itô's formula

Theorem 4.4.10 (Itô's theorem 3) *If Y is a Lévy-type stochastic integral of the form (4.13) then, for each $f \in C^2(\mathbb{R}^d)$, $t \geq 0$, with probability 1 we have*

$$f(Y(t)) - f(Y(0))$$
$$= \int_0^t \partial_i f(Y(s-)) \, dY^i(s) + \frac{1}{2} \int_0^t \partial_i \partial_j f(Y(s-)) \, d[Y_c^i, Y_c^j](s)$$
$$+ \sum_{0 \leq s \leq t} \left[f(Y(s)) - f(Y(s-)) - \Delta Y^i(s) \partial_i f(Y(s-)) \right].$$

Proof This is a straightforward but tedious exercise in algebraic manipulation of the result of Theorem 4.4.7, which we leave to the reader. We will show, however, that the random sum in the expression is almost surely finite. This follows from Proposition 4.4.8 since, by a Taylor series expansion and the Cauchy–Schwarz inequality, we have

$$\left| \sum_{0 \leq s \leq t} \left[f(Y(s)) - f(Y(s-)) - \Delta Y^i(s) \partial_i f(Y(s-)) \right] \right|$$

$$\leq \left| \int_0^t \int_{E^c} \left[f(Y(s-) + K(s,x)) - f(Y(s-)) \right. \right.$$
$$\left. \left. - K^i(s,x) \partial_i f(Y(s-)) \right] N(ds, dx) \right|$$

$$+ \left| \int_0^t \int_E \left[f(Y(s-) + H(s,x)) - f(Y(s-)) \right. \right.$$
$$\left. \left. + H^i(s,x) \partial_i f(Y(s-)) \right] N(ds, dx) \right|$$

$$\leq \int_0^t \int_{E^c} \left| K^i(s,x) K^j(s,x) \partial_i \partial_j f(Y(s-) + \theta(s,x) K(s,x)) \right| N(ds, dx)$$

$$+ \int_0^t \int_E \left| H^i(s,x) H^j(s,x) \partial_i \partial_j f(Y(s-) + \phi(s,x) H(s,x)) \right| N(ds, dx)$$

$$\leq \sup_{0 \leq s \leq t} |\partial_i \partial_j f(Y(s-))| \left(\int_0^t \int_{E^c} |K^i(s,x) K^j(s,x)| N(ds, dx) \right.$$
$$\left. + \int_0^t \int_E |H^i(s,x) H^j(s,x)| N(ds, dx) \right)$$

$$+ \int_0^t \int_{E^c} |K(s,x)|^2 N(ds, dx) + \int_0^t \int_E |H(s,x)|^2 N(ds, dx)$$

$$< \infty \quad \text{a.s.,}$$

where each $0 \leq s \leq t$ and $0 < \theta(s,x), \phi(s,x) < 1$. $\square$

The advantage of this formula over its predecessor is that the right-hand side is expressed entirely in terms of the process Y itself and its jumps. It is hence in a form that extends naturally to general semimartingales.

Exercise 4.4.11 Let $d = 1$ and apply Itô's formula to find the stochastic differential of $(e^{Y(t)}, t \geq 0)$, where Y is a Lévy-type stochastic integral. Can you find an adapted process Z that has a stochastic differential of the form

$$dZ(t) = Z(t-)dY(t)?$$

We will return to this question in the next chapter.

The investigation of stochastic integrals using Itô's formula is called *stochastic calculus* (or sometimes Itô calculus), and the remainder of this chapter and much of the next will be devoted to this topic.

4.4.3 Quadratic variation and Itô's product formula

We have already met the quadratic variation of a Brownian stochastic integral. We now extend this definition to the more general case of Lévy-type stochastic integrals $Y = (Y(t), t \geq 0)$ of the form (4.13). So for each $t \geq 0$ we define a $d \times d$ matrix-valued adapted process $[Y, Y] = ([Y, Y](t), t \geq 0)$ by the following prescription for its (i, j)th entry $(1 \leq i, j \leq d)$:

$$[Y^i, Y^j](t) = [Y_c^i, Y_c^j](t) + \sum_{0 \leq s \leq t} \Delta Y^i(s) \Delta Y^j(s). \quad (4.14)$$

By Corollary 4.4.9 each $[Y^i, Y^j](t)$ is almost surely finite, and we deduce that

$$\begin{aligned}[Y^i, Y^j](t) &= \sum_{k=1}^{m} \int_0^T F_k^i(s) F_k^j(s) ds + \int_0^t \int_{|x|<1} H^i(s, x) H^j(s, x) N(ds, dx) \\ &\quad + \int_0^t \int_{|x| \geq 1} K^i(s, x) K^j(s, x) N(ds, dx),\end{aligned} \quad (4.15)$$

so that we clearly have each $[Y^i, Y^j](t) = [Y^j, Y^i](t)$.

Exercise 4.4.12 Show that for each $\alpha, \beta \in \mathbb{R}$ and $1 \leq i, j, k \leq d$, $t \geq 0$,

$$[\alpha Y^i + \beta Y^j, Y^k](t) = \alpha [Y^i, Y^k](t) + \beta [Y^j, Y^k](t).$$

The importance of $[Y, Y]$ is that it measures the deviation in the stochastic differential of products from the usual Leibniz formula. The following result makes this precise.

Theorem 4.4.13 (Itô's product formula) *If Y^1 and Y^2 are real-valued Lévy-type stochastic integrals of the form (4.13) then, for all $t \geq 0$, with probability 1 we have that*

$$Y^1(t)Y^2(t) = Y^1(0)Y^2(0) + \int_0^t Y^1(s-)dY^2(s)$$
$$+ \int_0^t Y^2(s-)dY^1(s) + [Y^1, Y^2](t).$$

Proof We consider Y^1 and Y^2 as components of a vector $Y = (Y^1, Y^2)$ and take f in Theorem 4.4.10 to be the smooth mapping from $\mathbb{R}^2$ to $\mathbb{R}$ given by $f(x^1, x^2) = x^1 x^2$.

By Theorem 4.4.10 we then obtain, for each $t \geq 0$, with probability 1,

$$Y^1(t)Y^2(t) = Y^1(0)Y^2(0) + \int_0^t Y^1(s-)dY^2(s)$$
$$+ \int_0^t Y^2(s-)dY^1(s) + [Y_c^1, Y_c^2](t)$$
$$+ \sum_{0 \leq s \leq t} \{Y^1(s)Y^2(s) - Y^1(s-)Y^2(s-)$$
$$- [Y^1(s) - Y^1(s-)]Y^2(s-)$$
$$- [Y^2(s) - Y^2(s-)]Y^1(s-)\},$$

from which the required result easily follows. $\square$

Exercise 4.4.14 Extend this result to the case where Y^1 and Y^2 are d-dimensional.

We can learn much about the way our Itô formulae work by writing the product formula in differential form:

$$d(Y^1(t)Y^2(t)) = Y^1(t-)dY^2(t) + Y^2(t-)dY^1(t) + d[Y^1, Y^2](t).$$

By equation (4.15), we see that the term $d[Y^1, Y^2](t)$, which is sometimes called an *Itô correction*, arises as a result of the following formal product relations between differentials:

$$dB^i(t)dB^j(t) = \delta^{ij}dt, \quad N(dt, dx)N(dt, dy) = N(dt, dx)\delta(x - y)$$

for $1 \leq i, j \leq m$, where all other products of differentials vanish; if you have little previous experience of this game, these relations are a very valuable guide to intuition.

Note we have derived Itô's product formula as a corollary of Itô's theorem, but we could just as well have gone in the opposite direction, indeed the two results are equivalent; see e.g. Rogers and Williams [262], Chapter IV, Section 32.

Exercise 4.4.15 Consider the Brownian stochastic integrals given by $I^k(t) = \int_0^t F_j^k(s) dB^j(s)$ for each $t \geq 0$, $k = 1, 2$, and show that

$$[I^1, I^2](t) = \langle I^1, I^2 \rangle(t) = \sum_{j=1}^m \int_0^t F_j^1(s) F_j^2(s) ds.$$

Exercise 4.4.16 For the Poisson stochastic integrals

$$J^i(t) = \int_0^t \int_E H^i(s, x) \tilde{N}(dt, dx),$$

where $t \geq 0$ and $i = 1, 2$, deduce that

$$[J^1, J^2](t) - \langle J^1, J^2 \rangle(t) = \int_0^t \int_E H^1(s, x) H^2(s, x) \tilde{N}(dt, dx).$$

For completeness, we will give another characterisation of quadratic variation that is sometimes quite useful. We recall the sequence of partitions $(\mathcal{P}_n, n \in \mathbb{N})$ with mesh tending to zero that were introduced earlier.

Theorem 4.4.17 *If X and Y are real-valued Lévy-type stochastic integrals of the form (4.13), then, for each $t \geq 0$, with probability 1 we have*

$$[X, Y](t) = \lim_{n \to \infty} \sum_{j=0}^{m(n)} \left[X\left(t_{j+1}^{(n)}\right) - X\left(t_j^{(n)}\right) \right] \left[Y\left(t_{j+1}^{(n)}\right) - Y\left(t_j^{(n)}\right) \right],$$

where the limit is taken in probability.

Proof By polarisation, it is sufficient to consider the case $X = Y$. Using the identity

$$(x - y)^2 = x^2 - y^2 - 2y(x - y)$$

for $x, y \in \mathbb{R}$, we deduce that

$$\sum_{j=0}^{m(n)} \left[X\left(t_{j+1}^{(n)}\right) - X\left(t_j^{(n)}\right) \right]^2 = \sum_{j=0}^{m(n)} X\left(t_{j+1}^{(n)}\right)^2 - \sum_{j=0}^{m(n)} X\left(t_j^{(n)}\right)^2$$

$$- 2 \sum_{j=0}^{m(n)} X\left(t_j^{(n)}\right) \left[X\left(t_{j+1}^{(n)}\right) - X\left(t_j^{(n)}\right) \right],$$

4.4 Itô's formula

and the required result follows from Itô's product formula (Theorem 4.4.13) and Exercise 4.3.7. □

Many of the results of this chapter extend from Lévy-type stochastic integrals to arbitrary semimartingales, and full details can be found in Jacod and Shiryaev [151] and Protter [255]. In particular, if F is a simple process and X is a semimartingale we can again use Itô's prescription to define

$$\int_0^t F(s)dX(s) = \sum F(t_j)[X(t_{j+1}) - X(t_j)],$$

and then pass to the limit to obtain more general stochastic integrals. Itô's formula can be established in the form given in Theorem 4.4.10 and the quadratic variation of semimartingales defined as the correction term in the corresponding Itô product formula (or, equivalently, via the prescription of Theorem 4.4.17). In particular, if X and Y are semimartingales and X_c, Y_c denote their continuous parts then we have, for each $t \geq 0$,

$$[X, Y](t) = \langle X_c, Y_c \rangle(t) + \sum_{0 \leq s \leq t} \Delta X(s) \Delta Y(s); \qquad (4.16)$$

see Jacod and Shiryaev [151], p. 55.

Although stochastic calculus for general semimartingales is not the subject of this book, we do require one result, the famous Lévy characterisation of Brownian motion that we have already used in Chapter 2.

Theorem 4.4.18 (Lévy's characterisation) *Let* $M = (M(t), t \geq 0)$ *be a continuous centred martingale that is adapted to a given filtration* $(\mathcal{F}_t, t \geq 0)$. *If* $\langle M_i, M_j \rangle(t) = a_{ij}t$ *for each* $t \geq 0$, $1 \leq i, j \leq d$, *where* $a = (a_{ij})$ *is a positive definite symmetric matrix, then M is an $\mathcal{F}_t$-adapted Brownian motion with covariance* a.

Proof Fix $u \in \mathbb{R}^d$ and define the process $(Y_u(t), t \geq 0)$ by $Y_u(t) = e^{i\langle u, M(t) \rangle}$; then, by Itô's formula and incorporating (4.16), we obtain

$$dY_u(t) = iu^j Y_u(t) dM_j(t) - \tfrac{1}{2} u^i u^j Y_u(t) d\langle M_i, M_j \rangle(t)$$

$$= iu^j Y_u(t) dM_j(t) - \tfrac{1}{2}(u, au) Y_u(t) dt.$$

Upon integrating from s to t, we obtain

$$Y_u(t) = Y_u(s) + iu^j \int_s^t Y_u(\tau) dM_j(\tau) - \tfrac{1}{2}(u, au) \int_s^t Y_u(\tau) d\tau.$$

Now take conditional expectations of both sides with respect to $\mathcal{F}_s$, and use the conditional Fubini theorem (Theorem 1.1.7) to obtain

$$\mathbb{E}(Y_u(t)|\mathcal{F}_s) = Y_u(s) - \tfrac{1}{2}(u, au) \int_s^t \mathbb{E}(Y_u(\tau)|\mathcal{F}_s)\, d\tau.$$

Hence

$$\mathbb{E}(e^{i(u,M(t))}|\mathcal{F}_s) = e^{-(u,au)(t-s)/2}.$$

From here it is a straightforward exercise for the reader to confirm that M is a Brownian motion, as required. □

Note A number of interesting propositions that are equivalent to the Lévy characterisation can be found in Kunita [182], p. 67.

Exercise 4.4.19 Extend the Lévy characterisation to the case where M is a continuous local martingale.

Another classic and fairly straightforward application of Itô's formula for Brownian integrals is Burkholder's inequality. Let $M = (M(t), t \geq 0)$ be a (real-valued) Brownian integral of the form

$$M(t) = \int_0^t F^j(s) dB_j(s), \qquad (4.17)$$

where each $F^j \in \mathcal{H}_2(t)$, $1 \leq j \leq d$, $t \geq 0$. By Exercise 4.4.15,

$$[M, M](t) = \sum_{j=1}^m \int_0^t F_j(s)^2 ds$$

for each $t \geq 0$. Note that by Theorem 4.2.3(4), M is a square-integrable martingale.

Theorem 4.4.20 (Burkholder's inequality) *If $M = (M(t), t \geq 0)$ is a Brownian integral of the form (4.17), for which $\mathbb{E}([M, M](t)^{p/2}) < \infty$, then for any $p \geq 2$ there exists $C(p) > 0$ such that, for each $t \geq 0$,*

$$\mathbb{E}(|M(t)|^p) \leq C(p)\, \mathbb{E}([M, M](t)^{p/2}).$$

Proof We follow Kunita [182], p. 66. Assume first that each $M(t)$ is a bounded random variable. By Itô's formula we have, for each $t \geq 0$,

$$|M(t)|^p = p \int_0^t |M(s)|^{p-1} \operatorname{sgn}(M(s))\, F^j(s) dB_j(s)$$
$$+ \tfrac{1}{2} p(p-1) \int_0^t |M(s)|^{p-2} d[M, M](s),$$

4.4 Itô's formula

and, by the boundedness assumption, the stochastic integral is a martingale. Hence on taking expectations we obtain

$$\mathbb{E}(|M(t)|^p) = \frac{1}{2}p(p-1)\,\mathbb{E}\left(\int_0^t |M(s)|^{p-2} d[M,M](s)\right)$$

$$\leq \frac{1}{2}p(p-1)\,\mathbb{E}\left(\sup_{0\leq s\leq t} |M(s)|^{p-2}[M,M](t)\right).$$

By Hölder's inequality and Doob's martingale inequality, we obtain

$$\mathbb{E}\left(\sup_{0\leq s\leq t} |M(s)|^{p-2}[M,M](t)\right)$$

$$\leq \mathbb{E}\left(\sup_{0\leq s\leq t} |M(s)|^p\right)^{(p-2)/p} \mathbb{E}([M,M](t)^{p/2})^{2/p}$$

$$\leq \left(\frac{p}{p-1}\right)^{(p-2)/p} \mathbb{E}(|M(t)|^p)^{(p-2)/p}\,\mathbb{E}([M,M](t)^{p/2})^{2/p}.$$

Let $D(p) = \frac{1}{2}p(p-1)[p/(p-1)]^{(p-2)/p}$; then we have

$$\mathbb{E}(|M(t)|^p) \leq D(p)\,\mathbb{E}(|M(t)|^p)^{1-(2/p)}\,\mathbb{E}([M,M](t)^{p/2})^{2/p},$$

and the required result follows straightforwardly, with $C(p) = D(p)^{p/2}$.

For the general case, define a sequence of stopping times $(T_n, n \in \mathbb{N})$ by $T_n = \inf\{|M(t)| > n \text{ or } [M,M](t) > n\}$. Then the inequality holds for each process $(M(t \wedge T_n), t \geq 0)$, and we may use dominated convergence to establish the required result. □

Note By another more subtle application of Itô's formula, the inequality can be strengthened to show that there exists $c(p) > 0$ such that

$$c(p)\,\mathbb{E}([M,M](t)^{p/2}) \leq \mathbb{E}(|M(t)|^p) \leq C(p)\,\mathbb{E}([M,M](t)^{p/2});$$

see Kunita [182], pp. 66–7, for details. With more effort the inequality can be extended to arbitrary continuous local martingales M for which $M(0) = 0$ and also to all $p > 0$. This more extensive treatment can be found in Revuz and Yor [260], Chapter 4, Section 4. A further generalisation, where the pth power is replaced by an arbitrary convex function, is due to Burkholder, Davis and Gundy [62].

The inequalities still hold in the case where M is an arbitrary local martingale (so jumps are included), but we have been unable to find a direct proof using Itô's formula, as above, even in the case of Poisson integrals. Details of this general result may be found in Dellacherie and Meyer [78], pp. 303–4.

4.4.4 The Stratonovitch and Marcus canonical integrals

The Itô integral is a truly wonderful thing, and we will explore many more of its implications in the next chapter. Unfortunately it does have some disadvantages and one of the most important of these is – as Itô's formula has shown us – that it fails to satisfy the usual chain rule of differential calculus. This is the source of much beautiful mathematics, as we will see throughout this book, but if we examine stochastic differential equations and associated flows on smooth manifolds then we find that the Itô integral is not invariant under local co-ordinate changes and so is not a natural geometric object. Fortunately, there is a solution to this problem. We can define new 'integrals' as 'perturbations' of the Itô integral that have the properties we need.

The Stratonovitch integral

Let $M = (M(t), t \geq 0)$ be a Brownian integral of the form $M^i(t) = \int_0^t F_j^i(s) dB^j(s)$ and let $G = (G^1, \ldots, G^d)$ be a Brownian integral such that $G_i F_j^i \in \mathcal{P}_2(t)$ for each $1 \leq j \leq m, t \geq 0$. Then we define the *Stratonovitch integral* of G with respect to M by the prescription

$$\int_0^t G^i(s) \circ dM_i(s) = \int_0^t G^i(s) dM_i(s) + \frac{1}{2}[G^i, M_i](t).$$

The notation $\circ$ (sometimes called 'Itô's circle') clearly differentiates the Stratonovitch and Itô cases.

We also have the differential form

$$G^i(s) \circ dM_i(s) = G^i(s) dM_i(s) + \tfrac{1}{2} d[G^i, M_i](t).$$

Exercise 4.4.21 Establish the following relations, where $\alpha, \beta \in \mathbb{R}$ and X, Y, M_1 and M_2 are one-dimensional Brownian integrals:

(1) $(\alpha X + \beta Y) \circ dM = \alpha X \circ dM + \beta Y \circ dM$;
(2) $X \circ (dM_1 + dM_2) = X \circ dM_1 + X \circ dM_2$;
(3) $XY \circ dM = X \circ (Y \circ dM)$.

Find suitable extensions of these in higher dimensions.

The most important aspect of the Stratonovitch integral for us is that it satisfies a Newton–Leibniz-type chain rule.

Theorem 4.4.22 *If M is a Brownian integral and $f \in C^3(\mathbb{R}^d)$, then, for each $t \geq 0$, with probability 1 we have*

$$f(M(t)) - f(M(0)) = \int_0^t \partial_i f(M(s)) \circ dM^i(s).$$

Proof By the definition of the Stratonovitch integral, we have

$$\partial_i f(M(t)) \circ dM^i(t) = \partial_i f(M(t)) dM^i(t) + \tfrac{1}{2} d[\partial_i f(M(\cdot)), M^i](t)$$

and, by Itô's formula, for each $1 \leq i \leq d$,

$$d\{\partial_i f(M(t))\} = \partial_j \partial_i f(M(t)) \, dM^j(t) + \tfrac{1}{2} \partial_j \partial_k \partial_i f(M(t)) \, d[M^j, M^k](t),$$

giving

$$d[\partial_i f(M(\cdot)), M^i](t) = \partial_i \partial_j f(M(t)) \, d[M^i, M^j](t).$$

So, by using Itô's formula again, we deduce that

$$\int_0^t \partial_i f(M(s)) \circ dM^i(s)$$
$$= \partial_i f(M(s)) \, dM^i(s) + \frac{1}{2} \int_0^t \partial_i \partial_j f(M(s)) \, d[M^i, M^j](s)$$
$$= f(M(t)) - f(M(0)).$$

□

For those who hanker after a legitimate definition of the Stratonovitch integral as a limit of step functions, we consider again our usual sequence of partitions $(\mathcal{P}_n, n \in \mathbb{N})$.

Theorem 4.4.23

$$\int_0^t G_i(s) \circ dM^i(s)$$
$$= \lim_{n \to \infty} \sum_{j=0}^{m(n)} \frac{G_i\left(t_{j+1}^{(n)}\right) + G_i\left(t_j^{(n)}\right)}{2} \left[M^i\left(t_{j+1}^{(n)}\right) - M^i\left(t_j^{(n)}\right)\right],$$

where the limit is taken in probability.

Proof We suppress the indices i and n for convenience and note that, for each $0 \leq j \leq m$,

$$\frac{G(t_{j+1}) + G(t_j)}{2} [M(t_{j+1}) - M(t_j)]$$
$$= G(t_j)[M(t_{j+1}) - M(t_j)] + \tfrac{1}{2}[G(t_{j+1}) - G(t_j)][M(t_{j+1}) - M(t_j)],$$

and the result follows from the remark following Lemma 4.3.1 and Theorem 4.4.17.

□

The Marcus canonical integral

Now let Y be a Lévy-type stochastic integral; then you can check that the Stratonovich integral will no longer give us a chain rule of the Newton–Leibniz type and so we need a more sophisticated approach to take care of the jumps. The mechanism for doing this was developed by Marcus [213, 214].

We will define the *Marcus canonical integral* for integrands of the form $(G(s, Y(s-)), s \geq 0)$, where $G : \mathbb{R}^+ \times \mathbb{R}^d \to \mathbb{R}^d$ is such that $s \to G(s, Y(s-))$ is predictable and the Itô integrals $\int_0^t G_i(s, Y(s-))dY^i(s)$ exist for all $t \geq 0$.

We also need the following assumption.

There exists a measurable mapping $\Phi : \mathbb{R}^+ \times \mathbb{R} \times \mathbb{R}^d \times \mathbb{R}^d \to \mathbb{R}$ such that, for each $s \geq 0$, $x, y \in \mathbb{R}^d$:

(1) $u \to \Phi(s, u, x, y)$ is continuously differentiable;
(2) $\dfrac{\partial \Phi}{\partial u}(s, u, x, y) = y^i G_i(s, x + uy)$ for each $u \in \mathbb{R}$;
(3) $\Phi(s, 0, x, y) = \Phi(s, 0, x, 0)$.

Such a Φ is called a *Marcus mapping*.

Given such a mapping, we then define the *Marcus canonical integral* as follows: for each $t \geq 0$,

$$\int_0^t G_i(s, Y(s-)) \diamond dY^i(s)$$
$$= \int_0^t G_i(s, Y(s-)) \circ dY_c^i(s) + \int_0^t G_i(s, Y(s-))dY_d^i(s)$$
$$+ \sum_{0 \leq s \leq t} \big[\Phi(s, 1, Y(s-), \Delta Y(s)) - \Phi(s, 0, Y(s-), \Delta Y(s))$$
$$- \frac{\partial \Phi}{\partial u}(s, 0, Y(s-), \Delta Y(s))\big].$$

We consider two cases of interest.
(i) $G(s, y) = G(s)$ for all $s \geq 0$, $y \in \mathbb{R}^d$. In this case, we have

$$\Phi(s, u, x, y) = G_i(s)(x^i + uy^i).$$

We then have that

$$\int_0^t G_i(s) \diamond dY^i(s) = \int_0^t G_i(s) \circ dY_c^i(s) + \int_0^t G_i(s)dY_d^i(s);$$

so, if $Y_d \equiv 0$ then the Stratonovitch and Marcus integrals coincide while if $Y_c \equiv 0$ then the Marcus integral is the same as the Itô integral.

(ii) $G(s, y) = G(y)$ for all $s \geq 0$, $y \in \mathbb{R}^d$. We will consider this case within the context of the required Newton–Leibniz rule by writing $G_i(y) = \partial_i f(y)$ for each $1 \leq i \leq d$, where $f \in C^3(\mathbb{R}^d)$, $y \in \mathbb{R}^d$. We then have the following:

Theorem 4.4.24 *If Y is a Lévy-type stochastic integral of the form (4.13) and $f \in C^3(\mathbb{R}^d)$, then*

$$f(Y(t)) - f(Y(0)) = \int_0^t \partial_i f(Y(s-)) \diamond dY^i(s)$$

for each $t \geq 0$, with probability 1.

Proof Our Marcus map satisfies

$$\frac{\partial \Phi}{\partial u}(u, x, y) = y^i \partial_i f(x + uy),$$

and hence $\Phi(u, x, y) = f(x + uy)$.

We then find that

$$\int_0^t \partial_i f(Y(s-)) \diamond dY^i(s)$$
$$= \int_0^t \partial_i f(Y(s-)) \, dY^i(s) + \frac{1}{2} \int_0^t \partial_i \partial_j f(Y(s-)) \, d[Y_c^i, Y_c^j](s)$$
$$+ \sum_{0 \leq s \leq t} \left[f(Y(s)) - f(Y(s-)) - \Delta Y^j(s) \partial_i f(Y(s-)) \right]$$
$$= f(Y(t)) - f(Y(0)),$$

by Itô's formula. □

The probabilistic interpretation of the Marcus integral is as follows. The Marcus map introduces a fictitious time u with respect to which, at each jump time, the process travels at infinite speed along the straight line connecting the starting point $Y(s-)$ and the finishing point $Y(s)$. When we study stochastic differential equations later on, we will generalise the Marcus integral and replace the straight line by a curve determined by the geometry of the driving vector fields.

4.4.5 Backwards stochastic integrals

So far, in this book, we have fixed as forward the direction in time; all processes have started at time $t = 0$ and progressed to a later time t. For some applications it is useful to reverse the direction of time, so we fix a time T

and then proceed backwards to some earlier time s. In the discussion below, we will develop backwards notions of the concepts of filtration, martingale, stochastic integral etc. In this context, whenever we mention the more familiar notions that were developed earlier in this chapter, we will always prefix them by the word 'forward'.

We begin, as usual, with our probability space $(\Omega, \mathcal{F}, P)$. Let $(\mathcal{F}^s, 0 \leq s \leq T)$ be a family of sub σ-algebras of $\mathcal{F}$. We say that it is a *backwards filtration* if

$$\mathcal{F}^t \subseteq \mathcal{F}^s \text{ for all } 0 \leq s \leq t \leq T.$$

For an example of a backwards filtration, let $X = (X(s), 0 \leq s \leq T)$ be an $\mathbb{R}^d$-valued stochastic process on $(\Omega, \mathcal{F}, P)$ and, for each $0 \leq s \leq T$, define $\mathcal{G}_X^s = \sigma\{X(u); s \leq u \leq T\}$; $(\mathcal{G}_X^s, 0 \leq s \leq T)$ is then called the *natural backwards filtration* of X. Just as in the forward case, it is standard to impose the 'usual hypotheses' on backwards filtrations, these being

(1) (completeness) $\mathcal{F}^T$ contains all sets of P-measure zero in $\mathcal{F}$,
(2) (left continuity) for each $0 \leq s \leq T$, $\mathcal{F}^s = \mathcal{F}^{s-}$ where $\mathcal{F}^{s-} = \bigcap_{\epsilon > 0} \mathcal{F}^{s-\epsilon}$.

A process $X = (X(s), 0 \leq s \leq T)$ is said to be *backwards adapted* to a backwards filtration $(\mathcal{F}^s, 0 \leq s \leq T)$ if each $X(s)$ is $\mathcal{F}^s$-measurable, e.g. any process is backwards adapted to its own natural backwards filtration.

A backwards adapted process $(M(s), 0 \leq s \leq T)$ is called a *backwards martingale* if $\mathbb{E}(|M(s)|) < \infty$ and $\mathbb{E}(M(t)|\mathcal{F}^s) = M(s)$ whenever $0 \leq s \leq t \leq T$. Backwards versions of the supermartingale submartingale, local martingale and semimartingale are all obvious extensions (Note that some authors prefer to use 'reversed' rather than 'backwards').

Exercise 4.4.25 For each $0 \leq s \leq t \leq T$, let

$$M(s) = \sigma(B(T) - B(s)) + \lambda \int_s^T \int_{|x|<1} x \tilde{N}(ds, dx),$$

where $\sigma, \lambda \in \mathbb{R}$ and B and N are independent one-dimensional Brownian motions and Poisson random measures, respectively. Show that M is a backwards martingale with respect to its own natural backwards filtration.

Let $E \in \mathcal{B}(\mathbb{R}^d)$. We define $\mathbb{R}^d$-valued *backwards martingale-valued measures* on $[0, T] \times E$ analogously to the forward case, the only difference being that we replace the axiom (M2) (see the start of Section 4.1) by (M2)$_b$, where

(M2)$_b$ $M([s, t), A)$ is independent of $\mathcal{F}^t$ for all $0 \leq s \leq t \leq T$ and for all $A \in \mathcal{B}(E)$.

4.4 Itô's formula

Examples of the type of backwards martingale measure that will be of importance for us can be generated from Exercise 4.4.25, yielding

$$M([s, t], A) = \sigma(B(t) - B(s))\delta_0(A) + \lambda \int_s^t \int_A x \tilde{N}(ds, dx),$$

for each $0 \leq s \leq t \leq T$, $A \in \mathcal{B}(E)$, where $E = \hat{B} - \{0\}$.

We now want to carry out stochastic integration with respect to backwards martingale measures. First we need to consider appropriate spaces of integrands. Fix $0 \leq s \leq T$ and let $\mathcal{P}^-$ denote the smallest σ-algebra that contains all mappings $F : [s, T] \times E \times \Omega \to \mathbb{R}^d$ such that:

(1) for each $s \leq t \leq T$, the mapping $(x, \omega) \to F(t, x, \omega)$ is $\mathcal{B}(E) \otimes \mathcal{F}^t$ measurable;
(2) for each $x \in E$, $\omega \in \Omega$, the mapping $t \to F(t, x, \omega)$ is right-continuous.

We call $\mathcal{P}^-$ the *backwards predictable σ-algebra*. A $\mathcal{P}^-$-measurable mapping $G : [0, T] \times E \times \Omega \to \mathbb{R}^d$ is then said to be *backwards predictable*. Using the notion of $\mathcal{P}^-$ in place of $\mathcal{P}$, we can then form the backwards analogues of the spaces $\mathcal{H}_2(s, E)$ and $\mathcal{P}_2(s, E)$. We denote these by $\mathcal{H}_2^-(s, E)$ and $\mathcal{P}_2^-(s, E)$, respectively. The space of *backwards simple processes*, which we denote by $S^-(s, E)$, is defined to be the set of all $F \in \mathcal{H}_2(s, E)$ for which there exists a partition $s = t_1 < t_2 \cdots < t_m < t_{m+1} = T$ such that

$$F = \sum_{j=1}^m \sum_{k=1}^n F_k(t_{j+1}) \chi_{[t_j, t_{j+1})} \chi_{A_k},$$

where $A_1, \ldots, A_n$ are disjoint sets in $\mathcal{B}(E)$ with $\nu(A_k) < \infty$, for each $1 \leq k \leq n$, and each $F_k(t_j)$ is bounded and $\mathcal{F}^{t_j}$-measurable.

For such an F we can define its *Itô backwards stochastic integral* by

$$\int_s^t \int_E F(u, x) \cdot_b M(du, dx) = \sum_{j=1}^n \sum_{k=1}^n F_k(t_{j+1}) M([t_j, t_{j+1}), A_k).$$

We can then pass to the respective completions for $F \in \mathcal{H}_2^-(s, E)$ and $F \in \mathcal{P}_2^-(s, E)$, just as in Section 4.2. The reader should verify that backwards stochastic integrals with respect to backwards martingale measures are backwards local martingales. In particular, we can construct *Lévy-type backwards stochastic integrals*

$$Y^i(s) = Y^i(T) - \int_s^T G^i(u)du - \int_s^T F_j^i(u) \cdot_b dB^j(u)$$
$$- \int_s^T \int_{|x|<1} H^i(u,x) \cdot_b \tilde{N}(du, dx) - \int_s^T \int_{|x|\geq 1} K^i(u,x) N(du, dx),$$

where for each $1 \leq i \leq d$, $1 \leq j \leq m$, $t \geq 0$, we have $|G^i|^{1/2}, F_j^i \in \mathcal{P}_2^-(s)$, $H^i \in \mathcal{P}_2^-(s, E)$, and K is backwards predictable.

Exercise 4.4.26 Let $Y = (Y(s), 0 \leq s \leq T)$ be a backwards Lévy integral and suppose that $f \in C^3(\mathbb{R}^d)$. Derive the backwards Itô formula

$$f(Y(s)) = f(Y(T)) - \int_s^T \partial_i f(Y(u)) \cdot_b dY^i(u)$$
$$+ \frac{1}{2} \int_s^T \partial_i \partial_j f(Y(u)) \, d[Y_c^i, Y_c^j](u)$$
$$- \sum_{s \leq u \leq T} \left[f(Y(u)) - f(Y(u-)) - \Delta Y^i(u) \partial_i f(Y(u)) \right].$$

(Hint: Imitate the arguments for the forward case.)

As well as the backwards Itô integral, it is also useful to define backwards versions of the Stratonovitch and Marcus integrals.

Using the same notation as in Subsection 4.4.4, we define the backwards Stratonovitch integral by

$$\int_s^T G(u) \circ_b dM(u) = \int_s^T G(u) \cdot_b dM(u) + \tfrac{1}{2}[G,M](T) - \tfrac{1}{2}[G,M](s),$$

with the understanding that G is now backwards predictable. We again obtain a Newton–Leibniz-type chain rule,

$$f(M(T)) - f(M(s)) = \int_s^T \partial_i f(M(u)) \circ_b dM^i(u),$$

with probability 1, for each $f \in C^3(\mathbb{R}^d)$, $0 \leq s \leq T$. Taking the usual sequence of partitions $(\mathcal{P}_n, n \in \mathbb{N})$ of $[s, T]$, we have

$$\int_s^T G(u) \circ_b dM(u) = \lim_{n \to \infty} \sum_{j=0}^{m(n)} \frac{G\left(t_{j+1}^{(n)}\right) + G\left(t_j^{(n)}\right)}{2} \left[M\left(t_{j+1}^{(n)}\right) - M\left(t_j^{(n)}\right) \right],$$

where the limit is taken in probability.

4.4 Itô's formula

Again using the same notation as in Subsection 4.4.4, the *backwards Marcus canonical integral* is defined as

$$\int_s^T G_i(u, Y(u)) \diamond_b dY^i(u)$$

$$= \int_s^T G_i(u, Y(u)) \circ_b dY_c^i(u) + \int_s^T G_i(u, Y(u)) \cdot_b dY_d^i(u)$$

$$+ \sum_{s \leq t \leq T} \left[\Phi(t, 1, Y(t), \Delta Y(t)) - \Phi(t, 0, Y(t), \Delta Y(t)) \right.$$

$$\left. - \frac{\partial \Phi}{\partial u}(t, 0, Y(t), \Delta Y(t)) \right].$$

Sometimes we want to consider both forward and backwards stochastic integrals within the same framework. As usual, we fix $T > 0$. A *two-parameter filtration* of the σ-algebra $\mathcal{F}$ is a family $(\mathcal{F}_{s,t}; 0 \leq s < t \leq T)$ of sub σ-fields such that

$$\mathcal{F}_{s_1, t_1} \subseteq \mathcal{F}_{s_2, t_2} \text{ for all } 0 \leq s_2 \leq s_1 < t_1 \leq t_2.$$

If we now fix $s > 0$ then $(\mathcal{F}_{s,t}, t > s)$ is a forward filtration, while if we fix $t > 0$ then $(\mathcal{F}_{s,t}, 0 \leq s < t)$ is a backwards filtration. A martingale-valued measure on $[0, T] \times E$ is *localised* if $M((s, t), A)$ is $\mathcal{F}_{s,t}$-measurable for each $A \in \mathcal{B}(E)$ and each $0 \leq s < t \leq T$. Provided that both (M2) and (M2)$_b$ are satisfied, localised martingale measures can be used to define both forward and backwards stochastic integrals. Readers can check that examples of these are given by martingale measures built from processes with independent increments, as in Exercise 4.4.25.

4.4.6 Local times and extensions of Itô's formula

Here we sketch without proof some directions for extending Itô's formula beyond the case where f is a C^2-function, for $d = 1$. A far more comprehensive discussion can be found in Protter [255], pp. 162–80.

We begin by considering the case of a one-dimensional standard Brownian motion and we take $f(x) = |x|$ for $x \in \mathbb{R}$. Now f is not C^2; however, it is convex. We have $f'(x) = \text{sgn}(x)$ (for $x \neq 0$) but, in this case, f'' only makes sense as a distribution: $f''(x) = 2\delta(x)$ where δ is the Dirac delta function. We include a very swift proof of this to remind readers.

Proposition 4.4.27 *If $f(x) = |x|$, then $f''(x) = 2\delta(x)$, in the sense of distributions.*

Proof Let $g \in C_c^\infty(\mathbb{R})$ and, for convenience, assume that the support of g is the interval $[-a, b]$ where $a, b > 0$; then

$$\int_\mathbb{R} f''(x) g(x) dx = -\int_\mathbb{R} f'(x) g'(x) dx$$

$$= -\int_\mathbb{R} \text{sgn}(x)\, g'(x) dx$$

$$= \int_{-a}^0 g'(x) dx - \int_0^b g'(x) dx = 2g(0).$$

$\square$

Now let us naively apply Itô's formula to this set-up. So if B is our Brownian motion we see that, for each $t \geq 0$,

$$|B(t)| = \int_0^t \text{sgn}(B(s))\, dB(s) + \int_0^t \delta(B(s)) ds$$

$$= \int_0^t \text{sgn}(B(s))\, dB(s) + L(0, t),$$

where $(L(0, t), t \geq 0)$ is the local time of B at zero (see Subsection 1.5.3). In fact this result can be proved rigorously and is called *Tanaka's formula* in the literature.

Exercise 4.4.28 Show that $\int_0^t \text{sgn}(B(s))\, dB(s)$ is a Brownian motion.

We can push the idea behind Tanaka's formula a lot further. Let f be the difference of two convex functions; then f has a left derivative f_l' (see e.g. Dudley [84], pp. 158–9). If f_l'' is its second derivative (in the sense of distributions) then we have the following generalisation of Itô's formula for arbitrary real-valued semimartingales:

Theorem 4.4.29 (Meyer–Itô) *If $X = (X(t), t \geq 0)$ is a real-valued semimartingale and f is the difference of two convex functions, then, for each $x \in \mathbb{R}$, there exists an adapted process $(L(x, t), t \geq 0)$ such that, for each $t \geq 0$, with probability 1 we have*

$$f(X(t)) = f(X(0)) + \int_0^t f_l'(X(s-)) dX(s) + \frac{1}{2} \int_{-\infty}^\infty f_l''(x) L(x, t) dx$$

$$+ \sum_{0 \leq s \leq t} [f(X(s)) - f(X(s-)) - \Delta X(s) f_l'(X(s-))].$$

The quantity $(L(x, t), t \geq 0)$ is called the local time of the semimartingale X at the point x. In the case where X is a Lévy process, it clearly coincides with the notion introduced in Subsection 1.5.3.

4.5 Notes and further reading

Stochastic integration for adapted processes against Brownian motion was first developed by Itô [143] and his famous lemma was established in [144]. The extension of stochastic integration to square-integrable martingales is due to Kunita and Watanabe [181] while Meyer [224] took the next step in generalising to semimartingales. Any book with 'stochastic calculus' or 'stochastic differential equations' in the title contains an account of stochastic integration, with varying levels of difficulty. See e.g. Øksendal [241], Mikosch [228], Gihman and Skorohod [118], Liptser and Shiryaev [200] for Brownian motion; Karatzas and Shreve [167], Durrett [85], Krylov [178], Rogers and Williams [262] and Kunita [182] for continuous semimartingales; and Jacod and Shiryaev [151], Ikeda and Watanabe [140], Protter [255], Métivier [221] and Klebaner [170] for semimartingales with jumps. Millar's article [229] is interesting for Lévy stochastic integrals. For Wiener–Lévy stochastic integrals where the noise is a general infinitely divisible random measure, see Rajput and Rosiński [256].

Dinculeanu [80] utilises the concept of semivariation to unify Lebesgue–Stieltjes integration with stochastic integration for Banach-space-valued processes.

5

Exponential martingales, change of measure and financial applications

Summary We begin this chapter by studying two different types of 'exponential' of a Lévy-type stochastic integral Y. The first of these is the stochastic exponential, $dZ(t) = Z(t-)dY(t)$, and the second is the process e^Y. We are particularly interested in identifying conditions under which e^Y is a martingale. It can then be used to implement a change to an equivalent measure. This leads to Girsanov's theorem, and an important special case of this is the Cameron–Martin–Maruyama theorem, which underlies analysis in Wiener space. In Section 5.3 we prove the martingale representation theorem in the Brownian case and also discuss extensions to include jump processes. The final section of this chapter briefly surveys some applications to option pricing. We discuss the search for equivalent risk-neutral measures within a general 'geometric Lévy process' stock price model. In the Brownian case, we derive the Black–Scholes pricing formula for a European option. In the general case, where the market is incomplete, we discuss the Föllmer–Schweitzer minimal measure and Esscher transform approaches. The case where the market is driven by a hyperbolic Lévy process is discussed in some detail.

In this chapter, we will explore further important properties of stochastic integrals, particularly the implications of Itô's formula. Many of the developments which we will study here, although of considerable theoretical interest in their own right, are also essential tools in mathematical finance as we will see in the final section of this chapter. Throughout, we will for simplicity take $d = 1$ and deal with Lévy-type stochastic integrals $Y = (Y(t), t \geq 0)$ of the form (4.13) having the stochastic differential

$$dY(t) = G(t)dt + F(t)dB(t) + H(t,x)\tilde{N}(dt,dx)$$
$$+ K(t,x)N(dt,dx).$$

5.1 Stochastic exponentials

In this section we return to a question raised in Exercise 4.4.11, i.e. the problem of finding an adapted process $Z = (Z(t), t \geq 0)$ that has a stochastic differential

$$dZ(t) = Z(t-)dY(t).$$

The solution of this problem is obtained as follows. We take Z to be the *stochastic exponential* (sometimes called the *Doléans–Dade exponential* after its discoverer), which is denoted as $\mathcal{E}_Y = (\mathcal{E}_Y(t), t \geq 0)$ and defined as

$$\mathcal{E}_Y(t) = \exp\left\{Y(t) - \frac{1}{2}[Y_c, Y_c](t)\right\} \prod_{0 \leq s \leq t} [1 + \Delta Y(s)]e^{-\Delta Y(s)} \quad (5.1)$$

for each $t \geq 0$.

We will need the following assumption:

(SE) $\inf\{\Delta Y(t), t > 0\} > -1$ (a.s.).

Proposition 5.1.1 *If Y is a Lévy-type stochastic integral of the form (4.13) and (SE) holds, then each $\mathcal{E}_Y(t)$ is almost surely finite.*

Proof We must show that the infinite product in (5.1) converges almost surely. We write

$$\prod_{0 \leq s \leq t} [1 + \Delta Y(s)]e^{-\Delta Y(s)} = A(t) + B(t),$$

where

$$A(t) = \prod_{0 \leq s \leq t} [1 + \Delta Y(s)]e^{-\Delta Y(s)} \chi_{\{|\Delta Y(s)| \geq 1/2\}}$$

and

$$B(t) = \prod_{0 \leq s \leq t} [1 + \Delta Y(s)]e^{-\Delta Y(s)} \chi_{\{|\Delta Y(s)| < 1/2\}}.$$

Now, since Y is càdlàg, $\#\{0 \leq s \leq t; |\Delta Y(s)| \geq 1/2\} < \infty$ (a.s.), and so $A(t)$ is almost surely a finite product. Using the assumption (SE), we have

$$B(t) = \exp\left(\sum_{0 \leq s \leq t} \{\log[1 + \Delta Y(s)] - \Delta Y(s)\} \chi_{\{|\Delta Y(s)| < 1/2\}}\right).$$

We now employ Taylor's theorem to obtain the inequality

$$\log(1+y) - y \leq Ky^2$$

where $K > 0$, which is valid whenever $|y| < 1/2$. Hence

$$\left| \sum_{0 \leq s \leq t} \{\log[1 + \Delta Y(s)] - \Delta Y(s)\} \chi_{\{|\Delta Y(s)| < 1/2\}} \right|$$

$$\leq \sum_{0 \leq s \leq t} |\Delta Y(s)|^2 \chi_{\{|\Delta Y(s)| < 1/2\}} < \infty \quad \text{a.s.,}$$

by Corollary 4.4.9, and we have our required result. □

Of course (SE) ensures that $\mathcal{E}_Y(t) > 0$ (a.s.).

Note 1 In the next chapter we will see that the stochastic exponential is in fact the unique solution of the stochastic differential equation $dZ(t) = Z(t-)dY(t)$ with initial condition $Z(0) = 1$ (a.s.).

Note 2 The restrictions (SE) can be dropped and the stochastic exponential extended to the case where Y is an arbitrary (real-valued) càdlàg semimartingale, but the price we have to pay is that $\mathcal{E}_Y$ may then take negative values. See Jacod and Shiryaev [151], pp. 58–61, for details, and also for a further extension to the case of complex Y.

Exercise 5.1.2 Establish the following alternative form of (5.1):

$$\mathcal{E}_Y(t) = e^{S_Y(t)},$$

where

$$dS_Y(t) = F(t)dB(t) + \left[G(t) - \tfrac{1}{2}F(t)^2\right]dt$$

$$+ \int_{|x| \geq 1} \log[1 + K(t,x)] \, N(dt, dx)$$

$$+ \int_{|x| < 1} \log[1 + H(t,x)] \, \tilde{N}(dt, dx)$$

$$+ \int_{|x| < 1} \{\log[1 + H(t,x)] - H(t,x)\} \nu(dx) ds. \quad (5.2)$$

Theorem 5.1.3 *We have*

$$d\mathcal{E}_Y(t) = \mathcal{E}_Y(t-)dY(t).$$

Proof We apply Itô's formula to the result of Exercise 5.1.2 to obtain, for each $t \geq 0$,

$$\begin{aligned}
d\mathcal{E}_Y(t) = \mathcal{E}_Y(t-) \Bigg[& F(t)dB(t) + G(t)dt \\
& + \int_{|x|<1} \{\log[1 + H(t,x)] - H(t,x))\} \nu(dx) dt \\
& + \int_{|x|\geq 1} (\exp\{S_Y(t-) + \log[1 + K(t,x)]\} \\
& \quad - \exp[S_Y(t-)]) N(dt, dx) \\
& + \int_{|x|<1} (\exp\{S_Y(t-) + \log[1 + H(t,x)]\} \\
& \quad - \exp[S_Y(t-)]) \tilde{N}(dt, dx) \\
& + \int_{|x|<1} (\exp\{S_Y(t-) + \log[1 + H(t,x)]\} \\
& \quad - \exp[S_Y(t-)] \\
& \quad - \log[1 + H(t,x)][\exp S_Y(t-)] \nu(dx) dt \Bigg] \\
= \mathcal{E}_Y(t-) [& F(t)dB(t) + G(t)dt + K(t,x) N(dt,dx) \\
& + H(t,x) \tilde{N}(dt,dx)],
\end{aligned}$$

as required. □

Exercise 5.1.4 Let X and Y be Lévy-type stochastic integrals. Show that, for each $t \geq 0$,

$$\mathcal{E}_X(t) \mathcal{E}_Y(t) = \mathcal{E}_{X+Y+[X,Y]}(t).$$

Exercise 5.1.5 Let $Y = (Y(t), t \geq 0)$ be a compound Poisson process, so that each $Y(t) = X_1 + \cdots + X_{N(t)}$, where $(X_n, n \in \mathbb{N})$ are i.i.d. and N is an independent Poisson process. Deduce that, for each $t \geq 0$,

$$\mathcal{E}_Y(t) = \prod_{j=1}^{N(t)}(1 + X_j).$$

5.2 Exponential martingales

In this section our first goal is to find conditions under which $e^Y = (e^{Y(t)}, t \geq 0)$ is a martingale, where Y is as usual a Lévy-type stochastic integral. Such processes are an important source of Radon–Nikodým derivatives for changing the measure as described by Girsanov's theorem, and this leads to the Cameron–Martin–Maruyama formula, which underlies 'infinite-dimensional analysis' in Wiener space as well as being a vital tool in the derivation of the Black–Scholes formula in mathematical finance.

5.2.1 Lévy-type stochastic integrals as local martingales

Our first goal is to find necessary and sufficient conditions for a Lévy-type stochastic integral Y to be a local martingale. First we impose some conditions on K and G:

(LM1) for each $t > 0$, $K \in \mathcal{P}_2(t, E^c)$;
(LM2) $\mathbb{E}\left(\int_0^t \int_{|x| \geq 1} |K(s, x)| \nu(dx) ds\right) < \infty$;
(LM3) $G^{1/2} \in \mathcal{H}_2(t)$ for each $t > 0$.

(Note that $E = \hat{B} - \{0\}$ throughout this section.)

From (LM1) and (LM2), it follows that $\int_0^t \int_{|x| \geq 1} |K(s, x)| \nu(dx) ds < \infty$ (a.s.) and that we can write

$$\int_0^t \int_{|x| \geq 1} K(s, x) N(dx, ds)$$
$$= \int_0^t \int_{|x| \geq 1} K(s, x) \tilde{N}(dx, ds) + \int_0^t \int_{|x| \geq 1} K(s, x) \nu(dx) ds,$$

for each $t \geq 0$, the compensated integral being a local martingale.

Theorem 5.2.1 *If Y is a Lévy-type stochastic integral of the form (4.13) and the assumptions* (LM1) *to* (LM3) *are satisfied, then Y is a local martingale if and only if*

$$G(t) + \int_{|x| \geq 1} K(t, x) \nu(dx) = 0 \quad \text{a.s.,}$$

for (Lebesgue) almost all $t \geq 0$.

5.2 Exponential martingales

Proof First assume that Y is a local martingale with respect to the stopping times $(T_n, n \in \mathbb{N})$. Then, for each $n \in \mathbb{N}, 0 \leq s < t < \infty$,

$$Y(t \wedge T_n)$$
$$= Y(s \wedge T_n) + \int_{s \wedge T_n}^{t \wedge T_n} F(u) dB(u) + \int_{s \wedge T_n}^{t \wedge T_n} \int_{|x|<1} H(u,x) \tilde{N}(du, dx)$$
$$+ \int_{s \wedge T_n}^{t \wedge T_n} \int_{|x|>1} K(u,x) \tilde{N}(du, dx)$$
$$+ \int_{s \wedge T_n}^{t \wedge T_n} \left[G(u) + \int_{|x| \geq 1} K(u,x) v(dx) \right] du.$$

Now, for each $n \in \mathbb{N}$, $(Y(t \wedge T_n), t \geq 0)$ is a martingale, so we have

$$\mathbb{E}_s \left(\int_{s \wedge T_n}^{t \wedge T_n} \left[G(u) + \int_{|x| \geq 1} K(u,x) v(dx) \right] du \right) = 0.$$

We take the limit as $n \to \infty$ and, using the fact that by (LM1) and (LM3)

$$\left| \int_{s \wedge T_n}^{t \wedge T_n} \left[G(u) + \int_{|x| \geq 1} K(u,x) v(dx) \right] du \right|$$
$$\leq \int_0^t \left| G(u) + \int_{|x| \geq 1} K(u,x) v(dx) \right| du < \infty \quad \text{a.s.,}$$

together with the conditional version of dominated convergence (see e.g. Williams [304], p. 88), we deduce that

$$\mathbb{E}_s \left(\int_s^t \left[G(u) + \int_{|x| \geq 1} K(u,x) v(dx) \right] du \right) = 0.$$

Conditions (LM2) and (LM3) ensure that we can use the conditional Fubini theorem 1.1.7 to obtain

$$\int_s^t \mathbb{E}_s \left(G(u) + \int_{|x| \geq 1} K(u,x) v(dx) \right) du = 0.$$

It follows that

$$\lim_{h\to 0}\frac{1}{h}\int_s^{s+h}\mathbb{E}_s\left(G(u)+\int_{|x|\geq 1}K(u,x)\nu(dx)\right)du=0,$$

and hence by Lebesgue's differentiation theorem (see e.g. Cohn [73], p. 187) we have

$$\mathbb{E}_s\left(G(s)+\int_{|x|\geq 1}K(s,x)\nu(dx)\right)=0$$

for (Lebesgue) almost all $s \geq 0$. But $G(\cdot)+\int_{|x|\geq 1}K(\cdot,x)\nu(dx)$ is adapted, and the result follows. The converse is immediate. □

Note that, in particular, Y is a martingale if $F \in \mathcal{H}_2(t)$, $H \in \mathcal{H}_2(t, E)$ and $K \in \mathcal{H}_2(t, E^c)$ for all $t \geq 0$.

5.2.2 Exponential martingales

We continue to study Lévy-type stochastic integrals satisfying (LM1) to (LM3). We now turn our attention to the process $e^Y = (e^{Y(t)}, t \geq 0)$ (cf. Exercise 4.4.11).

By Itô's formula, we find, for each $t \geq 0$,

$$\begin{aligned}e^{Y(t)} &= 1 + \int_0^t e^{Y(s-)}F(s)dB(s) + \int_0^t\int_{|x|<1}e^{Y(s-)}\left(e^{H(s,x)}-1\right)\tilde{N}(ds,dx)\\ &+ \int_0^t\int_{|x|\geq 1}e^{Y(s-)}(e^{K(s,x)}-1)\tilde{N}(ds,dx)\\ &+ \int_0^t e^{Y(s-)}\bigg\{G(s)\frac{1}{2}F(s)^2+\int_{|x|<1}\left[e^{H(s,x)}-1-H(s,x)\right]\nu(dx)\\ &\quad+\int_{|x|\geq 1}(e^{K(s,x)}-1)\nu(dx)\bigg\}ds.\end{aligned} \quad (5.3)$$

Hence, by an immediate application of Theorem 5.2.1, we have:

5.2 Exponential martingales

Corollary 5.2.2 e^Y *is a local martingale if and only if*

$$G(s) + \tfrac{1}{2}F(s)^2 + \int_{|x|<1}\left[e^{H(s,x)} - 1 - H(s,x)\right]\nu(dx)$$
$$+ \int_{|x|\geq 1}(e^{K(s,x)} - 1)\nu(dx) = 0 \qquad (5.4)$$

almost surely and for (Lebesgue) almost all $s \geq 0$.

So e^Y is a local martingale if and only if for (Lebesgue) almost all $t \geq 0$,

$$e^{Y(t)} = 1 + \int_0^t e^{Y(s-)}F(s)dB(s)$$
$$+ \int_0^t \int_{|x|<1} e^{Y(s-)}(e^{H(s,x)} - 1)\tilde{N}(ds, dx)$$
$$+ \int_0^t \int_{|x|\geq 1} e^{Y(s-)}(e^{K(s,x)} - 1)\tilde{N}(ds, dx). \qquad (5.5)$$

We would like to go further and establish conditions under which e^Y is in fact a martingale. First we need the following general result about supermartingales.

Lemma 5.2.3 *If $M = (M(t), t \geq 0)$ is a supermartingale for which the mapping $t \to \mathbb{E}(M(t))$ is constant, then M is a martingale.*

Proof We follow the argument of Liptser and Shiryaev [200], p. 228.
Fix $0 < s < t < \infty$, let $A = \{\omega \in \Omega; \, \mathbb{E}_s(M(t))(\omega) < M(s)(\omega)\}$ and assume that $P(A) > 0$. Then

$$\mathbb{E}(M(t)) = \mathbb{E}(\mathbb{E}_s(M(t)))$$
$$= \mathbb{E}(\chi_A \, \mathbb{E}_s(M(t))) + \mathbb{E}((1 - \chi_A) \, \mathbb{E}_s(M(t)))$$
$$< \mathbb{E}(\chi_A M(s)) + \mathbb{E}((1 - \chi_A)M(s))$$
$$= \mathbb{E}(M(s)),$$

which contradicts the fact that $t \to \mathbb{E}(M(t))$ is constant. Hence $P(A) = 0$ and the result follows. □

From now on we assume that the condition (5.4) is satisfied for all $t \geq 0$, so that e^Y is a local martingale.

Theorem 5.2.4 *If Y is a Lévy-type stochastic integral of the form* (4.13) *which is such that e^Y is a local martingale, then e^Y is a martingale if and only if $\mathbb{E}(e^{Y(t)}) = 1$ for all $t \geq 0$.*

Proof Let $(T_n, n \in \mathbb{N})$ be the sequence of stopping times such that $(e^{Y(t \wedge T_n)}, t \geq 0)$ is a martingale; then, by the conditional form of Fatou's lemma (see e.g. Williams [304], p. 88), we have for each $0 \leq s < t < \infty$

$$\mathbb{E}_s(e^{Y(t)}) \leq \liminf_{n \to \infty} \mathbb{E}_s(e^{Y(t \wedge T_n)})$$
$$= \liminf_{n \to \infty} e^{Y(s \wedge T_n)} = e^{Y(s)},$$

so e^Y is a supermartingale. Now if we assume that the expectation is identically unity, it follows that e^Y is a martingale by Lemma 5.2.3. The converse is immediate from equation (5.5). □

For the remainder of this section, we will assume that the condition of Theorem 5.2.4 is valid (see also Kunita [185], Theorems 1.3 and 1.6). Under this constraint the process e^Y given by equation (5.5) is called an *exponential martingale*. Two important examples are:

Example 5.2.5 (The Brownian case) Here Y is a Brownian integral of the form

$$Y(t) = \int_0^t F(s) dB(s) + \int_0^t G(s) ds$$

for each $t \geq 0$. The unique solution to (5.4) is $G(t) = -\frac{1}{2}F(t)^2$ (a.e.). We then have, for each $t \geq 0$,

$$e^{Y(t)} = \exp\left(\int_0^t F(s) dB(s) - \frac{1}{2}\int_0^t F(s)^2 ds\right).$$

Example 5.2.6 (The Poisson case) Here Y is a Poisson integral driven by a Poisson process N of intensity λ and has the form

$$Y(t) = \int_0^t K(s) dN(s) + \int_0^t G(s) ds$$

for each $t \geq 0$.

The unique solution to (5.4) is $G(t) = -\lambda \int_0^t (e^{K(s)} - 1) ds$ (a.e.). For each $t \geq 0$, we obtain

$$e^{Y(t)} = \exp\left[\int_0^t K(s) dN(s) - \lambda \int_0^t (e^{K(s)} - 1) ds\right].$$

For the Brownian case, a more direct condition for e^Y to be a martingale than Theorem 5.2.4 is established in Liptser and Shiryaev [200], pp. 229–32. More precisely, it is shown that

$$\mathbb{E}\left(\exp\left(\int_0^\infty \tfrac{1}{2} F(s)^2 ds\right)\right) < \infty,$$

is a sufficient condition, called the *Novikov criterion*. More general results that establish conditions for

$$\left(\exp\left[M(t) - \tfrac{1}{2}\langle M, M\rangle(t)\right], t \geq 0\right)$$

to be a martingale, where M is an arbitrary continuous local martingale, can be found in, for example, Revuz and Yor [260], pp. 307–9, Durrett [85], pp. 108–9, and Chung and Williams [71], pp. 120–3.

Exercise 5.2.7 Let Y be a Lévy process with characteristics (b, a, ν) for which the generating function $\mathbb{E}(e^{uY(t)}) < \infty$ for all $t, u \geq 0$. Choose the parameter b to be such that condition (5.4) is satisfied, and hence show that e^Y is a martingale.

Exercise 5.2.8 Let Y be a Lévy-type stochastic integral. Show that e^Y coincides with the stochastic exponential $\mathcal{E}_Y$ if and only if Y is a Brownian integral.

5.2.3 Change of measure – Girsanov's theorem

If we are given two distinct probability measures P and Q on $(\Omega, \mathcal{F})$, we will write $\mathbb{E}_P$ ($\mathbb{E}_Q$) to denote expectation with respect to P (respectively, Q). We also use the terminology P-martingale, P-Brownian motion etc. when we want to emphasise that P is the operative measure. We remark that Q and P are each also probability measures on $(\Omega, \mathcal{F}_t)$, for each $t \geq 0$, and we will use the notation Q_t and P_t when the measures are restricted in this way. Suppose that $Q \ll P$; then each $Q_t \ll P_t$ and we sometimes write

$$\left.\frac{dQ}{dP}\right|_t = \frac{dQ_t}{dP_t}.$$

Lemma 5.2.9 $\left(\left.\dfrac{dQ}{dP}\right|_t, t \geq 0\right)$ is a P-martingale.

Proof For each $t \geq 0$, let $M(t) = \left.\dfrac{dQ}{dP}\right|_t$. For all $0 \leq s \leq t$, $A \in \mathcal{F}_s$,

$$\mathbb{E}_P\big(\chi_A\, \mathbb{E}_P(M(t)|\mathcal{F}_s)\big) = \mathbb{E}_P(\chi_A M(t))$$
$$= \mathbb{E}_{P_t}(\chi_A M(t)) = \mathbb{E}_{Q_t}(\chi_A)$$
$$= \mathbb{E}_{Q_s}(\chi_A) = \mathbb{E}_{P_s}(\chi_A M(s))$$
$$= \mathbb{E}_P(\chi_A M(s)).$$

$\square$

Now let e^Y be an exponential martingale. Then, since $\mathbb{E}_P(e^{Y(t)}) = \mathbb{E}_{P_t}(e^{Y(t)}) = 1$, we can define a probability measure Q_t on $(\Omega, \mathcal{F}_t)$ by

$$\frac{dQ_t}{dP_t} = e^{Y(t)}, \tag{5.6}$$

for each $t \geq 0$.

From now on, we will find it convenient to fix a time interval $[0, T]$. We write $P = P_T$ and $Q = Q_T$.

Before we establish Girsanov's theorem, which is the key result of this section, we need a useful lemma.

Lemma 5.2.10 $M = (M(t), 0 \leq t \leq T)$ *is a local Q-martingale if and only if* $Me^Y = (M(t)e^{Y(t)}, 0 \leq t \leq T)$ *is a local P-martingale.*

Proof We will establish a weaker result and show that M is a Q-martingale if and only if Me^Y is a P-martingale. We leave it to the reader to insert the appropriate stopping times.

Let $A \in \mathcal{F}_s$ and assume that M is a Q-martingale; then, for each $0 \leq s < t < \infty$,

$$\int_A M(t)e^{Y(t)} dP = \int_A M(t)e^{Y(t)} dP_t = \int_A M(t) dQ_t$$
$$= \int_A M(t) dQ = \int_A M(s) dQ = \int_A M(s) dQ_s$$
$$= \int_A M(s)e^{Y(s)} dP_s = \int_A M(s)e^{Y(s)} dP.$$

The converse is proved in the same way. $\square$

5.2 Exponential martingales

In the following we take Y to be a Brownian integral, so that for each $0 \leq t \leq T$

$$e^{Y(t)} = \exp\left[\int_0^t F(s)dB(s) - \frac{1}{2}\int_0^t F(s)^2 ds\right].$$

We define a new process $W = (W(t), 0 \leq t \leq T)$ by

$$W(t) = B(t) - \int_0^t F(s)ds,$$

for each $t \geq 0$.

Theorem 5.2.11 (Girsanov) *W is a Q-Brownian motion.*

Proof We follow the elegant proof given by Hsu in [134].

First we use Itô's product formula (Theorem 4.4.13) to find that, for each $0 \leq t \leq T$,

$$\begin{aligned}
d[W(t)e^{Y(t)}] &= dW(t)e^{Y(t)} + W(t)de^{Y(t)} + dW(t)de^{Y(t)} \\
&= e^{Y(t)}dB(t) - e^{Y(t)}F(t)dt + W(t)e^{Y(t)}F(t)dB(t) + e^{Y(t)}F(t)dt \\
&= e^{Y(t)}[1 - W(t)F(t)]dB(t).
\end{aligned}$$

Hence We^Y is a P-local martingale and so (by Lemma 5.2.10) W is a Q-local martingale. Moreover, since $W(0) = 0$ (a.s.), we see that W is centred (with respect to Q).

Now define $Z = (Z(t), 0 \leq t \leq T)$ by $Z(t) = W(t)^2 - t$; then, by another application of Itô's product formula, we find

$$dZ(t) = 2W(t)dW(t) - dt + dW(t)^2.$$

But $dW(t)^2 = dt$ and so Z is also a Q-local martingale. The result now follows from Lévy's characterisation of Brownian motion (Theorem 4.4.18 and Exercise 4.4.19). □

Exercise 5.2.12 Show that Girsanov's theorem continues to hold when e^Y is any exponential martingale with a Brownian component (so that F is not identically zero).

Exercise 5.2.13 Let $M = (M(t), 0 \leq t \leq T)$ be a local P-martingale of the form

$$M(t) = \int_0^t \int_{|x|<1} L(x,s) \tilde{N}(ds, dx),$$

where $L \in \mathcal{P}_2(t, E)$. Let e^Y be an exponential martingale. Use Lemma 5.2.10 to show that $N = (N(t), 0 \leq t \leq T)$ is a local Q-martingale, where

$$N(t) = M(t) - \int_0^t \int_{|x|<1} L(x,s)(e^{H(s,x)} - 1)\nu(dx)ds,$$

and we are assuming that the integral exists. A sufficient condition for this is that $\int_0^t \int_{|x|<1} |e^{H(s,x)} - 1|^2 \nu(dx) ds < \infty$. (Hint: Apply Lemma 5.2.10 and Itô's product formula.)

Quite abstract generalisations of Girsanov's theorem to general semimartingales can be found in Jacod and Shiryaev [151], pp. 152–66, and in Protter [255], pp. 108–10. The results established above, namely Theorem 5.2.11 and Exercises 5.2.12 and 5.2.13, will be adequate for all the applications we will consider.

Readers may be tempted to take the seemingly natural step of extending Girsanov's theorem to the whole of $\mathbb{R}^+$. Beware, this is fraught with difficulty! For a nice discussion of the pitfalls, see Bichteler [43], pp. 162–8.

5.2.4 Analysis on Wiener space

The Cameron–Martin–Maruyama theorem

In this subsection, we will continue to restrict all random motion to a finite time interval $I = [0, T]$.

We introduce the *Wiener space*,

$$\mathcal{W}_0(I) = \{\omega : I \to \mathbb{R}; \omega \text{ is continuous and } \omega(0) = 0\}.$$

Let $\mathcal{F}$ be the σ-algebra generated by cylinder sets and define a process $B = (B(t), t \in I)$ by $B(t)\omega = \omega(t)$ for each $t \geq 0, \omega \in \mathcal{W}_0(I)$. We have already mentioned Wiener's famous result, which asserts the existence of a probability measure P (usually called the *Wiener measure*) on $(\mathcal{W}_0(I), \mathcal{F})$ such that B is a standard Brownian motion; see [302, 301] for Wiener's justly celebrated original work on this).

Our first task is to consider a very important special case of the Girsanov theorem in this context, but first we need some preliminaries.

We define the *Cameron–Martin space* $\mathbb{H}(I)$ to be the set of all $h \in W_0(I)$ for which h is absolutely continuous with respect to Lebesgue measure and $\dot{h} \in L^2(I)$, where $\dot{h} = dh/dt$. Then $\mathbb{H}(I)$ is a Hilbert space with respect to the inner product

$$(h_1, h_2)_{\mathbb{H}} = \int_0^T \dot{h}_1(s)\dot{h}_2(s)ds,$$

and we denote the associated norm as $||h||_{\mathbb{H}}$.

We also need to consider translation in Wiener space so, for each $\phi \in \mathcal{W}_0(I)$, define $\tau_\phi : \mathcal{W}_0(I) \to \mathcal{W}_0(I)$ by

$$\tau_\phi(\omega) = \omega + \phi,$$

for each $\omega \in \mathcal{W}_0(I)$.

Since each τ_ϕ is measurable we can interpret it as a $\mathcal{W}_0(I)$-valued random variable with law P^ϕ, where $P^\phi(A) = P((\tau_\phi)^{-1}(A))$ for each $A \in \mathcal{F}$.

The final idea we will require is that of cylinder functions. Let $F : \mathcal{W}_0(I) \to \mathbb{R}$ be such that, for some $n \in \mathbb{N}$, there exists $f \in C^\infty(\mathbb{R}^n)$ and $0 < t_1 < \cdots < t_n \leq T$ such that

$$F(\omega) = f(\omega(t_1), \ldots, \omega(t_n)) \tag{5.7}$$

for each $\omega \in \mathcal{W}_0(I)$. We assume further that, for each $m \in \mathbb{N} \cup \{0\}$, $f^{(m)}$ is polynomially bounded, i.e., for each $x = (x_1, \ldots, x_n) \in \mathbb{R}^n$,

$$|f^{(m)}(x_1, \ldots, x_n)| \leq p_m(|x_1|, \ldots, |x_n|)$$

where p_m is a polynomial.

We call such an F a *cylinder function*. The set of all cylinder functions, which we denote as $\mathcal{C}(I)$, is dense in $L^p(\mathcal{W}_0(I), \mathcal{F}, P)$ for all $1 \leq p < \infty$; see e.g. Huang and Yan [135], pp. 62–3, for a proof of this.

Theorem 5.2.14 (Cameron–Martin–Maruyama) *If $h \in \mathbb{H}(I)$, then P^h is absolutely continuous with respect to P and*

$$\frac{dP^h}{dP} = \exp\left[\int_0^T \dot{h}(s)dB(s) - \tfrac{1}{2}||h||_{\mathbb{H}}^2\right].$$

Proof First note that by Lemma 4.3.11 we have

$$\mathbb{E}\left(\exp\left[\int_0^T \dot{h}(s)dB(s)\right]\right) = \exp\left(\tfrac{1}{2}||h||_{\mathbb{H}}^2\right).$$

By Theorem 5.2.4, $\left(\exp\left[\int_0^t \dot{h}(s)dB(s) - \tfrac{1}{2}\int_0^t \dot{h}(s)^2 ds\right], 0 \leq t \leq T\right)$ is a martingale, and so we can assert the existence of a probability measure Q on

($\mathcal{W}_0(I), \mathcal{F}$) such that

$$\frac{dQ}{dP} = \exp\left[\int_0^T \dot{h}(s)dB(s) - \tfrac{1}{2}\|h\|_{\mathbb{H}}^2\right].$$

Now, by Girsanov's theorem, W is a Q-Brownian motion where each $W(t) = B(t) - h(t)$. Let $F \in \mathcal{C}(I)$, and for ease of notation we will assume that $n = 1$, so that $F(\omega) = f(\omega(t))$ for some $0 < t \leq T$, for each $\omega \in \mathcal{W}_0(I)$. We then have

$$\mathbb{E}_Q(f(W(t))) = \mathbb{E}_P(f(B(t))).$$

Hence

$$\int_{\mathcal{W}_0(I)} f(B(t)(\omega - h))dQ(\omega) = \int_{\mathcal{W}_0(I)} f(B(t)(\omega))dQ(\omega + h)$$

$$= \int_{\mathcal{W}_0(I)} f(B(t)(\omega))dP(\omega),$$

and so

$$\int_{\mathcal{W}_0(I)} f(W(t)(\omega))dP^h(\omega) = \int_{\mathcal{W}_0(I)} f(B(t)(\omega - h))dP(\omega - h)$$

$$= \int_{\mathcal{W}_0(I)} f(B(t)(\omega))dP(\omega)$$

$$= \int_{\mathcal{W}_0(I)} f(B(t)(\omega - h))dQ(\omega)$$

$$= \int_{\mathcal{W}_0(I)} f(W(t)(\omega))\frac{dQ}{dP}(\omega)dP(\omega).$$

This extends to $f \in L^p(\mathcal{W}_0(I), \mathcal{F}, P)$ by a straightforward limiting argument and the required result follows immediately. □

In Stroock [291], pp. 287–8, it is shown that the condition $h \in \mathbb{H}(I)$ is both necessary and sufficient for P^h to be absolutely continuous with respect to h.

Directional derivative and integration by parts

An important development within the emerging field of infinite-dimensional analysis is the idea of differentiation of Wiener functionals. We give a brief insight into this, following the excellent exposition of Hsu [134].

5.2 Exponential martingales

Let $F \in \mathcal{C}(I)$ be of the form (5.7); then it is natural to define a directional derivative along $\phi \in \mathcal{W}_0(I)$ as

$$(D_\phi F)(\omega) = \lim_{\epsilon \to 0} \frac{F(\tau^{\epsilon\phi})(\omega) - F(\omega)}{\epsilon},$$

for each $\omega \in \mathcal{W}_0(I)$. It is then easy to verify that

$$(D_\phi F)(\omega) = \sum_{i=1}^n \partial_i F(\omega)\, \phi(t_i),$$

where

$$\partial_i F(\omega) = (\partial_i f)(\omega(t_1), \ldots, \omega(t_n)),$$

and ∂_i is the usual partial derivative with respect to the ith variable ($1 \le i \le n$). Hence, from an analytic point of view, D_ϕ is a densely defined linear operator in $L^p(\mathcal{W}_0(I), \mathcal{F}, P)$ for all $1 \le p < \infty$.

To be able to use the directional derivative effeceively, we need it to have some stronger properties. As we will see below, these become readily available when we make the requirement that $\phi \in \mathbb{H}(I)$. The key is the following result, which is usually referred to as *integration by parts in Wiener space*.

Theorem 5.2.15 (Integration by parts) *For all $h \in \mathbb{H}(I)$ and all $F, G \in \mathcal{C}(I)$,*

$$\mathbb{E}\big((D_h F)(G)\big) = \mathbb{E}(F(D_h^* G)),$$

where

$$D_h^* G = -D_h G + \int_0^T \dot{h}(s)\, dB(s).$$

Proof For each $\epsilon \in \mathbb{R}$ we have, by Theorem 5.2.14,

$$\mathbb{E}((F \circ \tau^{\epsilon h}) G) = \int_{\mathcal{W}_0(I)} F(\omega + \epsilon h) G(\omega)\, dP(\omega)$$

$$= \int_{\mathcal{W}_0(I)} F(\omega) G(\omega - \epsilon h)\, dP^{\epsilon h}(\omega)$$

$$= \int_{\mathcal{W}_0(I)} F(\omega) G(\omega - \epsilon h) \frac{dP^{\epsilon h}}{dP}(\omega)\, dP(\omega).$$

Now subtract $\mathbb{E}(FG)$ from both sides and pass to the limit as $\epsilon \to 0$. The required result follows when we use Theorem 5.2.14 to write

$$\frac{dP^{\epsilon h}}{dP} = \exp\left[\epsilon \int_0^T \dot{h}(s)dB(s) - \frac{\epsilon^2}{2}\|h\|_{\mathbb{H}}^2\right].$$

Note that the interchange of limit and integral is justified by dominated convergence, where we utilise the facts that cylinder functions are polynomially bounded and that Brownian paths are Gaussian and so have moments to all orders. □

Readers with a functional analysis background will be interested in knowing that each D_h is closable for each $1 < p < \infty$ and that $\mathcal{C}(I)$ is a core for the closure. Details can be found in Hsu [134].

Infinite-dimensional analysis based on the study of Wiener space is a deep and rapidly developing subject, which utilises techniques from probability theory, analysis and differential geometry. For further study, try Nualart [239], Stroock [292], Huang and Yan [135] or Malliavin [208].

5.3 Martingale representation theorems

Prior to the section on mathematical finance, it will be helpful to give a quick proof of the martingale representation theorem for the case of one-dimensional Brownian motion. In this section, we again fix $T > 0$ and all processes are adapted to a fixed filtration $(\mathcal{F}_t, 0 \le t \le T)$.

In the following we will find it convenient to take Ω to be Wiener space $\mathcal{W}_0([0, T])$ equipped with its usual filtration and the Wiener measure P. We will need to consider simultaneously several different Hilbert spaces. We define $L_0^2 = \{X \in L^2(\Omega, \mathcal{F}, P); \mathbb{E}(X) = 0\}$; then it is easy to verify that L_0^2 is a closed subspace of $L^2(\Omega, \mathcal{F}, P)$. Let $F \in \mathcal{H}_2(T)$, and recall the stochastic integral

$$I_T(F) = \int_0^T F(s)dB(s).$$

It was shown in Section 4.2 that I_T is a linear isometry from $\mathcal{H}_2(T)$ into L_0^2 and we denote the range of this isometry as h_T. Note that h_T is itself a closed subspace of L_0^2. Our main theorem will state that in fact $h_T = L_0^2$. This result has a very interesting probabilistic consequence. We follow the elegant exposition by Parthasarathy [248].

One of our key tools will be the martingales introduced in Chapter 2, $M_u = (M_u(t), 0 \le t \le T)$ where $u \in \mathbb{R}$ and each

$$M_u(t) = e^{uB(t) - u^2 t/2}.$$

5.3 Martingale representation theorems

Note that these are also stochastic exponentials $M_u(t) = \mathcal{E}_{uB}(t)$, so that we have $dM_u(t) = M_u(t)dB(t)$ and in integrated form, for each $0 \leq t \leq T$,

$$M_u(t) - 1 = \int_0^t M_u(s)dB(s). \tag{5.8}$$

Exercise 5.3.1 Show that for each $u, v \in \mathbb{R}, 0 \leq t \leq T$,

$$M_{u+v}(t) = e^{-uvt} M_u(t) M_v(t).$$

If $Y \in L_0^2$, we will denote by $H_Y = (H_Y(t), 0 \leq t \leq T)$ the closed martingale given by $H_Y(t) = \mathbb{E}(Y|\mathcal{F}_t)$ for each $0 \leq t \leq T$. A key role later will be played by the process $M_{F,Y} = (M_{F,Y}(t), t \geq 0)$ where, for each $0 \leq t \leq T$,

$$M_{F,Y}(t) = H_Y(t) I_t(F) = H_Y(t) \int_0^t F(s)dB(s).$$

Lemma 5.3.2 *If $Y \in L_0^2$ is orthogonal to every random variable in h_T, then, for each $F \in \mathcal{H}_2(T)$, $M_{F,Y}$ is a martingale.*

Proof Fix $0 \leq s < u \leq T$ and let X be a bounded $\mathcal{F}_s$-measurable random variable. Define an adapted process $F_X = (F_X(t), 0 \leq t \leq T)$ by

$$F_X(t) = \begin{cases} 0 & \text{for } 0 \leq t \leq s \text{ or } u < t \leq T, \\ XF(t) & \text{for } s < t \leq u; \end{cases}$$

then we easily see that $F_X \in \mathcal{H}_2(T)$. By repeated conditioning, we find

$$0 = \mathbb{E}\left(Y \int_0^T F_X(s)dB(s)\right) = \mathbb{E}(YX(I_u(F) - I_s(F)))$$
$$= \mathbb{E}(X(\mathbb{E}_u(Y)I_u(F) - YI_s(F)))$$
$$= \mathbb{E}(X(\mathbb{E}_s(H_Y(u)I_u(F)) - H_Y(s)I_s(F))).$$

Now, bounded $\mathcal{F}_s$-measurable random variables are dense in $L^2(\Omega, \mathcal{F}_s, P)$, so we deduce that

$$\mathbb{E}_s(H_Y(u)I_u(F)) = H_Y(s)I_s(F) \quad \text{a.s.,}$$

as required. □

As promised, our main result is the following.

Theorem 5.3.3

$$L_0^2 = h_T.$$

Proof Let $Z \in L_0^2$ be orthogonal to every vector in h_T. The result is established if we can show that $Z = 0$ (a.s.). We define $H_Z(t) = \mathbb{E}_t(Z)$ for each $0 \leq t \leq T$; then, by Lemma 5.3.2 and (5.8), $(H_Z(t)M_u(t), 0 \leq t \leq T)$ is a martingale for each $u \in \mathbb{R}$.

Now fix some $n \in \mathbb{N}$ and fix also $u_1, \ldots, u_n \in \mathbb{R}$ and $0 \leq t_1 < \cdots < t_n \leq T$. We define

$$\alpha = \mathbb{E}(M_{u_1}(t_1)M_{u_2}(t_2) \cdots M_{u_n}(t_n)Z); \quad (5.9)$$

then, by repeated conditioning, Lemma 5.3.2 and Exercise 5.3.1, we get

$$\alpha = \mathbb{E}(M_{u_1}(t_1)M_{u_2}(t_2) \cdots M_{u_n}(t_n)H_Z(t_n))$$
$$= \mathbb{E}\big(M_{u_1}(t_1)M_{u_2}(t_2) \cdots M_{u_{n-1}}(t_{n-1})\mathbb{E}_{t_{n-1}}(M_{u_n}(t_n)H_Z(t_n))\big)$$
$$= e^{u_n u_{n-1} t}\mathbb{E}(M_{u_1}(t_1)M_{u_2}(t_2) \cdots M_{u_n+u_{n-1}}(t_{n-1})H_Z(t_{n-1})).$$

Now iterate the above argument to see that there exist $a, b \in \mathbb{R}$ such that

$$\alpha = a\,\mathbb{E}(M_b(t_1)H_Z(t_1)).$$

By conditioning with respect to t_1, and the fact that $\mathbb{E}(Z) = 0$, we have

$$\alpha = a\,\mathbb{E}((M_b(t_1) - 1)Z);$$

but, by (5.8), $M_b(t_1) - 1$ is orthogonal to Z and so we have $\alpha = 0$. Applying this in (5.9) and using Exercise 5.3.1 repeatedly yields

$$\int \exp\{i[u_1 B(t_1) + \cdots + u_n B(t_n)]\} Z dP = 0.$$

From this we deduce that

$$\int \exp\big(i\{v_1 B(t_1) + v_2[B(t_2) - B(t_1)] + \cdots$$
$$+ v_n[B(t_n) - B(t_{n-1})]\}\big) Z dP = 0,$$

where each $v_r = u_r + \cdots + u_n$ ($1 \leq r \leq n$). Now $(B(t_1), B(t_2) - B(t_1), \ldots,$

5.3 Martingale representation theorems

$B(t_n) - B(t_{n-1}))$ is jointly Gaussian with density

$$f_n(x) = (2\pi)^{-n/2}\big[t_1(t_2 - t_1)\cdots(t_n - t_{n-1})\big]^{-1/2}$$
$$\times \exp\left\{-\frac{1}{2}\left[\frac{x_1^2}{t_1} + \frac{(x_2 - x_1)^2}{t_2 - t_1} + \cdots + \frac{(x_n - x_{n-1})^2}{t_n - t_{n-1}}\right]\right\}$$

for each $x = (x_1, \ldots, x_n) \in \mathbb{R}^n$. So we obtain

$$\int e^{i(v,x)}(Z \circ \Pi_{t_1,\ldots,t_n}^{-1})(x) f_n(x)\,dx = 0,$$

where $\Pi_{t_1,\ldots,t_n} : \Omega \to \mathbb{R}^n$ is given by

$$\Pi_{t_1,\ldots,t_n}(\omega) = \big(B(t_1)(\omega),\, B(t_2)(\omega) - B(t_1)(\omega),$$
$$\ldots, B(t_n)(\omega) - B(t_{n-1})(\omega)\big)$$

for each $\omega \in \Omega$. Recognising the left-hand side of the above identity as a Fourier transform, we deduce that

$$Z \circ \Pi_{t_1,\ldots,t_n}^{-1}(x) = 0$$

for (Lebesgue) almost all $x \in \mathbb{R}^n$. From this, we deduce that

$$Z \chi_{I_{t_1,\ldots,t_n}^{A_1,\ldots,A_n}} = 0 \quad \text{a.s.}$$

for cylinder sets $I_{t_1,\ldots,t_n}^{A_1,\ldots,A_n}$, where $A_1, \ldots A_n$ are arbitrary Borel sets in $\mathbb{R}$. Since the collection of such sets generates $\mathcal{F}$, it follows that $Z = 0$ (a.s.), as was required. □

The following corollary, sometimes called the *martingale representation theorem*, is of great probabilistic significance.

Corollary 5.3.4 (Martingale representation 1) *Fix $T > 0$. If $M = (M(t), t \geq 0)$ is a centred L^2-martingale, then, for each $0 \leq t \leq T$, there exists a unique $F_M \in \mathcal{H}_2(T)$ such that*

$$M(t) = \int_0^t F_M(s)\,dB(s).$$

Proof For each $0 \leq t \leq T$, we have $M(t) \in L_0^2$, so from the above theorem we deduce the existence of $F_{M,t} \in \mathcal{H}_2(t)$ such that $M(t) = \int_0^t F_{M,t}(s)\,dB(s)$. However, M is a martingale from which we easily deduce that, for each $0 < t_1 < t_2 \leq T$, $F_{M,t_1} = F_{M,t_2}$ (a.e.). Uniqueness follows from the easily proved fact that the Itô isometry is injective. □

Note that if M is not centred then the result continues to hold but with $M(t)$ in Corollary 5.3.4 replaced by $M(t) - \mu$, where $\mu = \mathbb{E}(M(t))$, for each $t \geq 0$. We emphasise that Corollary 5.3.4 only holds for martingales that are adapted to the filtration generated by Brownian motion.

The martingale representation theorem was originally established by Kunita and Watanabe in the classic paper [181]. In Proposition 5.2 of [181], they also obtained a generalisation for martingales adapted to the filtration of a Lévy process. Regrettably the proof is too complicated to present here. However, we state the result in a form due to Kunita [185]. In the following, $(\mathcal{F}_t, t \geq 0)$ is the augmented natural filtration of a d-dimensional Lévy process.

Theorem 5.3.5 (Martingale representation 2) *Fix $T \geq 0$. If $M = (M(t), t \geq 0)$ is a locally square-integrable (real-valued) martingale adapted to $(\mathcal{F}_t, t \geq 0)$ then there exist $\alpha \in \mathbb{R}$, $F \in \mathcal{H}_2(T)$ and $G \in \mathcal{H}_2(T, \mathbb{R}^d - \{0\})$ such that, for all $0 \leq t \leq T$,*

$$M(t) = \alpha + \int_0^t F_j(s) dB^j(s) + \int_0^t \int_{\mathbb{R}^d - \{0\}} G(s,x) \tilde{N}(dx, ds).$$

The triple (α, F, G) is uniquely (up to sets of measure zero) determined by M.

Using deep techniques, Jacod has extensively generalised the martingale representation theorem. The following result from Jacod [152] has been specialised to apply to the 'jump' part of a Lévy process. Let $\mathcal{G}_t = \sigma\{N([0,s) \times A); 0 \leq s \leq t; A \in \mathcal{B}(\mathbb{R}^d - \{0\})\}$ and assume that our filtration is such that $\mathcal{F}_t = \mathcal{G}_t \vee \mathcal{F}_0$.

Theorem 5.3.6 (Jacod) *If $M = (M(t), t \geq 0)$ is a càdlàg adapted process, there exists a sequence of stopping times $(S(n), n \in \mathbb{N})$ with respect to which $(M(t \wedge S(n)), t \geq 0)$ is a uniformly integrable martingale, for each $n \in \mathbb{N}$, if and only if there exists a predictable $H : \mathbb{R}^+ \times (\mathbb{R}^d - \{0\}) \times \Omega \to \mathbb{R}$ such that for each $t \geq 0$*

$$\int_0^t \int_{\mathbb{R}^d - \{0\}} |H(s,x)| \nu(dx) ds < \infty,$$

and then, with probability 1, we have the representation

$$M(t) = M(0) + \int_0^t \int_{\mathbb{R}^d - \{0\}} H(s,x) \tilde{N}(dx, ds).$$

Further, more extensive, results were obtained in Jacod [153] and, in particular, it is shown on p. 51 therein that any local martingale $M = (M(t), t \geq 0)$

adapted to the filtration of a Lévy process has a representation of the following type:

$$M(t) = M(0) + \int_0^t F_j(s) dB^j(s) + \int_0^t \int_{\mathbb{R}^d - \{0\}} H(s,x) \tilde{N}(dx, ds),$$

for each $t \geq 0$, where F and H satisfy suitable integrability conditions. A result of similar type will appear shortly in [186]. More on martingale representation can be found in Jacod and Shiryaev [151], pp. 179–91, Liptser and Shiryaev [201], Theorem 19.1, and Protter [255], pp. 147–57. In a recent interesting development, Nualart and Schoutens [237] established the martingale representation property (and a more general chaotic representation) for the Teugels martingales introduced in Exercise 2.4.19. This yields the Brownian and compensated Poisson representation theorems as special cases.

5.4 Stochastic calculus and mathematical finance

Beginning with the fundamental papers on option pricing by Black and Scholes [50] and Merton [219], there has been a revolution in mathematical finance in recent years, arising from the introduction of techniques based on stochastic calculus with an emphasis on Brownian motion and continuous semimartingales. Extensive accounts can be found in a number of specialist texts of varying rigour and difficulty; see e.g. Baxter and Rennie [32], Bingham and Kiesel [47], Etheridge [98], Lamberton and Lapeyre [188], Mel'nikov [218], Shiryaev [284] and Steele [288]. It is not our intention here to try to give a comprehensive account of such a huge subject. For a general introduction to financial derivatives see the classic text by Hull [137]. The short book by Shimko [283] is also highly recommended.

In recent years, there has been a growing interest in the use of Lévy processes and discontinuous semimartingales to model market behaviour (see e.g. Madan and Seneta [206], Eberlein and Keller [89], Barndorff-Nielsen [27], Chan [69], Geman, Madan and Yor [115] and articles on finance in [20]); not only are these of great mathematical interest but there is growing evidence that they may be more realistic models than those that insist on continuous sample paths. Our aim in this section is to give a brief introduction to some of these ideas.

5.4.1 Introduction to financial derivatives

Readers who are knowledgeable about finance can skip this first section.

We begin with a somewhat contrived example to set the scene. It is 1st April and the reader is offered the opportunity to buy shares in the Frozen Delight

Ice Cream Company (FDICC). These are currently valued at £1 each. Looking forward, we might envisage that a long hot summer will lead to a rise in value of these shares, while if there is a wet miserable one they may well crash. There are, of course, many other factors that can affect their value, such as advertising and trends in taste. Now suppose that as well as being able to buy shares, we might also purchase an 'option'. Specifically, for a cost of £0.20 we can buy a ticket that gives us the right to buy one share of FDICC for £1.20 on 1st August, irrespective of the actual market value of this share.

Now suppose I buy 1000 of these tickets and 1st August arrives. The summer has been hot and the directors of FDICC have wisely secured the franchise for merchandising for the summer's hit film with pre-teens – *The Infinitely Divisible Man*. Consequently shares are now worth £1.80 each. I then exercise my option to buy 1000 shares at £1.20 each and sell them immediately at their market value to make a profit of £600 (£400 if you include the cost of the options). Alternatively, suppose that the weather has been bad and the film nosedives, or competitors secure the franchise, and shares drop to £0.70 each. In this case, I simply choose not to exercise my option to purchase the shares and I throw all my tickets away to make an overall profit of £0 (or a loss of £200, if I include the cost of the tickets).

The fictional example that we have just described is an example of a *financial derivative*. The term 'derivative' is used to clarify that the value of the tickets depends on the behaviour of the stock, which is the primary financial object, sometimes called the 'underlying'. Such derivatives can be seen as a form of insurance, as they allow investors to spread risk over a range of options rather than being restricted to the primary stock and bond markets, and they have been gaining considerably in importance in recent years.

For now let us focus on the £0.20 that we paid for each option. Is this a fair price to pay? Does the market determine a 'rational price' for such options? These are questions that we will address in this section, using stochastic calculus.

We now introduce some general concepts and notations. We will work in a highly simplified context to make the fundamental ideas as transparent as possible[1]. Our market consists of stock of a single type and also a riskless investment such as a bank account. We model the value in time of a single unit of stock as a stochastic process $S = (S(t), t \geq 0)$ on some probability space $(\Omega, \mathcal{F}, P)$. We will also require S to be adapted to a given

---

[1] If you are new to option pricing then you should first study the theory in a discrete time setting, where it is much simpler. You can find this in the early chapters of any of the textbooks mentioned above.

5.4 Stochastic calculus and mathematical finance

filtration $(\mathcal{F}_t, t \geq 0)$, and indeed all processes discussed henceforth will be assumed to be $\mathcal{F}_t$-adapted. The bank account grows deterministically in accordance with the compound interest formula from a fixed initial value $A_0 > 0$, so that

$$A(t) = A_0 e^{rt}$$

for each $t \geq 0$, where $r > 0$ is the *interest rate*, which we will take to be constant (in practice, it is piecewise constant).

Now we will introduce our option. In this book, we will only be concerned with the simplest type and these are called *European call options*. In this scenario, one buys an option at time 0 to buy stock at a fixed later time T at a given price k. We call T the *expiration time* of the contract and k the *strike price* or *exercise price*. The *value* of the option at time T is the random variable

$$Z = \max\{S(T) - k, 0\} = (S(T) - k)^+.$$

Our contrived option for FDICC shares is a European call option with $T = 4$ months and $k = £1.20$ and we have already described two different scenarios, within which $Z = £0.60$ or $Z = 0$.

European call options are the simplest of a wide range of possible derivatives. Another common type is the *American call option*, where stocks may be purchased at any time within the interval $[0, T]$, not only at the endpoint. For every call option that guarantees you the right to buy at the exercise price there corresponds a *put option*, which guarantees owners of stock the right to sell at that price. Clearly a put option is only worth exercising when the strike price is below the current market value.

Exercise 5.4.1 Deduce that the value of a European put option is $Z = \max\{k - S(T), 0\}$.

To be able to consider all types of possible option in a unified framework, we define a *contingent claim*, with maturity date T, to be a non-negative $\mathcal{F}_T$-measurable random variable. So European and American options are examples of contingent claims.

A key concept is the notion of *arbitrage*. This is essentially 'free money' or risk-free profit and is forbidden in rational models of market behaviour. An *arbitrage opportunity* is the possibility of making a risk-free profit by the simultaneous purchase and sale of related securities. Here is an example of how arbitrage can take place, taken from Mel'nikov [218], p. 4. Suppose that a stock sells in Frankfurt for 150 euros and in New York for \$100 and that the dollar–euro exchange rate is 1.55. Then one can borrow 150 euros and buy the stock in Frankfurt to sell immediately in New York for $100. We then exchange this for 155 euros, which we use to immediately pay back the loan leaving a 5-euro profit. So, in this case, the disparity in pricing stocks in Germany and the USA has led to the availability of 'free money'. Of course this discussion is somewhat simplified as we have ignored all transaction costs. It is impossible to overestimate the importance of arbitrage in option pricing, as we will see shortly.

First we need to recall some basic ideas of compound interest. Suppose that a sum of money, called the principal and denoted P, is invested at a constant rate of interest r. After an amount of time t, it grows to Pe^{rt}. Conversely, if we want to obtain a given sum of money Q at time t then we must invest Qe^{-rt} at time zero. The process of obtaining Qe^{-rt} from Q is called *discounting*. In particular, if $(S(t), t \geq 0)$ is the stock price, we define the *discounted process* $\tilde{S} = (\tilde{S}(t), t \geq 0)$, where each $\tilde{S}(t) = e^{-rt}S(t)$.

At least in discrete time, we have the following remarkable result, which illustrates how the absence of arbitrage forces the mathematical modeller into the world of stochastic analysis.

Theorem 5.4.2 (Fundamental theorem of asset pricing 1) *If the market is free of arbitrage opportunities, then there exists a probability measure Q, which is equivalent to P, with respect to which the discounted process $\tilde{S}$ is a martingale.*

A similar result holds in the continuous case but we need to make more technical assumptions; see Bingham and Kiesel [47], pp. 176–7, or the fundamental paper by Delbaen and Schachermeyer [77]. The classic paper by Harrison and Pliska [127] is also valuable background for this topic. The philosophy of Theorem 5.4.2 will play a central role later.

Portfolios

An investor (which may be an individual or a company) will hold their investments as a combination of risky stocks and cash in the bank, say. Let $\alpha(t)$ and $\beta(t)$ denote the amount of each of these, respectively, that we hold at time t. The pair of adapted processes (α, β) where $\alpha = (\alpha(t), t \geq 0)$ and $\beta = (\beta(t), t \geq 0)$ is called a *portfolio* or *trading strategy*. The total value of all our investments at time t is denoted as $V(t)$, so

$$V(t) = \alpha(t)S(t) + \beta(t)A(t).$$

One of the key aims of the Black–Scholes approach to option pricing is to be able to *hedge* the risk involved in selling options, by being able to construct a portfolio whose value at the expiration time T is exactly that of the option. To be precise, a portfolio is said to be *replicating* if

$$V(T) = Z.$$

Clearly, replicating portfolios are desirable objects.

Another class of interesting portfolios are those that are *self-financing*, i.e. any change in wealth V is due only to changes in the values of stocks and bank accounts and not to any injections of capital from outside. We can model this using stochastic differentials if we make the assumption that the stock price process S is a semimartingale. We can then write

$$dV(t) = \alpha(t)dS(t) + \beta(t)dA(t) = \alpha(t)dS(t) + r\beta(t)A(t)dt,$$

so the infinitesimal change in V arises solely through those in S and A. Notice how we have sneakily slipped Itô calculus into the picture by the assumption that $dS(t)$ should be interpreted in the Itô sense. This is absolutely crucial. If we try to use any other type of integral (e.g. the Lebesgue–Stieltjes type) then certainly the theory that follows will no longer work.

A market is said to be *complete* if every contingent claim can be replicated by a self-financing portfolio. So, in a complete market, every option can be hedged by a portfolio that requires no injections of capital between its starting time and the expiration time. In discrete time, we have the following:

Theorem 5.4.3 (Fundamental theorem of asset pricing 2) *An arbitrage-free market is complete if and only if there exists a unique probability measure Q, which is equivalent to P, with respect to which the discounted process $\tilde{S}$ is a martingale.*

Once again, for the continuous-time version, see Bingham and Kiesel [47] and Delbaen and Schachermeyer [77].

Theorems 5.4.2 and 5.4.3 identify a key mathematical problem: to find a (unique, if possible) Q, which is equivalent to P, under which $\tilde{S}$ is a martingale. Such a Q is called a *martingale measure* or *risk-neutral measure*. If Q exists, but is not unique, then the market is said to be *incomplete*. We will address the problem of finding Q in the next two subsections.

5.4.2 Stock prices as a Lévy process

So far we have said little about the key process S that models the evolution of stock prices. As far back as 1900, Bachelier [16] in his Ph.D. thesis proposed that this should be a Brownian motion. Indeed, this can be intuitively justified on the basis of the central limit theorem if one perceives the movement of stocks as due to the 'invisible hand of the market', manifested as a very large number of independent, identically distributed, decisions. One immediate problem with this is that it is unrealistic, as stock prices cannot become negative but Brownian motion can. An obvious way out of this is to take exponentials, but let us be more specific.

Financial analysts like to study the *return* on their investment, which in a small time interval $[t, t + \delta t]$ will be

$$\frac{\delta S(t)}{S(t)} = \frac{S(t + \delta t) - S(t)}{S(t)};$$

it is then natural to introduce directly the noise at this level and write

$$\frac{\delta S(t)}{S(t)} = \sigma \delta X(t) + \mu \delta t,$$

where $X = (X(t), t \geq 0)$ is a semimartingale and σ, μ are parameters called the *volatility* and *stock drift* respectively. The parameter $\sigma > 0$ controls the strength of the coupling to the noise while $\mu \in \mathbb{R}$ represents deterministic effects; indeed if $\mathbb{E}(\delta X(t)) = 0$ for all $t \geq 0$ then μ is the logarithmic mean rate of return.

We now interpret this in terms of Itô calculus, by formally replacing all small changes that are written in terms of δ by Itô differentials. We then find that

$$dS(t) = \sigma S(t-)dX(t) + \mu S(t-)dt = S(t-)dZ(t), \quad (5.10)$$

where $Z(t) = \sigma X(t) + \mu t$.

We see immediately that $S(t) = \mathcal{E}_Z(t)$ is the stochastic exponential of the semimartingale Z, as described in Section 5.1. Indeed, when X is a standard Brownian motion $B = (B(t), t \geq 0)$ we obtain *geometric Brownian motion*, which is very widely used as a model for stock prices:

$$S(t) = S(0) \exp\left[\sigma B(t) + \left(\mu t - \tfrac{1}{2}\sigma^2 t\right)\right]. \quad (5.11)$$

There has been recently a great deal of interest in taking X to be a Lévy process. One argument in favour of this is that stock prices clearly do not move continuously, and a more realistic approach is one that allows small jumps in small time intervals. Moreover, empirical studies of stock prices indicate distributions with heavy tails, which are incompatible with a Gaussian model (see e.g. Akgiray and Booth [2]).

5.4 Stochastic calculus and mathematical finance

We will make the assumption from now on that X is indeed a Lévy process. Note immediately that in order for stock prices to be non-negative, (SE1) yields $\Delta X(t) > -\sigma^{-1}$ (a.s.) for each $t > 0$ and, for convenience, we will write $c = -\sigma^{-1}$ henceforth. We will also impose the following condition on the Lévy measure $\nu : \int_c^\infty (x^2 \vee x)\nu(dx) < \infty$. This means that each $X(t)$ has finite first and second moments, which would seem to be a reasonable assumption for stock returns.

By the Lévy–Itô decomposition (Theorem 2.4.16), for each $t \geq 0$,

$$X(t) = mt + \kappa B(t) + \int_c^\infty x \tilde{N}(t, dx) \qquad (5.12)$$

where $\kappa \geq 0$ and, in terms of the earlier parametrisation,

$$m = b + \int_{[c,-1] \cup [1,\infty)} x\nu(dx).$$

To keep the notation simple we assume in (5.12), and below, that 0 is omitted from the range of integration. Using Exercise 5.1.2, we obtain the following representation for stock prices:

$$d\big[\log(S(t))\big] = \kappa \sigma dB(t) + \big(m\sigma + \mu - \tfrac{1}{2}\kappa^2 \sigma^2\big) dt$$
$$+ \int_c^\infty \log[1 + \sigma x] \tilde{N}(dt, dx)$$
$$+ \int_c^\infty \big[\log(1 + \sigma x) - \sigma x\big] \nu(dx) dt. \qquad (5.13)$$

Note The use of Lévy processes in finance is at a relatively early stage of development and there seems to be some disagreement in the literature as to whether it is best to employ a stochastic exponential to model stock prices, as in (5.10), or to use *geometric Lévy motion*, $S(t) = e^{X(t)}$ (the reader can check that these are, more or less, equivalent when X is Gaussian). Indications are that the former is of greater theoretical interest while the latter may be more realistic in practical models. We will return to this point later on.

5.4.3 Change of measure

Motivated by the philosophy behind the fundamental theorems of asset pricing (Theorems 5.4.2 and 5.4.3), we seek to find measures Q, which are equivalent to P, with respect to which the discounted stock process $\tilde{S}$ is a martingale. Rather than consider all possible changes of measure, we work in a restricted context where we can exploit our understanding of stochastic calculus based on Lévy processes. In this respect, we will follow the exposition of Chan [69]; see also Kunita [185, 186].

Let Y be a Lévy-type stochastic integral that takes the form

$$dY(t) = G(t)dt + F(t)dB(t) + \int_{\mathbb{R}-\{0\}} H(t,x)\tilde{N}(dt,dx),$$

where in particular $H \in \mathcal{P}_2(t, \mathbb{R} - \{0\})$ for each $t \geq 0$. Note that we have deliberately chosen a restricted form of Y compatible with that of the Lévy process X, in order to simplify the discussion below.

We consider the associated exponential process e^Y and we assume that the conditions of Corollary 5.2.2 and Theorem 5.2.4 are satisfied, so that e^Y is a martingale (and G is determined by F and H). Hence we can define a new measure Q by the prescription $dQ/dP = e^{Y(T)}$. Furthermore, by Girsanov's theorem and Exercise 5.2.13, for each $0 \leq t \leq T$, $E \in \mathcal{B}([c, \infty))$,

$$B_Q(t) = B(t) - \int_0^t F(s)ds \quad \text{is a } Q\text{-Brownian motion}$$

and

$$\tilde{N}_Q(t, E) = \tilde{N}(t, E) - \nu_Q(t, E) \quad \text{is a } Q\text{-martingale,}$$

where

$$\nu_Q(t, E) = \int_0^t \int_E (e^{H(s,x)} - 1)\nu(dx)ds.$$

Note that

$$\mathbb{E}_Q(\tilde{N}_Q(t, E)^2) = \int_0^t \int_E \mathbb{E}_Q(e^{H(s,x)}) \, \nu(dx)ds;$$

see e.g. Ikeda and Watanabe [140], Chapter II, Theorem 3.1.

We rewrite the discounted stock price in terms of these new processes, to find

$$d\{\log[\tilde{S}(t)]\} = \kappa\sigma dB_Q(t) + \left(m\sigma + \mu - r - \tfrac{1}{2}\kappa^2\sigma^2 + \kappa\sigma F(t)\right.$$

$$\left. + \sigma \int_{\mathbb{R}-\{0\}} x(e^{H(t,x)} - 1)\nu(dx)\right)dt$$

$$+ \int_c^\infty \log(1 + \sigma x)\tilde{N}_Q(dt, dx)$$

$$+ \int_c^\infty [\log(1 + \sigma x) - \sigma x]\nu_Q(dt, dx). \qquad (5.14)$$

5.4 Stochastic calculus and mathematical finance

Now write $\tilde{S}(t) = \tilde{S}_1(t)\tilde{S}_2(t)$, where

$$d\{\log[\tilde{S}_1(t)]\} = \kappa\sigma dB_Q(t) - \frac{1}{2}\kappa^2\sigma^2 dt + \int_c^\infty \log(1+\sigma x)\tilde{N}_Q(dt, dx)$$
$$+ \int_c^\infty [\log(1+\sigma x) - \sigma x]\nu_Q(dt, dx)$$

and

$$d\{\log[\tilde{S}_2(t)]\}$$
$$= \left[m\sigma + \mu - r + \kappa\sigma F(t) + \sigma\int_{\mathbb{R}-\{0\}} x(e^{H(t,x)} - 1)\nu(dx)\right]dt.$$

On applying Itô's formula to $\tilde{S}_1$, we obtain

$$d\tilde{S}_1(t) = \kappa\sigma\tilde{S}_1(t)dB_Q(t) + \sigma\tilde{S}_1(t)x\tilde{N}_Q(dt, dx).$$

So $\tilde{S}_1$ is a Q-local martingale, and hence $\tilde{S}$ is a Q-local martingale if and only if

$$m\sigma + \mu - r + \kappa\sigma F(t) + \sigma\int_{\mathbb{R}-\{0\}} x(e^{H(t,x)} - 1)\nu(dx) = 0 \quad \text{a.s.}$$
(5.15)

In fact, if we impose the additional condition that

$$t \to \int_0^t \int_c^\infty x^2\, \mathbb{E}_Q(e^{H(s,x)})\,\nu(dx)ds$$

is locally bounded, then $\tilde{S}$ is a martingale. This follows from the representation of $\tilde{S}$ as the solution of a stochastic differential equation (see Exercise 6.2.5). Note that a sufficient condition for the above condition to hold is that H is uniformly bounded (in x and ω) on finite intervals $[0, t]$.

Now, equation (5.15) clearly has an infinite number of possible solution pairs (F, H). To see this, suppose that $f \in L^1(\mathbb{R} - \{0\}, \nu)$; then if (F, H) is a solution so too is

$$\left(F + \int_{\mathbb{R}-\{0\}} f(x)\nu(dx),\ \log\left(e^H - \frac{\kappa f}{x}\right)\right).$$

Consequently, there is an infinite number of possible measures Q with respect to which $\tilde{S}$ is a martingale. So the general Lévy-process model gives rise to incomplete markets. The following example is of considerable interest.

The Brownian case Here we have $\nu \equiv 0$, $\kappa \neq 0$, and the unique solution to (5.15) is

$$F(t) = \frac{r - \mu - m\sigma}{\kappa\sigma} \quad \text{a.s.}$$

So in this case the stock price is a geometric Brownian motion, and in fact we have a complete market; see e.g. Bingham and Keisel [47], pp. 189–90, for further discussion of this.

The only other example of a Levy process that gives rise to a complete market is that where the driving noise in (5.13) is a compensated Poisson process.

The Poisson case Here we take $\kappa = 0$ and $\nu = \lambda \delta_1$ for $\lambda > m + (\mu - r)/\sigma$. Writing $H(t, 1) = H(t)$, we find that

$$H(t) = \log\left[\frac{r - \mu + (\lambda - m)\sigma}{\lambda\sigma}\right] \quad \text{a.s.}$$

5.4.4 The Black–Scholes formula

We will follow the simple account given in Baxter and Rennie [32] of the classic Black–Scholes approach to pricing a European option. We will work with the geometric Brownian motion model for stock prices (5.11), so that the market is complete. Note that $\kappa = 1$ in (5.12). We will also make a slight change in the way we define the discounted stock price: in this section we will put $\tilde{S}(t) = A(t)^{-1}S(t)$ for each $0 \leq t \leq T$. We effect the change of measure as described above, and so by (5.14) and the condition (5.15) we obtain

$$d\{\log[\tilde{S}(t)]\} = \sigma dB_Q(t) - \tfrac{1}{2}\sigma^2 dt,$$

so that

$$d\tilde{S}(t) = \tilde{S}(t)\sigma dB_Q(t). \tag{5.16}$$

Let Z be a contingent claim and assume that it is square-integrable, i.e. $\mathbb{E}(|Z|^2) < \infty$. Define a martingale $(Z(t), t \geq 0)$ by discounting and conditioning as follows:

$$Z(t) = A(T)^{-1}\mathbb{E}(Z|\mathcal{F}_t)$$

for each $0 \leq t \leq T$. Then Z is an L^2-martingale, since, by the conditional form of Jensen's inequality, we have

$$\mathbb{E}(\mathbb{E}(Z|\mathcal{F}_t))^2) \leq \mathbb{E}(\mathbb{E}(Z^2|\mathcal{F}_t)) = \mathbb{E}(Z^2) < \infty.$$

Now we can appeal to the martingale representation theorem (Corollary 5.3.4)

5.4 Stochastic calculus and mathematical finance

to deduce that there exists a square-integrable process $\delta = (\delta(t), t \geq 0)$ such that, for all $0 \leq t \leq T$,

$$dZ(t) = \delta(t)dB_Q(t) = \gamma(t)d\tilde{S}(t), \tag{5.17}$$

where, by (5.16), each $\gamma(t) = \delta(t)/[\sigma \tilde{S}(t)]$.

The Black–Scholes strategy is to construct a portfolio V which is both self-financing and replicating and which effectively fixes the value of the option at each time t. We will show that the following prescription does the trick:

$$\alpha(t) = \gamma(t), \qquad \beta(t) = Z(t) - \gamma(t)\tilde{S}(t), \tag{5.18}$$

for all $0 \leq t \leq T$.

We call this the *Black–Scholes portfolio*. Its value is

$$V(t) = \gamma(t)S(t) + \big[Z(t) - \gamma(t)\tilde{S}(t)\big]A(t) \tag{5.19}$$

for each $0 \leq t \leq T$.

Theorem 5.4.4 *The Black–Scholes portfolio is self-financing and replicating.*

Proof First note that since for each $0 \leq t \leq T$ we have $\tilde{S}(t) = A(t)^{-1}S(t)$, (5.19) becomes

$$V(t) = A(t)\gamma(t)\tilde{S}(t) + \big[Z(t) - \gamma(t)\tilde{S}(t)\big]A(t) = Z(t)A(t). \tag{5.20}$$

To see that this portfolio is replicating, observe that

$$V(T) = A(T)Z(T) = A(T)A(T)^{-1}\,\mathbb{E}(Z|\mathcal{F}_T) = Z,$$

since Z is $\mathcal{F}_T$-measurable.

To see that the portfolio is self-financing, we apply the Itô product formula in (5.20) to obtain

$$dV(t) = dZ(t)A(t) + Z(t)dA(t) = \gamma(t)A(t)d\tilde{S}(t) + Z(t)dA(t),$$

by (5.17).

But, by (5.18), $Z(t) = \beta(t) + \gamma(t)\tilde{S}(t)$ and so

$$\begin{aligned}
dV(t) &= \gamma(t)A(t)d\tilde{S}(t) + \big[\beta(t) + \gamma(t)\tilde{S}(t)\big]dA(t) \\
&= \beta(t)dA(t) + \gamma(t)\big[A(t)d\tilde{S}(t) + \tilde{S}(t)dA(t)\big] \\
&= \beta(t)dA(t) + \gamma(t)d\big[A(t)\tilde{S}(t)\big] \\
&= \beta(t)dA(t) + \gamma(t)dS(t),
\end{aligned}$$

where again we have used the Itô product formula. $\square$

Using formula (5.20) in the above proof, we see that the value of the portfolio at any time $0 \le t \le T$ is given by

$$V(t) = A(t)\, \mathbb{E}_Q(A(T)^{-1}Z|\mathcal{F}_t) = e^{-r(T-t)}\, \mathbb{E}_Q(Z|\mathcal{F}_t) \qquad (5.21)$$

and, in particular,

$$V(0) = e^{-rT}\mathbb{E}_Q(Z). \qquad (5.22)$$

We note that $V(0)$ is the arbitrage price for the option, in that if the claim is priced higher or lower than this then there is the opportunity for risk-free profit for the seller or buyer, respectively. To see that this is true, suppose that the option sells for a price $P > V(0)$. If anyone is crazy enough to buy it at this price, then the seller can spend $V(0)$ to invest in $\gamma(0)$ units of stock and $\beta(0)$ units of the bank account. Using the fact that the portfolio is self-financing and replicating, we know that at time T it will deliver the value of the option $V(T)$ without any further injection of capital. Hence the seller has made $P - V(0)$ profit. A similar argument applies to the case where $P < V(0)$.

We can now derive the celebrated *Black–Scholes pricing formula* for a European option. First observe that, by (5.22), we have

$$V(0) = e^{-rT}\mathbb{E}_Q((S_T - k)^+).$$

Now, after the change of measure $\tilde{S}$ is a stochastic exponential driven by Brownian motion and so

$$\tilde{S}(T) = \tilde{S}(0)\exp\left[\sigma B_Q(T) - \tfrac{1}{2}\sigma^2 T\right],$$

hence

$$S(T) = A(0)\tilde{S}(0)\exp\left[\sigma B_Q(T) - \left(r - \tfrac{1}{2}\sigma^2\right)T\right]$$
$$= S(0)\exp\left(\sigma B_Q(T) - \left(r - \tfrac{1}{2}\sigma^2\right)T\right).$$

But $B_Q(T) \sim N(0, T)$, from which it follows that

$$V(0) = e^{-rT}\,\mathbb{E}_Q((se^{U+rT} - k)^+)$$

where $U \sim N(-\sigma^2 T/2, \sigma^2 T/2)$, and we have adopted the usual convention in finance of writing $S(0) = s$. Hence we have

$$V(0) = \frac{1}{\sigma\sqrt{2\pi T}} \int_{\log(k/s)}^{\infty} (se^x - ke^{-rT})\exp\left[-\frac{(x+\sigma^2 T/2)^2}{2\sigma^2 T}\right]dx.$$

Now write $\Phi(z) = P(Z \leq z)$, where $Z \sim N(0, 1)$ is a standard normal. Splitting the above formula into two summands and making appropriate substitutions (see for example Lamberton and Lapeyre [188], p. 70, if you need a hint) yields the celebrated *Black–Scholes pricing formula for European call options*:

$$V(0) = s\Phi\left(\frac{\log(s/k) + (r + \sigma^2/2)T}{\sigma\sqrt{T}}\right)$$
$$- ke^{-rT}\Phi\left(\frac{\log(s/k) + (r - \sigma^2/2)T}{\sigma\sqrt{T}}\right). \quad (5.23)$$

5.4.5 Incomplete markets

If the market is complete and if there is a suitable martingale representation theorem available, it is clear that the Black–Scholes approach described above can be applied in order to price contingent claims, in principle. However, if stock prices are driven by a general Lévy process as in (5.12), the market will be incomplete. Provided that there are no arbitrage opportunities, we know that equivalent measures Q exist with respect to which $\tilde{S}$ will be a martingale, but these will no longer be unique. In this subsection we examine briefly some approaches that have been developed for incomplete markets. These involve finding a 'selection principle' to reduce the class of all possible measures Q to a subclass within which a unique measure can be found. We again follow Chan [69]. An extensive discussion from a more general viewpoint can be found in Chapter 7 of Bingham and Kiesel [47].

The Föllmer–Schweizer minimal measure

In the Black–Scholes set-up, we have a unique martingale measure Q for which

$$\left.\frac{dQ}{dP}\right|_{\mathcal{F}_t} = e^{Y(t)},$$

where $d(e^{Y(t)}) = e^{Y(t)}F(t)dB(t)$ for $0 \leq t \leq T$. In the incomplete case, one approach to selecting Q would be simply to replace B by the martingale part of our Lévy process (5.12), so that we have

$$d(e^{Y(t)}) = e^{Y(t)}P(t)\left[\kappa dB(t) + \int_{(c,\infty)} x\tilde{N}(ds, dx)\right], \quad (5.24)$$

for some adapted process $P = (P(t), t \geq 0)$. If we compare this with the usual coefficients of exponential martingales in (5.3), we see that we have

$$\kappa P(t) = F(t), \qquad x P(t) = e^{H(t,x)} - 1$$

for each $t \geq 0$, $x > c$. Substituting these conditions into (5.15) yields

$$P(t) = \frac{r + \mu - m\sigma}{\sigma(\kappa^2 + \rho)}$$

where $\rho = \int_c^\infty x^2 \nu(dx)$, so this procedure selects a unique martingale measure under the constraint that we consider only measure changes of the type (5.24). Chan [69] demonstrates that this coincides with a general procedure introduced by Föllmer and Schweizer [104], which works by constructing a replicating portfolio of value $V(t) = \alpha(t)S(t) + \beta(t)A(t)$ and discounting it to obtain $\tilde{V}(t) = \alpha(t)\tilde{S}(t) + \beta(t)A(0)$. If we now define the cumulative cost $C(t) = \tilde{V}(t) - \int_0^t \alpha(s) d\tilde{S}(s)$ then Q minimises the risk $\mathbb{E}((C(T) - C(t))^2 | \mathcal{F}_t)$.

The Esscher transform

We will now make the additional assumption that

$$\int_{|x| \geq 1} e^{ux} \nu(dx) < \infty$$

for all $u \in \mathbb{R}$. In this case we can analytically continue the Lévy–Khintchine formula to obtain, for each $t \geq 0$,

$$\mathbb{E}(e^{-uX(t)}) = e^{-t\psi(u)}$$

where

$$\psi(u) = -\eta(iu)$$
$$= bu - \tfrac{1}{2}\kappa^2 u^2 + \int_c^\infty \left[1 - e^{-uy} - uy\chi_{\hat{B}}(y)\right] \nu(dy).$$

Now recall the martingales $M_u = (M_u(t), t \geq 0)$, where each $M_u(t) = e^{iuX(t) - t\eta(u)}$, which were defined in Chapter 2. Readers can check directly that the martingale property is preserved under analytic continuation, and we will write $N_u(t) = M_{iu}(t) = e^{-uX(t) + t\psi(u)}$. The key distinction between the martingales M_u and N_u is that the former are complex valued while the latter are strictly positive. For each $u \in \mathbb{R}$ we may thus define a new probability measure by the prescription

$$\left.\frac{dQ_u}{dP}\right|_{\mathcal{F}_t} = N_u(t),$$

for each $0 \leq t \leq T$. We call Q_u the *Esscher transform* of P by N_u. It has a long history of application within actuarial science (see Gerber and Shiu [116] and references therein). Applying Itô's formula to N_u, we obtain

$$dN_u(t) = N_u(t-)\left[-\kappa u B(t) + (e^{-ux} - 1)\tilde{N}(dt, dx)\right]. \quad (5.25)$$

On comparing this with our usual prescription (5.3) for exponential martingales e^Y, we find that

$$F(t) = -\kappa u, \qquad H(t, x) = -ux,$$

and so (5.15) yields the following condition for Q_u to be a martingale measure:

$$-\kappa^2 u\sigma + m\sigma + \mu - r + \sigma \int_c^\infty x(e^{-ux} - 1)\nu(dx) = 0.$$

Define $z(u) = \int_c^\infty x(e^{-ux} - 1)\nu(dx) - \kappa^2 u$ for each $u \in \mathbb{R}$. Then our condition takes the form

$$z(u) = \frac{r - \mu - m\sigma}{\sigma}.$$

Since $z'(u) \leq 0$, we see that z is monotonic decreasing and so is invertible. Hence this choice of u yields a martingale measure, under the constraint that we only consider changes of measure of the form (5.25).

Chan [69] showed that this Q_u minimises the relative entropy $H(Q|P)$, where

$$H(Q|P) = \int \frac{dQ}{dP} \log\left(\frac{dQ}{dP}\right) dP.$$

Further investigations of such *minimal entropy martingale measures* can be found in Fujiwara and Miyahara [110].

5.4.6 Hyperbolic Lévy processes in finance

So far we have concentrated our efforts in general discussions about Lévy processes as models of stock prices, without looking at any particular case other than Brownian motion. In fact, as far back as the 1960s Mandelbrot [209] proposed that α-stable processes might be a good model; see also Chapter 14 of his collected papers [210]. However, empirical studies appear to rule out these, as well as the classical Black–Scholes model (see e.g. Akgiray and Booth [2]). An example of a Lévy process that appears to be well suited to modelling stock price movements is the hyperbolic process, which we will now describe.

Hyperbolic distributions

Let $\Upsilon \in \mathcal{B}(\mathbb{R})$ and let $(g_\theta; \theta \in \Upsilon)$ be a family of probability density functions on $\mathbb{R}$ such that the mapping $(x, \theta) \to g_\theta(x)$ is jointly measurable from $\mathbb{R} \times \Upsilon$ to $\mathbb{R}$. Let ρ be another probability distribution on Υ, which we call the *mixing measure*; then, by Fubini's theorem, we see that the *probability mixture*

$$h(x) = \int_\Upsilon g_\theta(x) \rho(d\theta),$$

yields another probability density function h on $\mathbb{R}$. The hyperbolic distributions that we will now introduce arise in exactly this manner. First we need to describe the mixing measure ρ.

We begin with the following integral representation for Bessel functions of the third kind:

$$K_\nu(x) = \frac{1}{2} \int_0^\infty u^{\nu-1} \exp\left[-\frac{1}{2} x \left(u + \frac{1}{u}\right)\right] du$$

where $x, \nu \in \mathbb{R}$; see Section 5.6 for all the facts we need about Bessel functions in the present section.

From this, we see immediately that $f_\nu^{a,b}$ is a probability density function on $(0, \infty)$ for each $a, b > 0$, where

$$f_\nu^{a,b}(x) = \frac{(a/b)^{\nu/2}}{2 K_\nu(\sqrt{ab})} x^{\nu-1} \exp\left[-\frac{1}{2}\left(ax + \frac{b}{x}\right)\right].$$

The distribution that this represents is called a *generalised inverse Gaussian* and denoted $GIG(\nu, a, b)$. It clearly generalises the inverse Gaussian distribution discussed in Subsection 1.3.2. In our probability mixture, we now take ρ to be $GIG(1, a, b)$, $\Upsilon = (0, \infty)$, and g_{σ^2} to be the probability density function of an $N(\mu + \beta\sigma^2, \sigma^2)$, where $\mu, \beta \in \mathbb{R}$. A straightforward but tedious computation, in which we apply the beautiful result $K_{1/2}(x) = \sqrt{\pi/(2x)} e^{-x}$ (proved as Proposition 5.6.1 in Section 5.6), yields

$$h_{\delta,\mu}^{\alpha,\beta}(x) = \frac{\sqrt{\alpha^2 - \beta^2}}{2\alpha\delta K_1(\delta\sqrt{\alpha^2 - \beta^2})} \exp\left[-\alpha\sqrt{\delta^2 + (x-\mu)^2} + \beta(x-\mu)\right]$$
(5.26)

for all $x \in \mathbb{R}$, where we have, in accordance with the usual convention, introduced the parameters $\alpha^2 = a + \beta^2$ and $\delta^2 = b$.

The corresponding law is called a *hyperbolic distribution*, as $\log(h_{\delta,\mu}^{\alpha,\beta})$ is a hyperbola. These distributions were first introduced by Barndorff-Nielsen in [26], within models for the distribution of particle size in wind-blown sand deposits. In Barndorff-Nielsen and Halgreen [20], they were shown to be infinitely divisible. Halgreen [125] also established that they are self-decomposable.

All the moments of a hyperbolic distribution exist and we may compute the moment generating function $M_{\delta,\mu}^{\alpha,\beta}(u) = \int_{\mathbb{R}} e^{ux} h_{\delta,\mu}^{\alpha,\beta}(x) dx$, to obtain:

Proposition 5.4.5 For $|u + \beta| < \alpha$,

$$M_{\delta,\mu}^{\alpha,\beta}(u) = e^{\mu u} \frac{\sqrt{\alpha^2 - \beta^2}}{K_1(\delta\sqrt{\alpha^2 - \beta^2})} \frac{K_1(\delta\sqrt{\alpha^2 - (\beta+u)^2})}{\sqrt{\alpha^2 - (\beta+u)^2}}.$$

Proof Use straightforward manipulation (see Eberlein, Keller and Prause [88]). □

Note that, by analytic continuation, we get the characteristic function $\phi(u) = M(iu)$, which is valid for all $u \in \mathbb{R}$. Using this, Eberlein and Keller in [89] were able to show that the Lévy measure of the distribution is absolutely continuous with respect to Lebesgue measure, and they computed the exact form of the Radon–Nikodým derivative.

Exercise 5.4.6 Let X be a hyperbolically distributed random variable. Use Proposition 5.4.5 and (5.30) in Section 5.6 to establish

$$\mathbb{E}(X) = \mu + \frac{\delta\beta}{\sqrt{\alpha^2 - \beta^2}} \frac{K_2(\zeta)}{K_1(\zeta)}$$

and

$$\mathrm{Var}(X) = \delta^2 \left[\frac{K_2(\zeta)}{\zeta K_1(\zeta)} + \frac{\beta^2}{\alpha^2 - \beta^2} \left(\frac{K_3(\zeta)}{K_1(\zeta)} - \frac{K_2(\zeta)^2}{K_1(\zeta)^2} \right) \right],$$

where $\zeta = \delta\sqrt{\alpha^2 - \beta^2}$.

For simplicity, we will restrict ourselves to the symmetric case where $\mu = \beta = 0$. If we reparametrise, using $\zeta = \delta\alpha$, we obtain the two-parameter family of densities

$$h_{\zeta,\delta}(x) = \frac{1}{2\delta K_1(\zeta)} \exp\left[-\zeta\sqrt{1 + \left(\frac{x}{\delta}\right)^2} \right].$$

It is shown in Eberlein and Keller [89] that the corresponding Lévy process $X_{\zeta,\delta} = (X_{\zeta,\delta}(t), t \geq 0)$ has no Gaussian part and can be written

$$X_{\zeta,\delta}(t) = \int_0^t \int_{\mathbb{R}-\{0\}} x \tilde{N}(ds, dx)$$

for each $t \geq 0$.

Option pricing with hyperbolic Lévy processes

The hyperbolic Lévy process was first applied to option pricing by Eberlein and Keller in [89], following a suggestion by O. Barndorff-Nielsen. There is an intriguing analogy with sand production in that just as large rocks are broken down to smaller and smaller particles to create sand so, to quote Bingham and Kiesel in their review article [48], 'this "energy cascade effect" might be paralleled in the "information cascade effect", whereby price-sensitive information originates in, say, a global newsflash and trickles down through national and local level to smaller and smaller units of the economic and social environment.'

We may again model the stock price $S = (S(t), t \geq 0)$ as a stochastic exponential driven by a process $X_{\zeta,\delta}$, so that

$$dS(t) = S(t-)dX_{\zeta,\delta}(t)$$

for each $t \geq 0$ (we omit volatility for now and return to this point later). A drawback of this approach is that the jumps in $X_{\zeta,\delta}$ are not bounded below. Eberlein and Keller [89] suggested overcoming this problem by introducing a stopping time $\tau = \inf\{t > 0; \; \Delta X_{\zeta,\delta}(t) < -1\}$ and working with $\hat{X}_{\zeta,\delta}$ instead of $X_{\zeta,\delta}$, where for each $t \geq 0$

$$\hat{X}_{\zeta,\delta}(t) = X_{\zeta,\delta}(t)\chi_{\{t \leq \tau\}},$$

but this is clearly a somewhat contrived approach. An alternative point of view, also put forward by Eberlein and Keller [89], is to model stock prices by an exponential hyperbolic Lévy process and utilise

$$S(t) = S(0) \exp\left[X_{\zeta,\delta}(t)\right].$$

This has been found to be a highly successful approach from an empirical point of view. As usual we discount and consider

$$\hat{S}(t) = S(0) \exp\left[X_{\zeta,\delta}(t) - rt\right],$$

and we require a measure Q with respect to which $\hat{S} = (\hat{S}(t), t \geq 0)$ is a martingale. As expected, the market is incomplete, and we will follow Eberlein and Keller [89] and use the Esscher transform to price the option. Hence we seek a measure, of the form Q_u, that satisfies

$$\left.\frac{dQ_u}{dP}\right|_{\mathcal{F}_t} = N_u(t) = \exp\left\{-uX_{\zeta,\delta}(t) - t \, \log\left[M_{\zeta,\delta}(u)\right]\right\}.$$

5.4 Stochastic calculus and mathematical finance

Here $M_{\zeta,\delta}(u)$ denotes the moment generating function of $X_{\zeta,\delta}(1)$, as given by Proposition 5.4.5, for $|u| < \alpha$. Recalling Lemma 5.2.10, we see that $\hat{S}$ is a Q-martingale if and only if $\hat{S}N_u = (\hat{S}(t)N_u(t), t \geq 0)$ is a P-martingale. Now

$$\hat{S}(t)N_u(t) = \exp\left((1-u)X_{\zeta,\delta}(t) - t\{\log\left[M_{\zeta,\delta}(u)\right] + r\}\right).$$

But we know that $(\exp\left((1-u)X_{\zeta,\delta}(t) - t\{\log[M_{\zeta,\delta}(1-u)]\}\right), t \geq 0)$ is a martingale and, comparing the last two facts, we find that $\hat{S}$ is a Q-martingale if and only if

$$r = \log\left[M_{\zeta,\delta}(1-u)\right] - \log\left[M_{\zeta,\delta}(u)\right]$$
$$= \log\left[\frac{K_1\left(\sqrt{\zeta^2 - \delta^2(1-u)^2}\right)}{K_1\left(\sqrt{\zeta^2 - \delta^2 u^2}\right)}\right] - \frac{1}{2}\log\left[\frac{\zeta^2 - \delta^2(1-u)^2}{\zeta^2 - \delta^2 u^2}\right].$$

The required value of u can now be determined from this expression by numerical means.[1]

We can now price a European call option with strike price k and expiration time T. Writing $S(0) = s$ as usual, the price is

$$V(0) = \mathbb{E}_{Q_u}\left(e^{-rT}\left[S(T) - k\right]^+\right) = \mathbb{E}_{Q_u}\left(e^{-rT}\left\{s\exp\left[X_{\zeta,\delta}(t)\right] - k\right\}^+\right).$$

Exercise 5.4.7 Let $f_{\zeta,\delta}^{(t)}$ be the pdf of $X_{\zeta,\delta}(t)$ with respect to P. Use the Esscher transform to show that $X_{\zeta,\delta}(t)$ also has a pdf with respect to Q_u, which is given by

$$f_{\zeta,\delta}^{(t)}(x; u) = f_{\zeta,\delta}^{(t)}(x)\exp\left\{-ux - t\log[M_{\zeta,\delta}(u)]\right\}$$

for each $x \in \mathbb{R}, t \geq 0$. Hence obtain the pricing formula

$$V(0) = \int_{\log(k/s)}^{\infty} f_{\zeta,\delta}^{(T)}(x; 1-u)\, dx - e^{-rT} k \int_{\log(k/s)}^{\infty} f_{\zeta,\delta}^{(T)}(x; u)\, dx.$$

As shown in Eberlein and Keller [89], this model seems to give a more accurate description of stock prices than the usual Black–Scholes formula. An online programme for calculating stock prices directly can be found at the website http://www.fdm.uni-freiburg.de/groups/financial/UK.

Finally we discuss the volatility, as promised. Suppose that, instead of a hyperbolic process, we revert to a Brownian motion model of logarithmic stock price growth and write $S(t) = e^{Z(t)}$ where $Z(t) = \sigma B(t)$ for each $t \geq 0$; then

---

[1] Note that the equivalent expression in Eberlein and Keller [89], p. 297, is given in terms of the parameter $\theta = -u$.

the volatility is given by $\sigma^2 = \mathbb{E}(Z(1)^2)$. By analogy, we define the volatility in the hyperbolic case by $\sigma^2 = \mathbb{E}(X_{\zeta,\delta}(1)^2)$. Using the results of Exercise 5.4.6 we obtain

$$\sigma^2 = \frac{\delta^2 K_2(\zeta)}{\zeta K_1(\zeta)}.$$

Further discussions of pricing using hyperbolic models can be found in Eberlein, Keller and Prause [88] and Bingham and Kiesel [48]. Bibby and Sørenson [42] introduced a variation on this model in which the stock prices satisfy a stochastic differential equation driven by Brownian motion but the coefficients are chosen so that the stock price is approximately a geometric hyperbolic Lévy process for large time.

5.4.7 Other Lévy process models for stock prices

Hyperbolic processes are one of a number of different models that have been advocated to replace the Black–Scholes process by using Lévy processes. Here we briefly survey some others. One of the first of these was proposed by Merton [220] and simply interlaced the Brownian noise with the jumps of a compound Poisson process. So this model lets the stock price process $S = (S(t), t \geq 0)$ evolve as

$$S(t) = S(0) \exp\left[\beta t + \sigma B(t) - \frac{1}{2}\sigma^2 t\right] \prod_{j=1}^{N(t)} Y_j$$

for each $t \geq 0$, where the sequence $(Y_n, n \in \mathbb{N})$ of i.i.d. random variables, the Poisson process $(N(t), t \geq 0)$ and the Brownian motion $(B(t), t \geq 0)$ are all independent. There has recently been renewed interest in this approach; see Benhamou [33].

Although we have ruled out the use of stable noise to model stock prices on empirical grounds there is still some debate about this, and recent stable-law models are discussed by McCulloch [205] and Meerschaert and Scheffler [217].

One of the criticisms levelled at the classical Black–Scholes formula is that it assumes constant volatility σ. We could in practice test this by using knowledge of known option prices for fixed values of the other parameters to deduce the corresponding value of σ. Although the Black–Scholes pricing formula (5.23) is not invertible as a function of σ, we can use numerical methods to estimate σ, and the values so obtained are called *implied volatilities*. Rather than giving constant values, the graph of volatility against strike price produces a curve known as the *volatility smile*; see e.g. Hull [137], Chapter 7. To explain the volatility smile many authors have modified the Black–Scholes formalism to allow σ to be replaced by an adapted process $(\sigma(t), t \geq 0)$. Of particular interest to fans of Lévy processes is work by Barndorff-Nielsen and Shephard [25], wherein $(\sigma(t)^2, t \geq 0)$ is taken to be an Ornstein–Uhlenbeck process driven by a non-Gaussian Lévy process; see Subsection 4.3.5.

Recently, it has been argued in some fascinating papers by Geman, Madan and Yor [114, 115] that asset-price processes should be modelled as pure jump processes of finite variation. On the one hand, where the corresponding intensity measure is infinite the stock price manifests 'infinite activity', and this is the mathematical signature of the jitter arising from the interaction of pure supply shocks and pure demand shocks. On the other hand, where the intensity measure is finite we have 'finite activity', and this corresponds to sudden shocks that can cause unexpected movements in the market, such as a terrorist atrocity or a major earthquake.

By a remarkable result of Monroe [231] any such process (in fact, any semimartingale) can be realised as $(B(T(t)), t \geq 0)$, where B is a standard Brownian motion and $(T(t), t \geq 0)$ is a *time change*, i.e. a non-negative increasing process of stopping times. Of course, we obtain a Lévy process when T is an independent subordinator, and models of this type that had already been applied to option pricing are the variance gamma process (see Subsection 1.3.2) of Madan and Seneta [206] and its generalisations by Carr *et al.* [66], [68]. Another subordinated process, which we discussed in Subsection 1.3.2 and which has been applied to model option prices, is the normal inverse Gaussian process of Barndorff-Nielsen ([27, 28], see also Barndorff-Nielsen and Prause [24]), although this is of not of finite variation.

Barndorff-Nielsen and Levendorskiĭ [22] have proposed a model where the logarithm of the stock price evolves as a Feller process of Lévy type obtained by introducing a spatial dependence into the four parameters of the normal inverse Gaussian process. Their analysis relies upon the use of the pseudo-differential-operator techniques introduced in Chapter 3. A common criticism of Lévy-processes-driven models (and this of course includes Black–Scholes) is that it is unrealistic to assume that stock prices have independent increments. The use of more general Feller processes arising from stochastic differential equations driven by Lévy processes certainly overcomes this problem, and this is one of the main themes of the next chapter. Another interesting approach is to model the noise in the basic geometric model (5.10) by a more complicated process. For example, Rogers [263] proposed a Gaussian process that does not have independent increments. This process is related to fractional Brownian motion, which has also been proposed as a log-price process;

however, as is shown in [263], such a model is inadequate since it allows arbitrage opportunities.

There are a number of different avenues opening up in finance for the application of Lévy processes. For example, pricing American options is more complicated than the European case as the freedom in choosing any time in $[0, T]$ to trade the option is an optional stopping problem. For progress in using Lévy processes in this context see Avram, Chan and Usabel [15] and references therein. Boyarchenko and Levendorskiĭ [56] is a very interesting paper on the application of Lévy processes to pricing barrier and touch-and-out options. The pricing formula is obtained using Wiener–Hopf factorisation, and pseudo-differential operators also play a role in the analysis. The same authors have recently published a monograph [57], in which a wide range of problems in option pricing are tackled by using *Lévy processes of exponential type*, i.e. those for which there exist $\lambda_1 < 0 < \lambda_2$ such that

$$\int_{-\infty}^{-1} e^{-\lambda_2 x} \nu(dx) + \int_{1}^{\infty} e^{-\lambda_1 x} \nu(dx) < \infty.$$

A number of specific Levy processes used in financial modelling, such as the normal inverse Gaussian and hyperbolic processes, are of this type.

In addition to option pricing, Eberlein and Raible [91] considered a model of the bond market driven by the exponential of a Lévy stochastic integral; see also Eberlein and Özkan [90]. For other directions, see the articles on finance in the volume [23].

5.5 Notes and further reading

Stochastic exponentials were first introduced by C. Doléans-Dade in [81]. Although the order was reversed in the text, the Cameron–Martin–Maruyama formula as first conceived by Cameron and Martin [63] preceded the more general Girsanov theorem [120]. The first proof of the martingale representation theorem was given by Kunita and Watanabe in their ground-breaking paper [181].

We have already given a large number of references to mathematical finance in the text. The paper from which they all flow is Black and Scholes [50]. It was followed soon after by Merton [219], in which the theory was axiomatised and the key role of stochastic differentials was clarified. In recognition of this achievement, Merton and Scholes received the 1997 Bank of Sweden Prize in Economic Sciences in Memory of Alfred Nobel (which is often incorrectly referred to as the 'Nobel Prize for Economics'); sadly, Black was no longer alive at this time. See http://www.nobel.se/economics/laureates/1997/index.html for more information about this.

As this book was nearing completion, Schoutens [280] appeared in print; this book is sure to be of great interest to aficionados of Lévy processes in finance.

5.6 Appendix: Bessel functions

The material given here can be found in any reasonable book on special functions. We draw the reader's attention to the monumental treatise of Watson [300] in particular.

Let $\nu \in \mathbb{R}$. *Bessel's equation of order ν* is of great importance in classical mathematical physics. It takes the form

$$x^2 \frac{d^2 y}{dx^2} + x \frac{dy}{dx} + (x^2 - \nu^2)y = 0 \qquad (5.27)$$

for each $x \in \mathbb{R}$.

A series solution yields the general solution (for $\nu \notin \mathbb{Z}$)

$$y(x) = C_1 J_\nu(x) + C_2 J_{-\nu}(x),$$

where C_1, C_2 are arbitrary constants and J_ν is a *Bessel function of the first kind*,

$$J_\nu(x) = \sum_{n=0}^{\infty} \frac{(-1)^n (x/2)^{\nu+2n}}{n!\, \Gamma(\nu+n+1)}. \qquad (5.28)$$

An alternative representation of the general solution is

$$y(x) = C_1 J_\nu(x) + C_2 Y_\nu(x),$$

where Y_ν is a *Bessel function of the second kind*:

$$Y_\nu(x) = 2\pi e^{i\pi \nu} \frac{J_\nu(x) \cos(\nu \pi) - J_{-\nu}(x)}{\sin(2\nu \pi)}.$$

We now consider the *modified Bessel equation of order ν*,

$$x^2 \frac{d^2 y}{dx^2} + x \frac{dy}{dx} - (x^2 + \nu^2)y = 0. \qquad (5.29)$$

We could clearly write the general solution in the form

$$y(x) = C_1 J_\nu(ix) + C_2 J_{-\nu}(ix),$$

but it is more convenient to introduce the *modified Bessel functions*

$$I_\nu(x) = e^{-i\nu\pi/2} J_\nu(ix),$$

as these are real-valued.

Bessel functions of the third kind were introduced by H. M. Macdonald and are defined by

$$K_\nu(x) = \left(\frac{\pi}{2}\right) \frac{I_{-\nu}(x) - I_\nu(x)}{\sin(\nu\pi)}.$$

The most important results for us are the recurrence relations

$$K_{\nu+1}(x) = K_{\nu-1}(x) + \frac{2\nu}{x} K_\nu(x) \tag{5.30}$$

and the key integral formula,

$$K_\nu(x) = \frac{1}{2} \int_0^\infty u^{\nu-1} \exp\left[-\frac{1}{2}x\left(u + \frac{1}{u}\right)\right] du. \tag{5.31}$$

Note that a straightforward substitution yields, in particular,

$$K_{1/2}(x) = \int_0^\infty \exp\left[-\frac{1}{2}x\left(u^2 + \frac{1}{u^2}\right)\right] du.$$

The following result is so beautiful that we include a short proof. This was communicated to me by Tony Sackfield.

Proposition 5.6.1

$$K_{1/2}(x) = \sqrt{\frac{\pi}{2x}} e^{-x}.$$

Proof Using the various definitions given above we find that

$$K_{1/2}(x) = \frac{\pi}{2} e^{i\pi/4} \left[J_{-1/2}(ix) + i J_{1/2}(ix)\right].$$

But it follows from (5.28) that

$$J_{1/2}(x) = \sqrt{\frac{2}{\pi x}} \sin(x)$$

and
$$J_{-1/2}(x) = \sqrt{\frac{2}{\pi x}} \cos(x),$$
and so
$$K_{1/2}(x) = \frac{\pi}{2} e^{i\pi/4} \sqrt{\frac{2}{\pi x}} [\cos(ix) + i \, \sin(ix)] = \sqrt{\frac{\pi}{2ix}} \left(\frac{1+i}{\sqrt{2}}\right) e^{-x}.$$

The result follows from the fact that $\dfrac{1+i}{\sqrt{i}} = \sqrt{2}.$ □

6

Stochastic differential equations

Summary After a review of first-order differential equations and their associated flows, we investigate stochastic differential equations (SDEs) driven by Brownian motion and an independent Poisson random measure. We establish the existence and uniqueness of solutions under the standard Lipschitz and growth conditions, using the Picard iteration technique. We then turn our attention to investigating properties of the solution. These are exhibited as stochastic flows and as multiplicative cocycles. The interlacing structure is established, and we prove the continuity of solutions as a function of their initial conditions. We then show that solutions of SDEs are Feller processes and compute their generators. Perturbations are studied via the Feynman–Kac formula. We briefly survey weak solutions and associated martingale problems. Finally, we study solutions of Marcus canonical equations and discuss the respective conditions under which these yield stochastic flows of homeomorphisms and diffeomorphisms.

One of the most important applications of Itô's stochastic integral is in the construction of *stochastic differential equations* (SDEs). These are important for a number of reasons.

(1) Their solutions form an important class of Markov processes where the infinitesimal generator of the corresponding semigroup can be constructed explicitly. Important subclasses that can be studied in this way include diffusion and jump-diffusion processes.
(2) Their solutions give rise to stochastic flows, and hence to interesting examples of random dynamical systems.
(3) They have many important applications to, for example, filtering, control, finance and physics.

Before we begin our study of SDEs, it will be useful to remind ourselves of some of the key features concerning the construction and elementary properties of ordinary differential equations (ODEs).

6.1 Differential equations and flows

Our main purpose in this section is to survey some of those aspects of ODEs that recur in the study of SDEs. We aim for a simple pedagogic treatment that will serve as a useful preparation and we do not attempt to establish optimal results. We mainly follow Abraham et al. [1], Section 4.1.

Let $b : \mathbb{R}^d \to \mathbb{R}^d$, so that $b = (b^1, \ldots, b^d)$ where $b^i : \mathbb{R}^d \to \mathbb{R}$ for $1 \leq i \leq d$.

We study the vector-valued differential equation

$$\frac{dc(t)}{dt} = b(c(t)), \tag{6.1}$$

with fixed initial condition $c(0) = c_0 \in \mathbb{R}^d$, whose solution, if it exists, is a curve $(c(t), t \in \mathbb{R})$ in $\mathbb{R}^d$.

Note that (6.1) is equivalent to the system of ODEs

$$\frac{dc^i(t)}{dt} = b^i(c(t))$$

for each $1 \leq i \leq d$.

To solve (6.1), we need to impose some structure on b. We say that b is *(globally) Lipschitz* if there exists $K > 0$ such that, for all $x, y \in \mathbb{R}^d$,

$$|b(x) - b(y)| \leq K|x - y|. \tag{6.2}$$

The expression (6.2) is called a *Lipschitz condition* on b and the constant K appearing therein is called a *Lipschitz constant*. Clearly if b is Lipschitz then it is continuous.

Exercise 6.1.1 Show that if b is differentiable with bounded partial derivatives then it is Lipschitz.

Exercise 6.1.2 Deduce that if b is Lipschitz then it satisfies a linear growth condition

$$|b(x)| \leq L(1 + |x|)$$

for all $x \in \mathbb{R}^d$, where $L = \max\{K, |b(0)|\}$.

The following existence and uniqueness theorem showcases the important technique of *Picard iteration*. We first rewrite (6.1) as an integral equation,

$$c(t) = c(0) + \int_0^t b(c(s))ds,$$

for each $t \in \mathbb{R}$. Readers should note that we are adopting the convention whereby $\int_0^t$ is understood to mean $\int_t^0$ when $t < 0$.

Theorem 6.1.3 *If $b : \mathbb{R}^d \to \mathbb{R}^d$ is (globally) Lipschitz, then there exists a unique solution $c : \mathbb{R} \to \mathbb{R}^d$ of the initial value problem* (6.1).

Proof Define a sequence $(c_n, n \in \mathbb{N} \cup \{0\})$, where $c_n : \mathbb{R} \to \mathbb{R}^d$ is defined by

$$c_0(t) = c_0, \qquad c_{n+1}(t) = c_0 + \int_0^t b(c_n(s))ds,$$

for each $n \geq 0, t \in \mathbb{R}$. Using induction and Exercise 6.1.2, it is straightforward to deduce that each c_n is integrable on $[0, t]$, so that the sequence is well defined.

Define $\alpha_n = c_n - c_{n-1}$ for each $n \in \mathbb{N}$. By Exercise 6.1.2, for each $t \in \mathbb{R}$ we have

$$|\alpha_1(t)| \leq |b(c_0)| \, |t| \leq Mt \tag{6.3}$$

where $M = L(1 + |c_0|)$.

Using the Lipschitz condition (6.2), for each $t \in \mathbb{R}, n \in \mathbb{N}$, we obtain

$$|\alpha_{n+1}(t)| \leq \int_0^t |b(c_n(s)) - b(c_{n-1}(s))|ds \leq K \int_0^t |\alpha_n(s)|ds \tag{6.4}$$

and a straightforward inductive argument based on (6.3) and (6.4) yields the estimate

$$|\alpha_n(t)| \leq \frac{MK^{n-1}|t|^n}{n!}$$

for each $t \in \mathbb{R}$. Hence for all $t > 0$ and $n, m \in \mathbb{N}$ with $n > m$, we have

$$\sup_{0 \leq s \leq t} |c_n(s) - c_m(s)| \leq \sum_{r=m+1}^n \sup_{0 \leq s \leq t} |\alpha_r(s)| \leq \sum_{r=m+1}^n \frac{MK^{r-1}|t|^r}{r!}.$$

Hence $(c_n, n \in \mathbb{N})$ is uniformly Cauchy and so uniformly convergent on finite intervals $[0, t]$ (and also on intervals of the form $[-t, 0]$ by a similar argument.) Define $c = (c(t), t \in \mathbb{R})$ by

$$c(t) = \lim_{n \to \infty} c_n(t) \qquad \text{for each } t \in \mathbb{R}.$$

To see that c solves (6.1), note first that by (6.2) and the uniformity of the

6.1 Differential equations and flows

convergence we have, for each $t \in \mathbb{R}$, $n \in \mathbb{N}$,

$$\left| \int_0^t b(c(s))ds - \int_0^t b(c_n(s))ds \right| \leq \int_0^t |b(c(s)) - b(c_n(s))|ds$$
$$\leq Kt \sup_{0 \leq s \leq t} |c(s) - c_n(s)|$$
$$\to 0 \quad \text{as} \quad n \to \infty.$$

Hence, for each $t \in \mathbb{R}$,

$$c(t) - c(0) + \int_0^t b(c(s))ds = \lim_{n \to \infty} \left[c_{n+1}(t) - c(0) + \int_0^t b(c_n(s))ds \right]$$
$$= 0.$$

Finally, we show that the solution is unique. Assume that c' is another solution of (6.1) and, for each $n \in \mathbb{N}$, $t \in \mathbb{R}$, define

$$\beta_n(t) = c_n(t) - c'(t),$$

so that $\beta_{n+1}(t) = \int_0^t b(\beta_n(s))ds$.

Arguing as above, we obtain the estimate

$$|\beta_n(t)| \leq \frac{MK^{n-1}|t|^n}{n!},$$

from which we deduce that each $\lim_{n \to \infty} \beta_n(t) = 0$, so that $c(t) = c'(t)$ as required. $\square$

Note that by the uniformity of the convergence in the proof of Theorem 6.1.3 the map $t \to c(t)$ is continuous from $\mathbb{R}$ to $\mathbb{R}^d$.

Now that we have constructed unique solutions to equations of the type (6.1), we would like to explore some of their properties. A useful tool in this regard is *Gronwall's inequality*, which will also play a major role in the analysis of solutions to SDEs.

Proposition 6.1.4 (Gronwall's inequality) *Let $[a, b]$ be a closed interval in $\mathbb{R}$ and $\alpha, \beta : [a, b] \to \mathbb{R}$ be non-negative with α locally bounded and β integrable. If there exists $C \geq 0$ such that, for all $t \in [a, b]$,*

$$\alpha(t) \leq C + \int_a^t \alpha(s)\beta(s)ds, \tag{6.5}$$

then we have

$$\alpha(t) \leq C \exp\left[\int_a^t \beta(s)ds \right]$$

for all $t \in [a, b]$.

Proof First assume that $C > 0$ and let $h : [a, b] \to (0, \infty)$ be defined by

$$h(t) = C + \int_a^t \alpha(s)\beta(s)ds$$

for all $t \in [a, b]$. By Lebesgue's differentiation theorem (see e.g. Cohn [73], p. 187), h is differentiable on (a, b), with

$$h'(t) = \alpha(t)\beta(t) \leq h(t)\beta(t)$$

by (6.5), for (Lebesgue) almost all $t \in (a, b)$.

Hence $h'(t)/h(t) \leq \beta(t)$ (a.e.) and the required result follows on integrating both sides between a and b.

Now suppose that $C = 0$; then, by the above analysis, for each $t \in [a, b]$ we have $\alpha(t) \leq (1/n) \exp\left[\int_a^b \beta(s)ds\right]$ for each $n \in \mathbb{N}$, hence $\alpha(t) = 0$ as required. □

Note that in the case where equality holds in (6.5), Gronwall's inequality is (essentially) just the familiar integrating factor method for solving first-order linear differential equations.

Now let us return to our consideration of the solutions to (6.1). There are two useful perspectives from which we can regard these.

- If we fix the initial condition $c_0 = x \in \mathbb{R}$ then the solution is a curve $(c(t), t \in \mathbb{R})$ in $\mathbb{R}^d$ passing through x when $t = 0$.
- If we allow the initial condition to vary, we can regard the solution as a function of two variables $(c(t, x), t \in \mathbb{R}, x \in \mathbb{R}^d)$ that generates a family of curves.

It is fruitful to introduce some notation that allows us to focus more clearly on our ability to vary the initial conditions. To this end we define for each $t \in \mathbb{R}, x \in \mathbb{R}^d$,

$$\xi_t(x) = c(t, x),$$

so that each $\xi_t : \mathbb{R}^d \to \mathbb{R}^d$.

Lemma 6.1.5 *For each $t \in \mathbb{R}$, $x, y \in \mathbb{R}^d$,*

$$|\xi_t(x) - \xi_t(y)| \leq e^{K|t|}|x - y|,$$

so that, in particular, each $\xi_t : \mathbb{R}^d \to \mathbb{R}^d$ is continuous.

6.1 Differential equations and flows

Proof Fix t, x and y and let $\gamma_t = |\xi_t(x) - \xi_t(y)|$. By (6.1) and (6.2) we obtain

$$\gamma_t \leq |x - y| + \int_0^t |b(\xi_s(x)) - b(\xi_s(y))| ds \leq |x - y| + K \int_0^t \gamma_s ds,$$

and the result follows by Gronwall's inequality. □

Suppose now that b is C^1; then we may differentiate b at each $x \in \mathbb{R}^d$, and its derivative $Db(x) : \mathbb{R}^d \to \mathbb{R}^d$ is the Jacobian matrix of b. We will now investigate the implications of the smoothness of b for the solution $(\xi_t, t \in \mathbb{R})$.

Exercise 6.1.6 Let $(\xi_t, t \geq 0)$ be the solution of (6.1) and suppose that $b \in C_b^1(\mathbb{R}^d)$. Deduce that for each $x \in \mathbb{R}^d$ there is a unique solution to the $d \times d$-matrix-valued differential equation

$$\frac{d}{dt}\gamma(t, x) = Db(\xi_t(x))\gamma(t, x)$$

with initial condition $\gamma(0, x) = I$.

Theorem 6.1.7 If $b \in C_b^k(\mathbb{R}^d)$ for some $k \in \mathbb{N}$, then $\xi_t \in C^k(\mathbb{R}^d)$ for each $t \in \mathbb{R}$.

Proof We begin by considering the case $k = 1$.

Let γ be as in Exercise 6.1.6. We will show that ξ_t is differentiable and that $D\xi_t(x) = \gamma(t, x)$ for each $t \in \mathbb{R}$, $x \in \mathbb{R}^d$.

Fix $h \in \mathbb{R}^d$ and let $\theta(t, h) = \xi_t(x + h) - \xi_t(x)$. Then, by (6.1),

$$\theta(t, h) - \gamma(t, x)(h) = \int_0^t \left[b(\xi_s(x + h)) - b(\xi_s(x))\right] ds$$

$$- \int_0^t Db(\xi_s(x))\gamma(s, x)(h) ds$$

$$= I_1(t) + I_2(t) \tag{6.6}$$

where

$$I_1(t) = \int_0^t \left[b(\xi_s(x + h)) - b(\xi_s(x)) - Db(\xi_s(x))\theta(s, h)\right] ds$$

and

$$I_2(t) = \int_0^t Db(\xi_s(x))\big(\theta(s, h) - \gamma(s, x)(h)\big) ds.$$

By the mean value theorem,

$$|b(\xi_s(x+h)) - b(\xi_s(x))| \leq C|\theta(s,h)|$$

where $C = d \sup_{y \in \mathbb{R}^d} \max_{1 \leq i,j \leq d} |Db(y)_{ij}|$. Hence, by Lemma 6.1.5,

$$|I_1(t)| \leq 2Ct \sup_{0 \leq s \leq t} |\theta(s,h)| \leq 2Ct|h|e^{K|t|}, \tag{6.7}$$

while

$$|I_2(t)| \leq C' \int_0^t |\theta(s,h) - \gamma(s,x)(h)|ds,$$

where $C' = Cd^{1/2}$.

Substitute (6.7) and (6.1) in (6.6) and apply Gronwall's inequality to deduce that

$$|\theta(t,h) - \gamma(t,x)(h)| \leq 2Ct|h|e^{(K+C')|t|},$$

from which the required result follows. From the result of Exercise 6.1.6, we also have the 'derivative flow' equation

$$\frac{dD\xi_t(x)}{dt} = Db(\xi_t(x))D\xi_t(x).$$

The general result is proved by induction using the argument given above. $\square$

Exercise 6.1.8 Under the conditions of Theorem 6.1.7, show that, for all $x \in \mathbb{R}^d$, the map $t \to \xi_t(x)$ is C^{k+1}.

We recall that a bijection $\phi : \mathbb{R}^d \to \mathbb{R}^d$ is a *homeomorphism* if ϕ and ϕ^{-1} are both continuous and a C^k-*diffeomorphism* if ϕ and ϕ^{-1} are both C^k.

A family $\phi = \{\phi_t, t \in \mathbb{R}\}$ of homeomorphisms of $\mathbb{R}^d$ is called a *flow* if

$$\phi_0 = I \quad \text{and} \quad \phi_s \phi_t = \phi_{s+t} \tag{6.8}$$

for all $s, t \in \mathbb{R}$. If each ϕ_t is a C^k-diffeomorphism, we say that ϕ is a *flow of C^k-diffeomorphisms*.

Equation (6.8) is sometimes called the *flow property*. Note that an immediate consequence of it is that

$$\phi_t^{-1} = \phi_{-t}$$

for all $t \in \mathbb{R}$, so that (6.8) tells us that ϕ is a one-parameter group of homeomorphisms of $\mathbb{R}^d$.

6.1 Differential equations and flows

Lemma 6.1.9 *If $\phi = \{\phi_t, t \geq 0\}$ is a family of C^k-mappings from $\mathbb{R}^d$ to $\mathbb{R}^d$ such that $\phi_0 = I$ and $\phi_s \phi_t = \phi_{s+t}$ for all $s, t \in \mathbb{R}$ then ϕ is a flow of C^k-diffeomorphisms.*

Proof It is enough to observe that, for all $t \in \mathbb{R}$, we have $\phi_{-t}\phi_t = \phi_t \phi_{-t} = I$, so that each ϕ_t has a two-sided C^k-inverse and thus is a C^k-diffeomorphism. $\square$

Theorem 6.1.10 *Let $\xi = (\xi_t, t \in \mathbb{R})$ be the unique solution of (6.1). If $b \in C_b^k(\mathbb{R}^d)$, then ξ is a flow of C^k-diffeomorphisms.*

Proof We seek to apply Lemma 6.1.9. By Theorem 6.1.7 we see that each $\xi_t \in C^k(\mathbb{R}^d)$, so we must establish the flow property.

The fact that $\xi_0 = I$ is immediate from (6.1). Now, for each $x \in \mathbb{R}^d$ and $s, t \in \mathbb{R}$,

$$\xi_{t+s}(x) = x + \int_0^{t+s} b(\xi_u(x))du$$
$$= x + \int_0^s b(\xi_u(x))du + \int_s^{t+s} b(\xi_u(x))du$$
$$= \xi_s(x) + \int_s^{t+s} b(\xi_u(x))du$$
$$= \xi_s(x) + \int_0^t b(\xi_{u+s}(x))du.$$

However, we also have

$$\xi_t(\xi_s(x)) = \xi_s(x) + \int_0^t b(\xi_u(\xi_s(x)))du,$$

and it follows that $\xi_{t+s}(x) = \xi_t(\xi_s(x))$ by the uniqueness of solutions to (6.1). $\square$

Exercise 6.1.11 Deduce that if b is Lipschitz then the solution $\xi = (\xi(t), t \in \mathbb{R})$ is a flow of homeomorphisms.

Exercise 6.1.12 Let ξ be the solution of (6.1) and let $f \in C^k(\mathbb{R}^d)$; show that

$$\frac{df}{dt}(\xi_t(x)) = b^i(\xi_t(x))\frac{\partial f}{\partial x_i}(\xi_t(x)). \tag{6.9}$$

If $b \in C^k(\mathbb{R}^d)$, it is convenient to consider the linear mapping $Y : C^{k+1}(\mathbb{R}^d) \to C^k(\mathbb{R}^d)$ defined by

$$(Yf)(x) = b^i(x)\frac{\partial f}{\partial x_i}(x)$$

for each $f \in C^k(\mathbb{R}^d)$, $x \in \mathbb{R}^d$. The mapping Y is called a C^k-*vector field*. We denoted as $\mathcal{L}_k(\mathbb{R}^d)$ the set of all C^k-vector fields on $\mathbb{R}^d$.

Exercise 6.1.13 Let X, Y and $Z \in \mathcal{L}_k(\mathbb{R}^d)$, where $k > 2$.

(1) Show that $\alpha X + \beta Y \in \mathcal{L}_k(\mathbb{R}^d)$ for all $\alpha, \beta \in \mathbb{R}$.
(2) Show that the commutator $[X, Y] \in \mathcal{L}_k(\mathbb{R}^d)$, where

$$([X, Y]f)(x) = (X(Y(f))(x) - (Y(X(f)))(x)$$

for each $f \in C^k(\mathbb{R}^d)$, $x \in \mathbb{R}^d$.
(3) Establish the *Jacobi identity*

$$[X, [Y, Z]] + [Y, [Z, X]] + [Z, [X, Y]] = 0.$$

We saw in the last exercise that $\mathcal{L}_k(\mathbb{R}^d)$ is a *Lie algebra*, i.e. it is a real vector space equipped with a binary operation $[\cdot, \cdot]$ that satisfies the Jacobi identity and the condition $[X, X] = 0$ for all $X \in \mathcal{L}_k(\mathbb{R}^d)$. Note that a Lie algebra is not an 'algebra' in the usual sense since the commutator bracket is not associative (we have the Jacobi identity instead).

In general, a vector field $Y = b^i \partial_i$ is said to be *complete* if the associated differential equation (6.1) has a unique solution $(\xi(t)(x), t \in \mathbb{R})$ for all initial conditions $x \in \mathbb{R}^d$. The vector field Y fails to be complete if the solution only exists locally, e.g. for all $t \in (a, b)$ where $-\infty < a < b < \infty$, and 'blows up' at a and b.

Exercise 6.1.14 Let $d = 1$ and $Y(x) = x^2 d/dx$ for each $x \in \mathbb{R}$. Show that Y is not complete.

If Y is complete, each $(\xi(t)(x), t \in \mathbb{R})$ is called the *integral curve* of Y through the point x, and the notation $\xi(t)(x) = \exp(Y)(x)$ is often employed to emphasise that, from an infinitesimal viewpoint, Y is the fundamental object from which all else flows. We call 'exp' the *exponential map*. These ideas all extend naturally to the more general set-up where $\mathbb{R}^d$ is replaced by a differentiable manifold.

6.2 Stochastic differential equations – existence and uniqueness

We now turn to the main business of this chapter. Let $(\Omega, \mathcal{F}, P)$ be a probability space equipped with a filtration $\{\mathcal{F}_t, t \geq 0\}$ that satisfies the usual hypotheses. Let $B = (B(t), t \geq 0)$ be an r-dimensional standard Brownian motion and N an independent Poisson random measure on $\mathbb{R}^+ \times (\mathbb{R}^d - \{0\})$ with associated compensator $\tilde{N}$ and intensity measure ν, where we assume that ν is a Lévy measure. We always assume that B and N are independent of $\mathcal{F}_0$.

In the last section, we considered ODEs of the form

$$\frac{dy(t)}{dt} = b(y(t)), \tag{6.10}$$

whose solution $(y(t), t \in \mathbb{R})$ is a curve in $\mathbb{R}^d$.

We begin by rewriting this 'Itô-style' as

$$dy(t) = b(y(t))dt. \tag{6.11}$$

Now restrict the parameter t to the non-negative half-line $\mathbb{R}^+$ and consider $y = (y(t), t \geq 0)$ as the evolution in time of the state of a system from some initial value $y(0)$. We now allow the system to be subject to random noise effects, which we introduce additively in (6.11). In general, these might be described in terms of arbitrary semimartingales (see e.g. Protter [255]), but in line with the usual philosophy of this book, we will use the 'noise' associated with a Lévy process.

We will focus on the following SDE:

$$dY(t) = b(Y(t-))dt + \sigma(Y(t-))dB(t)$$
$$+ \int_{|x|<c} F(Y(t-), x)\tilde{N}(dt, dx)$$
$$+ \int_{|x|\geq c} G(Y(t-), x)N(dt, dx), \tag{6.12}$$

which is a convenient shorthand for the system of SDEs

$$dY^i(t) = b^i(Y(t-))dt + \sigma^i_j(Y(t-))dB^j(t)$$
$$+ \int_{|x|<c} F^i(Y(t-), x)\tilde{N}(dt, dx)$$
$$+ \int_{|x|\geq c} G^i(Y(t-), x)N(dt, dx), \tag{6.13}$$

where each $1 \leq i \leq d$. Here the mappings $b^i : \mathbb{R}^d \to \mathbb{R}, \sigma^i_j : \mathbb{R}^d \to \mathbb{R}, F^i : \mathbb{R}^d \times \mathbb{R}^d \to \mathbb{R}$ and $G^i : \mathbb{R}^d \times \mathbb{R}^d \to \mathbb{R}$ are all assumed to be measurable for $1 \leq i \leq d, 1 \leq j \leq r$. Further conditions on these mappings will follow later. The convenient parameter $c \in [0, \infty]$ allows us to specify what we mean by 'large' and 'small' jumps in specific applications. Quite often, it will be convenient to take $c = 1$. If we want to put both 'small' and 'large' jumps on the same footing we take $c = \infty$ (or 0), so that the term involving G (or F, respectively) is absent in (6.12)).

We will always consider (6.12), or equivalently (6.13), as a random initial-value problem with a fixed initial condition $Y(0) = Y_0$, where Y_0 is a given $\mathbb{R}^d$-valued random vector. Sometimes we may want to fix $Y_0 = y_0$ (a.s.), where $y_0 \in \mathbb{R}^d$.

In order to give (6.13) a rigorous meaning we rewrite it in integral form, for each $t \geq 0$, $1 \leq i \leq d$, as

$$Y^i(t) = Y^i(0) + \int_0^t b^i(Y(t-))dt + \int_0^t \sigma^i_j(Y(t-))dB^j(t)$$
$$+ \int_0^t \int_{|x|<c} F^i(Y(t-), x)\tilde{N}(dt, dx)$$
$$+ \int_0^t \int_{|x|\geq c} G^i(Y(t-), x)N(dt, dx) \quad \text{a.s.} \quad (6.14)$$

The solution to (6.14), when it exists, will be an $\mathbb{R}^d$-valued stochastic process $(Y(t), t \geq 0)$ with each $Y(t) = (Y^1(t), \ldots, Y^d(t))$. Note that we are implicitly assuming that Y has left-limits in our formulation of (6.14), and we will in fact be seeking càdlàg solutions so that this is guaranteed.

As we have specified the noise B and N in advance, any solution to (6.14) is sometimes called a *strong solution* in the literature. There is also a notion of a *weak solution*, which we will discuss in Subsection 6.7.3. We will require solutions to (6.14) to be unique, and there are various notions of uniqueness available. The strongest of these, which we will look for here, is to require our solutions to be *pathwise unique*, i.e. if $Y_1 = (Y_1(t), t \geq 0)$ and $Y_2 = (Y_2(t), t \geq 0)$ are both solutions to (6.14) then $P(Y_1(t) = Y_2(t)$ for all $t \geq 0) = 1$.

The term in (6.14) involving large jumps is that controlled by G. This is easy to handle using interlacing, and it makes sense to begin by omitting this term and concentrate on the study of the equation driven by continuous noise interspersed with small jumps. To this end, we introduce the *modified* SDE

$$dZ(t) = b(Z(t-))dt + \sigma(Z(t-))dB(t) + \int_{|x|<c} F(Z(t-), x)\tilde{N}(dt, dx),$$
$$(6.15)$$

with initial condition $Z(0) = Z_0$.

6.2 Stochastic differential equations

We now impose some conditions on the mappings b, σ and F that will enable us to solve (6.15). First, for each $x, y \in \mathbb{R}^d$ we introduce the $d \times d$ matrix

$$a(x, y) = \sigma(x)\sigma(y)^{\mathrm{T}},$$

so that $a^{ik}(x, y) = \sum_{j=1}^{r} \sigma_j^i(x)\sigma_j^k(y)$ for each $1 \leq i, k \leq d$.

We will have need of the matrix seminorm on $d \times d$ matrices, given by

$$||a|| = \sum_{i=1}^{d} |a_i^i|.$$

We impose the following two conditions.

(C1) Lipschitz condition There exists $K_1 > 0$ such that, for all $y_1, y_2 \in \mathbb{R}^d$,

$$|b(y_1) - b(y_2)|^2 + ||a(y_1, y_1) - 2a(y_1, y_2) + a(y_2, y_2)||$$
$$+ \int_{|x|<c} |F(y_1, x) - F(y_2, x)|^2 \nu(dx) \leq K_1 |y_1 - y_2|^2. \quad (6.16)$$

(C2) Growth condition There exists $K_2 > 0$ such that, for all $y \in \mathbb{R}^d$,

$$|b(y)|^2 + ||a(y, y)|| + \int_{|x|<c} |F(y, x)|^2 \nu(dx) \leq K_2(1 + |y|^2).$$
$$(6.17)$$

We make some comments on these.

First, the condition $||a(y_1, y_1) - 2a(y_1, y_2) + a(y_2, y_2)|| \leq L|y_1 - y_2|^2$, for some $L > 0$, is sometimes called *bi-Lipschitz continuity*. It may seem at odds with the other terms on the left-hand side of (6.16) but this is an illusion. A straightforward calculation yields

$$||a(y_1, y_1) - 2a(y_1, y_2) + a(y_2, y_2)|| = \sum_{i=1}^{d} \sum_{j=1}^{r} \left[\sigma_j^i(y_1) - \sigma_j^i(y_2)\right]^2,$$

and if you take $d = r = 1$ then

$$|a(y_1, y_1) - 2a(y_1, y_2) + a(y_2, y_2)| = |\sigma(y_1) - \sigma(y_2)|^2.$$

Exercise 6.2.1 If a is bi-Lipschitz continuous, show that there exists $L_1 > 0$ such that

$$||a(y, y)|| \leq L_1(1 + ||y||^2)$$

for all $y \in \mathbb{R}^d$.

Our second comment on the conditions is this: if you take $F = 0$, it follows from Exercises 6.1.2 and 6.2.1 that the growth condition (C2) is a consequence of the Lipschitz condition (C1). Hence in the case of non-zero F, in the presence of (C1), (C2) is equivalent to the requirement that there exists $M > 0$ such that, for all $y \in \mathbb{R}^d$,

$$\int_{|x|<c} |F(y, x)|^2 \nu(dx) \leq M(1 + |y|^2).$$

Exercise 6.2.2

(1) Show that if ν is finite, then the growth condition is a consequence of the Lipschitz condition.
(2) Show that if $F(y, x) = H(y) f(x)$ for all $y \in \mathbb{R}^d$, $|x| \leq c$, where H is Lipschitz continuous and $\int_{|x|\leq c} |f(x)|^2 \nu(dx) < \infty$, then the growth condition is a consequence of the Lipschitz condition.

Having imposed conditions on our coefficients, we now discuss the initial condition. Throughout this chapter, we will always deal with the *standard initial condition* $Y(0) = Y_0$ (a.s.), for which Y_0 is $\mathcal{F}_0$-measurable. Hence $Y(0)$ is independent of the noise B and N.

Finally we note that, throughout the remainder of this chapter, we will frequently employ the following inequality for $n \in \mathbb{N}$ and $x_1, x_2, \ldots x_n \in \mathbb{R}$:

$$|x_1 + x_2 + \cdots + x_n|^2 \leq n(|x_1|^2 + |x_2| + \cdots + |x_n|^2). \qquad (6.18)$$

This is easily verified by using induction and the Cauchy–Schwarz inequality.

Our existence and uniqueness theorem will employ the technique of Picard iteration, which served us well in the ODE case (Theorem 6.1.3); cf. Ikeda and Watanabe [140], Chapter 4, Section 9.

Theorem 6.2.3 *Assume the Lipschitz and growth conditions. There exists a unique solution $Z = (Z(t), t \geq 0)$ to the modified SDE (6.15) with the standard initial condition. The process Z is adapted and càdlàg.*

Our strategy is to first carry out the proof of existence and uniqueness in the case $\mathbb{E}(|Z_0|^2) < \infty$ and then consider the case $\mathbb{E}(|Z_0|^2) = \infty$.

6.2 Stochastic differential equations

Proof of existence for $\mathbb{E}(|Z_0|^2) < \infty$ Define a sequence of processes $(Z_n, n \in \mathbb{N} \cup \{0\})$ by $Z_0(t) = Z_0$ and, for all $n \in \mathbb{N} \cup \{0\}, t \geq 0$,

$$dZ_{n+1}(t) = b(Z_n(t-))dt + \sigma(Z_n(t-))dB(t) + \int_{|x|<c} F(Z_n(t-), x)\tilde{N}(dt, dx).$$

A simple inductive argument and use of Theorem 4.2.12 demonstrates that each Z_n is adapted and càdlàg.

For each $1 \leq i \leq d, n \in \mathbb{N} \cup \{0\}, t \geq 0$, we have

$$Z_{n+1}^i(t) - Z_n^i(t)$$
$$= \int_0^t \left[b^i(Z_n(s-)) - b^i(Z_{n-1}(s-)) \right] ds$$
$$+ \int_0^t \left[\sigma_j^i(Z_n(s-)) - \sigma_j^i(Z_{n-1}(s-)) \right] dB^j(s)$$
$$+ \int_0^t \int_{|x|<c} \left[F^i(Z_n(s-), x) - F^i(Z_{n-1}(s-), x) \right] \tilde{N}(ds, dx).$$

We need to obtain some inequalities, and we begin with the case $n = 0$.

First note that on using the inequality (6.18), with $n = 3$, we have

$$|Z_1(t) - Z_0(t)|^2$$
$$= \sum_{i=1}^d \left[\int_0^t b^i(Z(0))ds + \int_0^t \sigma_j^i(Z(0))dB^j(s) \right.$$
$$\left. + \int_0^t \int_{|x|<c} F^i(Z(0), x)\tilde{N}(ds, dx) \right]^2$$
$$\leq 3 \sum_{i=1}^d \left\{ \left[\int_0^t b^i(Z(0))ds \right]^2 + \left[\int_0^t \sigma_j^i(Z(0))dB^j(s) \right]^2 \right.$$
$$\left. + \left[\int_0^t \int_{|x|<c} F^i(Z(0), x)\tilde{N}(ds, dx) \right]^2 \right\}$$
$$= 3 \sum_{i=1}^d \left\{ t^2 \left[b^i(Z(0)) \right]^2 + \left[\sigma_j^i(Z(0))B^j(t) \right]^2 \right.$$
$$\left. + \left[\int_{|x|<c} F^i(Z(0), x)\tilde{N}(t, dx) \right]^2 \right\}$$

for each $t \geq 0$. We now take expectations and apply Doob's martingale inequality to obtain

$$\mathbb{E}\left(\sup_{0 \leq s \leq t} |Z_1(s) - Z_0(s)|^2\right)$$
$$\leq 3t^2 \, \mathbb{E}(|b(Z(0))|^2) + 12t \, \mathbb{E}(\|a(Z(0), Z(0))\|)$$
$$+ 12t \int_{|x|<c} \mathbb{E}(|F(Z(0), x)|)^2 \, \nu(dx).$$

On applying the growth condition (C2), we can finally deduce that

$$\mathbb{E}\left(\sup_{0 \leq s \leq t} |Z_1(s) - Z_0(s)|^2\right) \leq C_1(t) t K_2 (1 + \mathbb{E}(|Z(0)|^2)), \quad (6.19)$$

where $C_1(t) = \max\{3t, 12\}$.

We now consider the case for general $n \in \mathbb{N}$. Arguing as above, we obtain

$$\mathbb{E}\left(\sup_{0 \leq s \leq t} |Z_{n+1}(s) - Z_n(s)|^2\right)$$
$$\leq \sum_{i=1}^{d} \Bigg[3 \, \mathbb{E}\left(\sup_{0 \leq s \leq t} \left\{\int_0^s [b^i(Z_n(u-)) - b^i(Z_{n-1}(u-))]du\right\}^2\right)$$
$$+ 12 \, \mathbb{E}\left(\left\{\int_0^t [\sigma_j^i(Z_n(s-)) - \sigma_j^i(Z_{n-1}(s-))]dB^j(s)\right\}^2\right)$$
$$+ 12 \, \mathbb{E}\left(\left\{\int_0^t \int_{|x|<c} \big[F^i(Z_n(s-), x)$$
$$- F^i(Z_{n-1}(s-), x)\big]\tilde{N}(ds, dx)\right\}^2\right)\Bigg].$$

By the Cauchy–Schwarz inequality, for all $s \geq 0$,

$$\left\{\int_0^s [b^i(Z_n(u-)) - b^i(Z_{n-1}(u-))]du\right\}^2$$
$$\leq s \int_0^s [b^i(Z_n(u-)) - b^i(Z_{n-1}(u-))]^2 du$$

6.2 Stochastic differential equations

and so, by Itô's isometry, we obtain

$$\mathbb{E}\left(\sup_{0\leq s\leq t} |Z_{n+1}(s) - Z_n(s)|^2\right)$$
$$\leq C_1(t)\left[\int_0^t \mathbb{E}\big(|b(Z_n(s-)) - b(Z_{n-1}(s-))|^2\big)ds \right.$$
$$+ \int_0^t \mathbb{E}\big(||a(Z_n(s-), Z_n(s-)) - 2a(Z_n(s-), Z_{n-1}(s-))$$
$$+ a(Z_{n-1}(s-), Z_{n-1}(s-))||\big)ds$$
$$\left. + \int_0^t\!\!\int_{|x|<c} \mathbb{E}\big(|F(Z_n(t-), x) - F(Z_{n-1}(t-), x)|^2\big)v(dx)\right].$$

We now apply the Lipschitz condition (C1) to find that

$$\mathbb{E}\left(\sup_{0\leq s\leq t} |Z_{n+1}(s) - Z_n(s)|^2\right)$$
$$\leq C_1(t)K_1 \int_0^t \mathbb{E}\left(\sup_{0\leq u\leq s} |Z_n(u) - Z_{n-1}(u)|^2\right)ds \quad (6.20)$$

By induction based on (6.19) and (6.20), we thus deduce the key estimate

$$\mathbb{E}\left(\sup_{0\leq s\leq t} |Z_n(s) - Z_{n-1}(s)|^2\right) \leq \frac{C_2(t)^n K_3^n}{n!} \quad (6.21)$$

for all $n \in \mathbb{N}$, where $C_2(t) = tC_1(t)$ and

$$K_3 = \max\{K_1, K_2[1 + \mathbb{E}(|Z(0)|^2)]\}.$$

Our first observation is that $(Z_n(t), t \geq 0)$ is convergent in L^2 for each $t \geq 0$. Indeed, for each $m, n \in \mathbb{N}$ we have (using $||\cdot||_2 = [\mathbb{E}(|\cdot|^2)]^{1/2}$ to denote the L^2-norm), for each $0 \leq s \leq t$,

$$||Z_n(s) - Z_m(s)||_2 = \sum_{r=m+1}^n ||Z_r(s) - Z_{r-1}(s)||_2 \leq \sum_{r=m+1}^n \frac{C_2(t)^{r/2} K_3^{r/2}}{(r!)^{1/2}},$$

and, since the series on the right converges, we have that each $(Z_n(s), n \in \mathbb{N})$ is Cauchy and hence convergent to some $Z(s) \in L^2(\Omega, \mathcal{F}, P)$. We denote as Z the process $(Z(t), t \geq 0)$. A standard limiting argument yields the useful estimate

$$||Z(s) - Z_n(s)||_2 \leq \sum_{r=n+1}^\infty \frac{C_2(t)^{r/2} K_3^{r/2}}{(r!)^{1/2}}, \quad (6.22)$$

for each $n \in \mathbb{N} \cup \{0\}$, $0 \leq s \leq t$.

We also need to establish the almost sure convergence of $(Z_n, n \in \mathbb{N})$. Applying the Chebyshev–Markov inequality in (6.21), we deduce that

$$P\left(\sup_{0 \leq s \leq t} |Z_n(s) - Z_{n-1}(s)| \geq \frac{1}{2^n}\right) \leq \frac{[4K_3 C_2(t)]^n}{n!},$$

from which we see that

$$P\left(\limsup_{n \to \infty} \sup_{0 \leq s \leq t} |Z_n(s) - Z_{n-1}(s)| \geq \frac{1}{2^n}\right) = 0,$$

by Borel's lemma. Arguing as in Theorem 2.5.2, we deduce that $(Z_n, n \in \mathbb{N} \cup \{0\})$ is almost surely uniformly convergent on finite intervals $[0, t]$ to Z, from which it follows that Z is adapted and càdlàg.

Now we must verify that Z really satisfies the SDE. Define a stochastic process $\tilde{Z} = (\tilde{Z}(t), t \geq 0)$ by

$$\tilde{Z}^i(t) = Z_0^i + \int_0^t b^i(Z(s-))ds + \int_0^t \sigma_j^i(Z(s-))dB^j(s)$$
$$+ \int_0^t \int_{|x|<c} F^i(Z(s-), x)\tilde{N}(ds, dx)$$

for each $1 \leq i \leq d, t \geq 0$. Hence, for each $n \in \mathbb{N} \cup \{0\}$,

$$\tilde{Z}^i(t) - Z_n^i(t) = \int_0^t \left[b^i(Z(s-)) - b^i(Z_n(s-))\right]ds$$
$$+ \int_0^t \left[\sigma_j^i(Z(s-)) - \sigma_j^i(Z_n(s-))\right]dB^j(s)$$
$$+ \int_0^t \int_{|x|<c} \left[F^i(Z(s-), x) - F^i(Z_n(s-), x)\right]\tilde{N}(ds, dx).$$

Now using the same argument with which we derived (6.20) and then applying (6.22), we obtain for all $0 \leq s \leq t < \infty$,

$$\mathbb{E}(|\tilde{Z}(s) - Z_n(s)|^2) \leq C_1(t)K_1 \int_0^t \mathbb{E}(|Z(u) - Z_n(u)|^2)du$$
$$\leq C_2(t)K_1 \sup_{0 \leq u \leq t} \mathbb{E}(|Z(u) - Z_n(u)|^2)$$
$$\leq C_2(t)K_1 \left(\sum_{r=n+1}^{\infty} \frac{C_2(t)^{r/2} K_3^{r/2}}{(r!)^{1/2}}\right)^2$$
$$\to 0 \quad \text{as} \quad n \to \infty.$$

6.2 Stochastic differential equations

Hence each $Z(s) = L^2 - \lim_{n\to\infty} Z_n(s)$ and so, by uniqueness of limits, $\tilde{Z}(s) = Z(s)$ (a.s.) as required.

Proof of uniqueness for $\mathbb{E}(|Z_0|^2) < \infty$ Let Z_1 and Z_2 be two distinct solutions to (6.15). Hence, for each $t \geq 0$, $1 \leq i \leq d$,

$$Z_1^i(t) - Z_2^i(t)$$
$$= \int_0^t [b^i(Z_1(s-)) - b^i(Z_2(s-))]ds$$
$$+ \int_0^t [\sigma_j^i(Z_1(s-)) - \sigma_j^i(Z_2(s-))]dB^j(s)$$
$$+ \int_0^t \int_{|x|<c} [F^i(Z_1(s-), x) - F^i(Z_2(s-), x)]\tilde{N}(ds, dx).$$

We again follow the same line of argument as used in deducing (6.20), to find that

$$\mathbb{E}\left(\sup_{0\leq s\leq t} |Z_1(s) - Z_2(s)|^2\right)$$
$$\leq C_1(t)K_1 \int_0^t \mathbb{E}\left(\sup_{0\leq u\leq s} |Z_1(u) - Z_2(u)|^2\right) ds.$$

Thus, by Gronwall's inequality, $\mathbb{E}\left(\sup_{0\leq s\leq t} |Z_1(s) - Z_2(s)|^2\right) = 0$. Hence $Z_1(s) = Z_2(s)$ for all $0 \leq s \leq t$ (a.s.). By continuity of probability, we obtain, as required,

$$P(Z_1(t) = Z_2(t) \text{ for all } t \geq 0)$$
$$= P\left(\bigcap_{N\in\mathbb{N}} (Z_1(t) = Z_2(t) \text{ for all } 0 \leq t \leq N)\right) = 1.$$

Proof of existence and uniqueness for $\mathbb{E}(|Z_0|^2) = \infty$ (cf. Itô [145]). For each $n \in \mathbb{N}$, define $\Omega_N = \{\omega \in \Omega; |Z_0| \leq N\}$. Then $\Omega_N \subseteq \Omega_M$ whenever $n \leq M$ and $\Omega = \bigcup_{n\in\mathbb{N}} \Omega_N$. Let $Z_0^N = Z_0 \chi_{\Omega_N}$. By the above analysis, the equation (6.15) with initial condition Z_0^N has a unique solution $(Z_N(t), t \geq 0)$. Clearly, for $M > N$, $Z_M(t)(\omega) = Z_{M-1}(t)(\omega) = \cdots = Z_N(t)(\omega)$ for all $t \geq 0$ and almost all $\omega \in \Omega_N$.

By continuity of probability, given any $\epsilon > 0$ there exists $n \in \mathbb{N}$ such that $n > N \Rightarrow P(\Omega_n) > 1 - \epsilon$. Then given any $\delta > 0$, for all $m, n > N$,

$$P\left(\sup_{t \geq 0} |Z_n(t) - Z_m(t)| > \delta\right) < \epsilon.$$

Hence $(Z_n, n \in \mathbb{N})$ is uniformly Cauchy in probability and so is uniformly convergent in probability to a process $Z = (Z(t), t \geq 0)$. We can extract a subsequence for which the convergence holds uniformly (a.s.) and from this it follows that Z is adapted, càdlàg and solves (6.15).

For uniqueness, suppose that $Z' = (Z'(t), t \geq 0)$ is another solution to (6.15); then, for all $M \geq N$, $Z'(t)(\omega) = Z_M(t)(\omega)$ for all $t \geq 0$ and almost all $\omega \in \Omega_N$. For, suppose this fails to be true for some $M \geq N$. Define $Z''_M(t)(\omega) = Z'_M(t)(\omega)$ for $\omega \in \Omega_N$ and $Z''_M(t)(\omega) = Z_M(t)(\omega)$ for $\omega \in \Omega_N^c$. Then Z''_M and Z_M are distinct solutions to (6.15) with the same initial condition Z_0^M, and our earlier uniqueness result gives the required contradiction. That $P(Z(t) = Z'(t)$ for all $t \geq 0) = 1$ follows by a straightforward limiting argument, as above. $\square$

Corollary 6.2.4 *Let Z be the unique solution of (6.15) as constructed in Theorem 6.2.3. If $\mathbb{E}(|Z_0|^2) < \infty$ then $\mathbb{E}(|Z(t)|^2) < \infty$ for each $t \geq 0$ and there exists a constant $D(t) > 0$ such that*

$$\mathbb{E}(|Z(t)|^2) \leq D(t)\left[1 + \mathbb{E}(|Z_0|^2)\right].$$

Proof By (6.22) we see that, for each $t \geq 0$, there exists $C(t) \geq 0$ such that

$$||Z(t) - Z_0||_2 \leq \sum_{n=0}^{\infty} ||Z_n(t) - Z_{n-1}(t)||_2 \leq C(t).$$

Now

$$\mathbb{E}(|Z(t)|^2) \leq 2\,\mathbb{E}(|Z(t) - Z(0)|^2) + 2\,\mathbb{E}(|Z(0)|^2),$$

and the required result follows with $D(t) = 2\max\{1, C(t)^2\}$. $\square$

Exercise 6.2.5 Consider the SDE

$$dZ(t) = \sigma(Z(t-))dB(t) + \int_{|x|<c} F(Z(t-), x)\tilde{N}(dt, dx)$$

satisfying all the conditions of Corollary 6.2.4. Deduce that Z is a square-integrable martingale. Hence deduce that the discounted stock price $\tilde{S}_1$ discussed in Subsection 5.4.3 is indeed a martingale, as was promised.

6.2 Stochastic differential equations

Exercise 6.2.6 Deduce that $Z = (Z(t), t \geq 0)$ has continuous sample paths, where

$$dZ(t) = b(Z(t))dt + \sigma(Z(t))dB(t).$$

(Hint: Use the uniformity of the convergence in Theorem 6.2.3 and recall the discussion of Subsection 4.3.1.)

Exercise 6.2.7 Show that the following Lipschitz condition on the matrix-valued function $\sigma(\cdot)$ is a sufficient condition for the bi-Lipschitz continuity of $a(\cdot) = \sigma(\cdot)\sigma(\cdot)^{\mathrm{T}}$: there exists $K > 0$ such that, for each $1 \leq i \leq d, 1 \leq j \leq r$, $y_1, y_2 \in \mathbb{R}^d$,

$$\left|\sigma_j^i(y_1) - \sigma_j^i(y_2)\right| \leq K|y_1 - y_2|.$$

Having dealt with the modified equation, we can now apply a standard interlacing procedure to construct the solution to the original equation (6.12). We impose the following assumption on the coefficient G, which ensures that the integrands in Poisson integrals are predictable.

Assumption 6.2.8 We require that the mapping $y \to G(y, x)$ is continuous for all $x \geq c$.

Theorem 6.2.9 *There exists a unique càdlàg adapted solution to* (6.12).

Proof Let $(\tau_n, n \in \mathbb{N})$ be the arrival times for the jumps of the compound Poisson process $(P(t), t \geq 0)$, where each $P(t) = \int_{|x| \geq c} x N(t, dx)$. We then construct a solution to (6.12) as follows:

$$\begin{aligned}
Y(t) &= Z(t) & \text{for } 0 \leq t < \tau_1, \\
Y(\tau_1) &= Z(\tau_1-) + G(Z(\tau_1-), \Delta P(\tau_1)) & \text{for } t = \tau_1, \\
Y(t) &= Y(\tau_1) + Z_1(t) - Z_1(\tau_1) & \text{for } \tau_1 < t < \tau_2, \\
Y(\tau_2) &= Z(\tau_2-) + G(Z(\tau_2-), \Delta P(\tau_2)) & \text{for } t = \tau_1,
\end{aligned}$$

and so on, recursively. Here Z_1 is the unique solution to (6.15) with initial condition $Z_1(0) = Y(\tau_1)$. Y is clearly adapted, càdlàg and solves (6.12). Uniqueness follows by the uniqueness in Theorem 6.2.3 and the interlacing structure. □

Note Theorem 6.2.9 may be generalised considerably. More sophisticated techniques were developed by Protter [255], pp. 193–201, and Jacod [154], pp. 451ff., in the case where the driving noise is a general semimartingale with jumps.

In some problems we might require time-dependent coefficients and so we study the inhomogeneous SDE

$$dY(t) = b(t, Y(t-))dt + \sigma(t, Y(t-))dB(t)$$
$$+ \int_{|x|<c} F(t, Y(t-), x)\tilde{N}(dt, dx)$$
$$+ \int_{|x|>c} G(t, Y(t-), x)N(dt, dx). \quad (6.23)$$

We can again reduce this problem by interlacing to the study of the modified SDE with small jumps. In order to solve the latter we can impose the following (crude) Lipschitz and growth conditions.

For each $t > 0$, there exists $K_1(t) > 0$ such that, for all $y_1, y_2 \in \mathbb{R}^d$,

$$|b(t, y_1) - b(t, y_2)| + ||a(t, y_1, y_1) - 2a(t, y_1, y_2) + a(t, y_2, y_2)||$$
$$+ \int_{|x|<c} |F(t, y_1, x) - F(t, y_2, x)|^2 \nu(dx) \leq K_1(t)|y_1 - y_2|^2.$$

There exists $K_2(t) > 0$ such that, for all $y \in \mathbb{R}^d$,

$$|b(t, y)|^2 + ||a(t, y, y)|| + \int_{|x|<c} |F(t, y, x)|^2 \leq K_2(t)(1 + |y|^2),$$

where $a(t, y_1, y_2) = \sigma(t, y_1)\sigma(t, y_2)^T$ for each $t \geq 0$, $y_1, y_2 \in \mathbb{R}^d$.

Exercise 6.2.10 Show that (6.23) has a unique solution under the above conditions.

The final variation which we will examine in this chapter involves local solutions. Let T_∞ be a stopping time and suppose that $Y = (Y(t), 0 \leq t < T_\infty)$ is a solution to (6.12). We say that Y is a *local solution* if $T_\infty < \infty$ (a.s.) and a *global solution* if $T_\infty = \infty$ (a.s.). We call T_∞ the *explosion time* for the SDE (6.12). So far in this chapter we have looked at global solutions. If we want to allow local solutions to (6.12) we can weaken our hypotheses to allow *local* Lipschitz and growth conditions on our coefficients. More precisely we impose:

(C3) Local Lipschitz condition For all $n \in \mathbb{N}$ and $y_1, y_2 \in \mathbb{R}^d$ with $\max\{|y_1|, |y_2|\} \leq n$, there exists $K_1(n) > 0$ such that

$$|b(y_1) - b(y_2)| + ||a(y_1, y_1) - 2a(y_1, y_2) + a(y_2, y_2)||$$
$$+ \int_{|x|<c} |F(y_1, x) - F(y_2, x)|^2 \nu(dx) \leq K_1(n)|y_1 - y_2|^2.$$

6.2 Stochastic differential equations

(C4) Local Growth condition For all $n \in \mathbb{N}$ and for all $y \in \mathbb{R}^d$ with $|y| \leq n$, there exists $K_2(n) > 0$ such that

$$|b(y)|^2 + \|a(y, y)\| + \int_{|x|<c} |F(y, x)|^2 \leq K_2(n)(1 + |y|^2).$$

We then have

Theorem 6.2.11 *If we assume* (C3) *and* (C4) *and impose the standard initial condition, then there exists a unique local solution* $Y = (Y(t), 0 \leq t < T_\infty)$ *to the SDE* (6.12).

Proof Once again we can reduce the problem by interlacing to the solution of the modified SDE. The proof in this case is almost identical to the case of equations driven by Brownian motion, and we refer the reader to the account of Durrett [85] for the details. $\square$

We may also consider *backwards stochastic differential equations* on a time interval $[0, T]$. We write these as follows (in the time-homogeneous case):

$$dY(t) = -b(Y(t))dt - \sigma(Y(t)) \cdot_b dB(t)$$
$$- \int_{|x|<c} F(t, Y(t-), x) \cdot_b \tilde{N}(dt, dx)$$
$$- \int_{|x|>c} G(t, Y(t-), x) N(dt, dx),$$

where $\cdot_b$ denotes the backwards stochastic integral. The solution (when it exists) is a backwards adapted process $(Y(s); 0 \leq s \leq T)$. Instead of an initial condition $Y(0) = Y_0$ (a.s.) we impose a final condition $Y(T) = Y_T$ (a.s.). We then have the integral form

$$Y^i(s) = Y_T^i - \int_s^T b^i(Y(u))du - \int_s^T \sigma_j^i(Y(u)) \cdot_b dB^j(u)$$
$$- \int_s^T \int_{|x|<c} F^i(Y(u), x) \cdot_b \tilde{N}(dt, dx)$$
$$- \int_s^T \int_{|x|\geq c} G^i(Y(u)), x) N(dt, dx) \quad \text{a.s.}$$

for each $0 \leq s \leq T$, $1 \leq i \leq d$.

The theory of backwards SDEs can now be developed just as in the forward case, so we can obtain the existence and uniqueness of solutions by imposing Lipschitz and growth conditions on the coefficients and the usual independence condition on Y_T.

Backwards SDEs with discontinuous noise have not been developed so thoroughly as the forward case. This may change in the future as more applications are found; see e.g. Nualart and Schoutens [238], where backwards SDEs driven by the Teugels martingales of Exercise 2.4.24 are applied to option pricing. Other articles on backwards SDEs include Situ [286] and Ouknine [243].

6.3 Examples of SDEs

SDEs driven by Lévy processes

Let $X = (X(t), t \geq 0)$ be a Lévy process taking values in $\mathbb{R}^m$. We denote its Lévy–Itô decomposition as

$$X^i(t) = \lambda^i t + \tau^i_j B^j(t) + \int_0^t \int_{|x|<1} x^i \tilde{N}(ds, dx) + \int_0^t \int_{|x|\geq 1} x^i N(ds, dx)$$

for each $1 \leq i \leq m$, $t \geq 0$. Here, as usual, $\lambda \in \mathbb{R}^m$ and (τ^i_j) is a real-valued $m \times r$ matrix.

For each $1 \leq i \leq d$, $1 \leq j \leq m$, let $L^i_j : \mathbb{R}^d \to \mathbb{R}^d$ be measurable, and form the $d \times m$ matrix $L(x) = (L^i_j(x))$ for each $x \in \mathbb{R}^d$. We consider the SDE

$$dY(t) = L(Y(t-))dX(t), \tag{6.24}$$

with standard initial condition $Y(0) = Y_0$ (a.s.), so that, for each $1 \leq i \leq d$,

$$dY^i(t) = L^i_j(Y(t-))dX^j(t).$$

This is of the same form as (6.12), with coefficients given by $b(\cdot) = L(\cdot)\lambda$, $\sigma(\cdot) = L(\cdot)\tau$, $F(\cdot, x) = L(\cdot)x$ for $|x| < 1$ and $G(\cdot, x) = L(\cdot)x$ for $|x| \geq 1$.

To facilitate discussion of the existence and uniqueness of solutions of (6.24), we introduce two new matrix-valued functions, N, a $d \times d$ matrix given by $N(x) = L(x)\tau\tau^T L(x)^T$ for each $x \in \mathbb{R}^d$ and M, a $m \times m$ matrix defined by $M(x) = L(x)^T L(x)$.

We impose the following Lipschitz-type conditions on M and N: there exist $D_1, D_2 > 0$ such that, for all $y_1, y_2 \in \mathbb{R}^d$,

$$\|N(y_1, y_1) - 2N(y_1, y_2) + N(y_2, y_2)\| \leq D_1 |y_1 - y_2|^2, \tag{6.25}$$

$$\max_{1 \leq p, q \leq m} |M^p_q(y_1, y_1) - 2M^p_q(y_1, y_2) + M^p_q(y_2, y_2)| \leq D_2 |y_1 - y_2|^2. \tag{6.26}$$

Note that (6.25) is just the usual bi-Lipschitz condition, which allows control of the Brownian integral terms within SDEs.

Tedious but straightforward algebra then shows that (6.25) and (6.26) imply the Lipschitz and growth conditions (C1) and (C2) and hence, by Theorems 6.2.3 and 6.2.9, the equation (6.24) has a unique solution. In applications, we often meet the case $m = d$ and $L = \text{diag}(L_1, \ldots, L_d)$. In this case, readers can check that a sufficient condition for (6.25) and (6.26) is the single Lipschitz condition that there exists $D_3 > 0$ such that, for all $y_1, y_2 \in \mathbb{R}^d$,

$$|L(y_1) - L(y_2)| \leq D_3 |y_1 - y_2|, \tag{6.27}$$

where we are regarding L as a vector-valued function.

Another class of SDEs that are often considered in the literature take the form

$$dY(t) = b(Y(t-))dt + L(Y(t-))dX(t),$$

and these clearly have a unique solution whenever L is as in (6.27) and b is globally Lipschitz. The important case where X is α-stable was studied by Janicki, Michna and Weron [156].

Stochastic exponentials

We consider the equation

$$dY(t) = Y(t-)dX(t),$$

so that, for each $1 \leq i \leq d$, $dY^i(t) = Y^i(t-)dX^i(t)$.

This trivially satisfies the Lipschitz condition (6.27) and so has a unique solution. In the case $d = 1$ with $Y_0 = 1$ (a.s.), we saw in Section 5.1 that the solution is given by the *stochastic exponential*

$$Y(t) = \mathcal{E}_X(t) = \exp\left\{X(t) - \tfrac{1}{2}[X_c, X_c](t)\right\} \prod_{0 \leq s \leq t} [1 + \Delta X(s)] e^{-\Delta X(s)},$$

for each $t \geq 0$.

The Langevin equation and Ornstein–Uhlenbeck process revisited

The process $B = (B(t), t \geq 0)$ that we have been calling 'Brownian motion' throughout this book is not the best possible description of the physical phenomenon of Brownian motion.

A more realistic model was proposed by Ornstein and Uhlenbeck [242] in the 1930s; see also Chapter 9 of Nelson [236] and Chandrasekar [70]. Let

$x = (x(t), t \geq 0)$, where $x(t)$ is the displacement after time t of a particle of mass m executing Brownian motion, and let $v = (v(t), t \geq 0)$, where $v(t)$ is the velocity of the particle. Ornstein and Uhlenbeck argued that the total force on the particle should arise from a combination of random bombardments by the molecules of the fluid and also a macroscopic frictional force, which acts to dampen the motion. In accordance with Newton's laws, this total force equals the rate of change of momentum and so we write the formal equation

$$m\frac{dv}{dt} = -\beta m v + m \frac{dB}{dt},$$

where β is a positive constant (related to the viscosity of the fluid) and the formal derivative 'dB/dt' describes random velocity changes due to molecular bombardment. This equation acquires a meaning as soon as we interpret it as an Itô-style SDE. We thus obtain the *Langevin equation*, named in honour of the French physicist Paul Langevin,

$$dv(t) = -\beta v(t) dt + dB(t). \tag{6.28}$$

It is more appropriate for us to generalise this equation and replace B by a Lévy process $X = (X(t), t \geq 0)$, to obtain

$$dv(t) = -\beta v(t) dt + dX(t), \tag{6.29}$$

which we continue to call the Langevin equation. It has a unique solution by Theorem 6.2.9. We can in fact solve (6.29) by multiplying both sides by the integrating factor $e^{-\beta t}$ and using Itô's product formula. This yields our old friend the *Ornstein–Uhlenbeck process* (4.9),

$$v(t) = e^{-\beta t} v_0 + \int_0^t e^{-\beta(t-s)} dX(s)$$

for each $t \geq 0$. Recall from Exercise 4.3.18 that when X is a Brownian motion, v is Gaussian. In this latter case, the integrated Ornstein–Uhlenbeck process also has a physical interpretation. It is nothing but the displacement of the Brownian particle

$$x(t) = \int_0^t v(s) ds,$$

for each $t \geq 0$.

An interesting generalisation of the Langevin equation is obtained when the number β is replaced by a matrix Q, all of whose eigenvalues have a positive real part. We thus obtain the equation

$$dY(t) = -QY(t)dt + dX(t),$$

whose unique solution is the *generalised Ornstein–Uhlenbeck process*, $Y = (Y(t), t \geq 0)$, where, for each $t \geq 0$,

$$Y(t) = e^{-Qt}Y_0 + \int_0^t e^{-Q(t-s)} dX(s).$$

For further details see Sato and Yamazoto [273] and Barndorff-Nielsen, Jensen and Sørensen [21].

Diffusion processes

The most intensively studied class of SDEs is the class of those that lead to diffusion processes. These generalise the Ornstein–Uhlenbeck process for Brownian motion, but now the aim is to describe all possible random motions that are due to 'diffusion'. A hypothetical particle that diffuses should move continuously and be characterised by two functions, a 'drift coefficient' b that describes the deterministic part of the motion and a 'diffusion coefficient a' that corresponds to the random part. Generalising the Langevin equation, we model diffusion as a stochastic process $Y = (Y(t), t \geq 0)$, starting at $Y(0) = Y$ (a.s.) and solving the SDE

$$dY(t) = b(Y(t))dt + \sigma(Y(t))dB(t), \tag{6.30}$$

where $a(\cdot) = \sigma(\cdot)\sigma(\cdot)^{\mathrm{T}}$. We impose the usual Lipschitz conditions on b and the stronger one given in Exercise 6.2.7 on σ; these ensure that (6.30) has a unique strong solution. In this case, $Y = (Y(t), t \geq 0)$ is sometimes called an *Itô diffusion*.

A more general approach was traced back to Kolmogorov [172] by David Williams in [303]. A *diffusion process* in $\mathbb{R}^d$ is a path-continuous Markov process $Y = (Y(t), t \geq 0)$ starting at $Y_0 = x$ (a.s) for which there exist continuous functions $\beta : \mathbb{R}^d \to \mathbb{R}^d$ and $\alpha : \mathbb{R}^d \to M_d(\mathbb{R})$ such that

$$\left.\frac{d}{dt}\mathbb{E}(Y(t))\right|_{t=0} = \beta(x) \quad \text{and} \quad \left.\frac{d}{dt}\text{Cov}(Y(t), Y(t))_{ij}\right|_{t=0} = \alpha_{ij}(x). \tag{6.31}$$

We call β and α the *infinitesimal mean* and *infinitesimal covariance*, respectively. The link between this more general definition and SDEs is given in the following result.

Theorem 6.3.1 *Every Itô diffusion is a diffusion with $\beta = b$ and $\alpha = a$.*

Proof The Markov property will be discussed later in this chapter. Continuity of b and a follows from the Lipschitz conditions. For the explicit calculations below, we follow Durrett [85], pp. 178–9.

Writing the Itô diffusion in integral form we have, for each $t \geq 0$,

$$Y(t) = x + \int_0^t b(Y(s))ds + \int_0^t \sigma(Y(s))dB(s).$$

Since the Brownian integral is a centred L^2-martingale, we have that

$$\mathbb{E}(Y(t)) = x + \int_0^t \mathbb{E}(b(X(s))\, ds,$$

and $\beta(x) = b(x)$ now follows on differentiating.

For each $1 \leq i, j \leq d, t \geq 0$,

$$\text{Cov}\,(Y(t), Y(t))_{ij} = \mathbb{E}(Y_i(t)Y_j(t)) - \mathbb{E}(Y_i(t))\,\mathbb{E}(Y_j(t)).$$

By Itô's product formula,

$$\begin{aligned}d(Y_i(t)Y_j(t)) &= dY_i(t)Y_j(t) + Y_i(t)dY_j(t) + d[Y_i, Y_j](t) \\ &= dY_i(t)Y_j(t) + Y_i(t)dY_j(t) + a_{ij}(Y(t))dt.\end{aligned}$$

Hence

$$\mathbb{E}(Y_i(t)Y_j(t)) = x_i x_j + \int_0^t \mathbb{E}\bigl(Y_i(s)b_j(Y(s)) + Y_j(s)b_i(Y(s)) \\ + a_{ij}(Y(s))\bigr)ds$$

and so

$$\left.\frac{d}{dt}\mathbb{E}(Y_i(t)Y_j(t))\right|_{t=0} = x_i b_j(x) + x_j b_i(x) + a_{ij}(x).$$

We can easily verify that

$$\left.\frac{d}{dt}\bigl[\mathbb{E}(Y_i(t))\,\mathbb{E}(Y_j(t))\bigr]\right|_{t=0} = x_i b_j(x) + x_j b_i(x),$$

and the required result follows. $\square$

Diffusion processes have a much wider scope than physical models of diffusing particles; for example, the Black–Scholes model for stock prices $(S(t), t \geq 0)$ is an Itô diffusion taking the form

$$dS(t) = \beta S(t)dt + \sigma S(t)dt,$$

where $\beta \in \mathbb{R}$ and $\sigma > 0$ denote the usual stock drift and volatility parameters.

6.4 Stochastic flows, cocycle and Markov properties of SDEs

We will not make a detailed investigation of diffusions in this book. For more information on this extensively studied topic, see e.g. Durrett [85], Ikeda and Watanabe [140], Itô and McKean [141], Krylov [178], Rogers and Williams [261], [262] and Stroock and Varadhan [289].

When a particle diffuses in accordance with Brownian motion, its standard deviation at time t is $\sqrt{t}$. In *anomalous diffusion*, particles diffuse through a non-homogeneous medium that either slows the particles down (*subdiffusion*) or speeds them up (*superdiffusion*). The standard deviation behaves like t^ν, where $\nu < 1/2$ for subdiffusion and $\nu > 1/2$ for superdiffusion. A survey of some of these models is given in Chapter 12 of Uchaikin and Zolotarev [297]; compound Poisson processes and symmetric stable laws play a key role in the analysis.

Jump-diffusion processes

By a *jump-diffusion* process, we mean the strong solution $Y = (Y(t), t \geq 0)$ of the SDE

$$dY(t) = b(Y(t-))dt + \sigma(Y(t-))dB(t)$$
$$+ \int_{\mathbb{R}^d - \{0\}} G(Y(t-), x) N(dt, dx),$$

where N is a Poisson random measure that is independent of the Brownian motion B having finite intensity measure ν (so we have taken $c = 0$ in (6.12)). It then follows, by the construction in the proof of Theorem 6.2.9, that the paths of Z simply consist of that of an Itô diffusion process interlaced by jumps at the arrival times of the compound Poisson process $P = (P(t), t \geq 0)$, where each $P(t) = \int_0^t \int_{\mathbb{R}^d - \{0\}} x N(dt, dx)$.

We note that there is by no means universal agreement about the use of the phrase 'jump-diffusion process', and some authors use it to denote the more general processes arising from the solution to (6.12). The terminology may also be used when N is the random measure counting the jumps of a more general point process.

6.4 Stochastic flows, cocycle and Markov properties of SDEs

6.4.1 Stochastic flows

Let $Y_y = (Y_y(t), t \geq 0)$ be the strong solution of the SDE (6.12) with fixed deterministic initial condition $Y_0 = y$ (a.s.). Just as in the case of ordinary differential equations, we would like to study the properties of $Y_y(t)$ as y varies. Imitating the procedure of Section 6.1, we define $\Phi_t : \mathbb{R}^d \times \Omega \to \mathbb{R}^d$ by

$$\Phi_t(y, \omega) = Y_y(t)(\omega)$$

for each $t \geq 0$, $y \in \mathbb{R}^d$, $\omega \in \Omega$. We will find it convenient below to fix $y \in \mathbb{R}^d$ and regard these mappings as random variables. We then employ the notation $\Phi_t(y)(\cdot) = \Phi_t(y, \cdot)$.

Based on equation (6.8), we might expect that

$$`\Phi_{s+t}(y, \omega) = \Phi_t(\Phi_s(y, \omega), \omega),'$$

for each $s, t \geq 0$. In fact this is not the case, as the following example shows.

Example 6.4.1 (Random translation) Consider the simplest SDE driven by a Lévy process $X = (X(t), t \geq 0)$,

$$dY_y(t) = dX(t), \quad Y_y(0) = y \quad \text{a.s.},$$

whose solution is the random translation $\Phi_t(y) = y + X(t)$. Then

$$\Phi_{t+s}(y) = y + X(t+s).$$

But $\Phi_t(\Phi_s(y)) = y + X(t) + X(s)$ and these are clearly not the same (except in the trivial case where $X(t) = mt$, for all $t \geq 0$, with $m \in \mathbb{R}$). However, if we define the two-parameter motion

$$\Phi_{s,t}(y) = y + X(t) - X(s),$$

where $0 \leq s \leq t < \infty$, then it is easy to check that, for all $0 \leq r < s < t < \infty$,

$$\Phi_{r,t}(y) = \Phi_{s,t}(\Phi_{r,s}(y)),$$

and this gives us a valuable clue as to how to proceed in general.

Example 6.4.1 suggests that if we want to study the flow property for random dynamical systems then we need a two-parameter family of motions. The interpretation of the random mapping $\Phi_{s,t}$ is that it describes motion commencing at the 'starting time' s and ending at the 'finishing time' t. We now give some general definitions.

Let $\Phi = \{\Phi_{s,t}, 0 \leq s \leq t < \infty\}$ be a family of measurable mappings from $\mathbb{R}^d \times \Omega \to \mathbb{R}^d$. For each $\omega \in \Omega$, we have associated mappings $\Phi_{s,t}^\omega : \mathbb{R}^d \to \mathbb{R}^d$, given by $\Phi_{s,t}^\omega(y) = \Phi_{s,t}(\omega, y)$ for each $y \in \mathbb{R}^d$.

We say that Φ is a *stochastic flow* if there exists $\mathcal{N} \subset \Omega$, with $P(\mathcal{N}) = 0$, such that for all $\omega \in \Omega - \mathcal{N}$:

(1) $\Phi_{r,t}^\omega = \Phi_{s,t}^\omega \circ \Phi_{r,s}^\omega$ for all $0 \leq r < s < t < \infty$;
(2) $\Phi_{s,s}^\omega(y) = y$ for all $s \geq 0$, $y \in \mathbb{R}^d$.

If, in addition, each $\Phi_{s,t}^\omega$ is a homeomorphism (C^k-diffeomorphism) of $\mathbb{R}^d$, for all $\omega \in \Omega - \mathcal{N}$, we say that Φ is a *stochastic flow of homeomorphisms* (C^k-*diffeomorphisms*, respectively).

If, in addition to properties (1) and (2), we have that

(3) for each $n \in \mathbb{N}, 0 \le t_1 < t_2 < \cdots < t_n < \infty, y \in \mathbb{R}^d$, the random variables $\{\Phi_{t_j, t_{j+1}}(y); 1 \le j \le n-1\}$ are independent,
(4) the mappings $t \to \Phi_{s,t}(y)$ are càdlàg for each $y \in \mathbb{R}^d, 0 \le s < t$,

we say that Φ is a *Lévy flow*.

If (4) can be strengthened from 'càdlàg' to 'continuous', we say that Φ is a *Brownian flow*.

The reason for the terminology 'Lévy flow' and 'Brownian flow' is that when property (3) holds we can think of Φ as a Lévy process on the group of all diffeomorphisms from $\mathbb{R}^d$ to itself (see Baxendale [30], Fujiwara and Kunita [108] and Applebaum and Kunita [5] for more about this viewpoint).

Brownian flows of diffeomorphisms were studied extensively by Kunita in [182]. It was shown in Section 4.2 therein that they all arise as solutions of SDEs driven by (a possibly infinite number of) standard Brownian motions. The programme for Lévy flows is less complete – see Section 3 of Fujiwara and Kunita [108] for some partial results.

Here we will study flows driven by the SDEs studied in Section 6.2. We consider two-parameter versions of these, i.e.

$$d\Phi_{s,t}(y) = b(\Phi_{s,t-}(y))dt + \sigma(\Phi_{s,t-}(y))dB(t)$$
$$+ \int_{|x|<c} F(\Phi_{s,t-}(y), x)\tilde{N}(dt, dx)$$
$$+ \int_{|x|\ge c} G(\Phi_{s,t-}(y), x)N(dt, dx) \quad (6.32)$$

with initial condition $\Phi_{s,s}(y) = y$ (a.s.), so that, for each $1 \le i \le d$,

$$\Phi_{s,t}(y)^i = y^i + \int_0^t b^i(\Phi_{s,u-}(y))du + \int_0^t \sigma^i_j(\Phi_{s,u-}(y))dB^j(u)$$
$$+ \int_0^t \int_{|x|<c} F^i(\Phi_{s,u-}(y), x)\tilde{N}(du, dx)$$
$$+ \int_0^t \int_{|x|\ge c} G^i(\Phi_{s,u-}(y), x)N(du, dx).$$

The fact that (6.32) has a unique strong solution under the usual Lipschitz and growth conditions is achieved by a minor modification to the proofs of Theorems 6.2.3 and 6.2.9.

Theorem 6.4.2 Φ *is a Lévy flow.*

Proof The measurability of each $\Phi_{s,t}$ and the càdlàg property (4) follow from the constructions of Theorems 6.2.3 and 6.2.9. Property (2) is immediate. To establish the flow property (1), we follow similar reasoning to that in the proof of Theorem 6.1.10.

To simplify the form of expressions appearing below, we will omit, without loss of generality, all except the compensated Poisson terms in (6.32).

For all $0 \leq r < s < t < \infty$, $1 \leq i \leq d$, $y \in \mathbb{R}^d$, we have

$$\begin{aligned}\Phi_{r,t}(y)^i &= y^i + \int_r^t \int_{|x|<c} F^i(\Phi_{r,u-}(y), x)\tilde{N}(du, dx) \\ &= y^i + \int_r^s \int_{|x|<c} F^i(\Phi_{r,u-}(y), x)\tilde{N}(du, dx) \\ &\quad + \int_s^t \int_{|x|<c} F^i(\Phi_{r,u-}(y), x)\tilde{N}(du, dx) \\ &= \Phi_{r,s}(y)^i + \int_s^t \int_{|x|<c} F^i(\Phi_{r,u-}(y), x)\tilde{N}(du, dx).\end{aligned}$$

However,

$$\Phi_{s,t}(\Phi_{r,s}(y))^i = \Phi_{r,s}(y)^i + \int_s^t \int_{|x|<c} F^i(\Phi_{s,u-}(\Phi_{r,s}(y)), x)\tilde{N}(du, dx),$$

and the required result follows by the uniqueness of solutions to SDEs.

For the independence (3), consider the sequence of Picard iterates $(\Phi_{s,t}^{(n)}, n \in \mathbb{N} \cup \{0\})$ constructed in the proof of Theorem 6.2.3. Using induction and arguing as in the proof of Lemma 4.3.12, we see that each $\Phi_{s,t}^{(n)}$ is measurable with respect to $\sigma\{N(v, A) - N(u, A), 0 \leq s \leq u < v \leq t, A \in \mathcal{B}(B_c(0))\}$, from which the required result follows. $\square$

Exercise 6.4.3 Extend Theorem 6.4.2 to the case of the general standard initial condition.

Example 6.4.4 (Randomising deterministic flows) We assume that $b \in C_b^k(\mathbb{R})$ and consider the one-dimensional ODE

$$\frac{d\xi(a)}{da} = b(\xi(a)).$$

6.4 Stochastic flows, cocycle and Markov properties of SDEs

By Theorem 6.1.10, its unique solution is a flow of C^k-diffeomorphisms $\xi = (\xi(a), a \in \mathbb{R})$. We randomise the flow ξ by defining

$$\Phi_{s,t}(y) = \xi(X(t) - X(s))(y)$$

for all $0 \leq s \leq t < \infty$, $y \in \mathbb{R}^d$, where X is a one-dimensional Lévy process with characteristics (m, σ^2, ν). It is an easy exercise to check that Φ is a Lévy flow of C^k-diffeomorphisms. It is of interest to find the SDE satisfied by Φ. Thanks to Exercise 6.1.8, we can use Itô's formula to obtain

$$\begin{aligned}
d\Phi_{s,t}(y) &= mb(\Phi_{s,t-}(y))dt + \sigma b(\Phi_{s,t-}(y))dB(t) \\
&+ \tfrac{1}{2}\sigma^2 b'(\Phi_{s,t-}(y))b(\Phi_{s,t-}(y))dt \\
&+ \int_{|x|<1} \left[\xi(x)(\Phi_{s,t-}(y)) - \Phi_{s,t-}(y)\right]\tilde{N}(dt,dx) \\
&+ \int_{|x|\geq 1} \left[\xi(x)(\Phi_{s,t-}(y)) - \Phi_{s,t-}(y)\right]N(dt,dx) \\
&+ \int_{|x|<1} \left[\xi(x)(\Phi_{s,t-}(y)) - \Phi_{s,t-}(y) - xb(\Phi_{s,t-}(y))\right]\nu(dx)dt.
\end{aligned}$$
(6.33)

Here we have used the flow property for ξ in the jump term and the fact that

$$\frac{d^2}{da^2}\xi(a) = b'(\xi(a))b(\xi(a)).$$

The SDE (6.33) is the simplest example of a *Marcus canonical equation*. We will return to this theme in Section 6.8.

6.4.2 The Markov property

Here we will apply the flow property established above to prove that solutions of SDEs give rise to Markov processes.

Theorem 6.4.5 *The strong solution to (6.12) is a Markov process.*

Proof Let $t \geq 0$. Following Exercise 6.4.3, we can consider the solution $Y = (Y(t), t \geq 0)$ as a stochastic flow with random initial condition Y_0, and we will abuse notation to the extent of writing each

$$Y(t) = \Phi_{0,t}(Y_0) = \Phi_{0,t}.$$

Our aim is to prove that

$$\mathbb{E}(f(\Phi_{0,t+s})|\mathcal{F}_s) = \mathbb{E}(f(\Phi_{0,t+s})|\Phi_{0,s})$$

for all $s, t \geq 0$, $f \in B_b(\mathbb{R}^d)$.

Now define $G_{f,s,t} \in B_b(\mathbb{R}^d)$ by

$$G_{f,s,t}(y) = \mathbb{E}(f(\Phi_{s,s+t}(y))),$$

for each $y \in \mathbb{R}^d$. By Theorem 6.4.2, and Exercise 6.4.3, we have that $\Phi_{0,t+s} = \Phi_{s,s+t} \circ \Phi_{0,s}$ (a.s.) and that $\Phi_{s,s+t}$ is independent of $\mathcal{F}_s$. Hence, by Lemma 1.1.8,

$$\mathbb{E}(f(\Phi_{0,t+s})|\mathcal{F}_s) = \mathbb{E}(f(\Phi_{s,s+t} \circ \Phi_{0,s})|\mathcal{F}_s) = \mathbb{E}(G_{f,s,t}(\Phi_{0,s})).$$

By the same argument, we also get $\mathbb{E}(f(\Phi_{0,t+s})|\Phi_{0,s}) = \mathbb{E}(G_{f,s,t}(\Phi_{0,s}))$, and the required result follows. $\square$

As in Section 3.1, we can now define an associated stochastic evolution $(T_{s,t}, 0 \leq s \leq t < \infty)$, by the prescription

$$(T_{s,t}f)(y) = \mathbb{E}\big(f(\Phi_{s,t})|\Phi_{0,s} = y\big)$$

for each $f \in B_b(\mathbb{R}^d)$, $y \in \mathbb{R}^d$. We will now strengthen Theorem 6.4.5.

Theorem 6.4.6 *The strong solution to* (6.12) *is a homogeneous Markov process.*

Proof We must show that $T_{s,s+t} = T_{0,t}$ for all $s, t \geq 0$.

Without loss of generality, we just consider the compensated Poisson terms in (6.12). Using the stationary increments property of Lévy processes, we obtain for each $f \in B_b(\mathbb{R}^d)$, $y \in \mathbb{R}^d$,

$$(T_{s,s+t}f)(y) = \mathbb{E}\big(f(\Phi_{s,s+t}(y))\big|\Phi_{0,s} = y\big)$$
$$= \mathbb{E}\left(f\left(y + \int_s^{s+t}\int_{|x|<c} F(\Phi_{0,s+u-}(y), x)\tilde{N}(ds, du)\right)\right)$$
$$= \mathbb{E}\left(f\left(y + \int_0^t \int_{|x|<c} F(\Phi_{0,u-}(y), x)\tilde{N}(ds, du)\right)\right)$$
$$= \mathbb{E}\big(f(\Phi_{0,t}(y))\big|\Phi_{0,0}(y) = y\big)$$
$$= (T_{0,t}f)(y).$$

$\square$

Referring again to Section 3.1, we see that we have a semigroup $(T_t, t \geq 0)$ on $B_b(\mathbb{R}^d)$, which is given by

$$(T_t f)(y) = \mathbb{E}\left(f(\Phi_{0,t}(y)) \big| \Phi_{0,0}(y) = y\right) = \mathbb{E}(f(\Phi_{0,t}(y)))$$

for each $t \geq 0$, $f \in B_b(\mathbb{R}^d)$, $y \in \mathbb{R}^d$. We would like to investigate the Feller property for this semigroup, but first we need to probe deeper into the properties of solution flows.

Exercise 6.4.7 Establish the strong Markov property for SDEs, i.e. show that

$$\mathbb{E}(f(\Phi_{0,t+S})|\mathcal{F}_S) = \mathbb{E}(f(\Phi_{0,t+S})|\Phi_{0,S})$$

for any $t \geq 0$, where S is a stopping time with $P(S < \infty) = 1$.

(Hint: Imitate the proof of Theorem 6.4.5, or see Theorem 31 in Protter [255], Chapter 6, pp. 237–9.)

6.4.3 Cocycles

As we will see below, the cocycle property of SDEs is quite closely related to the flow property. In this subsection we will work throughout with the canonical Lévy process constructed in Subsection 1.4.1. So Ω is the path space $\{\omega : \mathbb{R}^+ \to \mathbb{R}; \omega(0) = 0\}$, $\mathcal{F}$ is the σ-algebra generated by the cylinder sets and P is the unique probability measure given by Kolmogorov's existence theorem from the recipe (1.27) on cylinder sets. Hence $X = (X(t), t \geq 0)$ is a Lévy process on $(\Omega, \mathcal{F}, P)$, where $X(t)\omega = \omega(t)$ for each $\omega \in \Omega$, $t \geq 0$.

The space Ω comes equipped with a *shift* $\theta = (\theta_t, t \geq 0)$, each $\theta_t : \Omega \to \Omega$ being defined as follows. For each $s, t \geq 0$,

$$(\theta_t \omega)(s) = \omega(t+s) - \omega(t). \qquad (6.34)$$

Exercise 6.4.8 Deduce the following:
(1) θ is a one-parameter semigroup, i.e. $\theta_{t+s} = \theta_t \theta_s$ for all $s, t \geq 0$;
(2) the measure P is θ-invariant, i.e. $P(\theta_t^{-1}(A)) = P(A)$ for all $A \in \mathcal{F}$, $t \geq 0$.
 (Hint: First establish this on cylinder sets, using (1.27).)

Lemma 6.4.9 *X is an additive cocycle for θ, i.e., for all $s, t \geq 0$,*

$$X(t+s) = X(s) + (X(t) \circ \theta(s)).$$

Proof For each $s, t \geq 0$, $\omega \in \Omega$,

$$X(t)(\theta_s(\omega)) = (\theta_s \omega)(t) = \omega(s+t) - \omega(s) = X(t+s)(\omega) - X(s)(\omega).$$

$\square$

Additive cocycles were introduced into probability theory by Kolmogorov [173], who called them *helices*; see also de Sam Lazaro and Meyer [272] and Arnold and Scheutzow [12].

We now turn to the Lévy flow Φ that arises from solving (6.32).

Lemma 6.4.10 *For all* $0 \leq s \leq t < \infty$, $y \in \mathbb{R}^d$, $\omega \in \Omega$,

$$\Phi_{s,s+t}(y, \omega) = \Phi_{0,t}(y, \theta_s\omega) \qquad \text{a.s.}$$

Proof (See Proposition 24 of Arnold and Scheutzow [12].) We use the sequence of Picard iterates $(\Phi^{(n)}_{s,s+t}, n \in \mathbb{N} \cup \{0\})$ constructed in the proof of Theorem 6.4.2 and aim to show that

$$\Phi^{(n)}_{s,s+t}(y, \omega) = \Phi^{(n)}_{0,t}(y, \theta_s\omega) \qquad \text{a.s.}$$

for all $n \in \mathbb{N} \cup \{0\}$, from which the result follows on taking limits as $n \to \infty$.

We proceed by induction. Clearly the result is true when $n = 0$. Suppose that it holds for some $n \in \mathbb{N}$. Just as in the proof of Theorem 6.4.2 we will consider a condensed SDE, without loss of generality, and this time we will retain only the Brownian motion terms. Using our usual sequence of partitions and the result of Lemma 6.4.9, for each $1 \leq i \leq d$, $s, t \geq 0$, $y \in \mathbb{R}^d$, $\omega \in \Omega$, the following holds with probability 1:

$$\Phi^{i,(n+1)}_{s,s+t}(y, \omega)$$

$$= y^i + \int_s^{t+s} \sigma^i_j(\Phi^{(n)}_{s,s+u}(y, \omega))dB^j(u)(\omega)$$

$$= y^i + \lim_{n\to\infty} \sum_{k=0}^{m(n)} \sigma^i_j(\Phi^{(n)}_{s,s+t_k}(y, \omega))(B^j(s + t_{k+1})$$
$$\qquad\qquad\qquad\qquad\qquad\qquad - B^j(s + t_k))(\omega)$$

$$= y^i + \lim_{n\to\infty} \sum_{k=0}^{m(n)} \sigma^i_j(\Phi^{(n)}_{0,t_k}(y, \theta_s\omega))(B^j(t_{k+1}) - B^j(t_k))(\theta_s\omega)$$

$$= y^i + \int_0^t \sigma^i_j(\Phi^{(n)}_{0,u}(y, \theta_s\omega))dB^j(u)(\theta_s\omega)$$

$$= \left[y^i + \int_0^t \sigma^i_j(\Phi^{(n)}_{0,u}(y))dB^j(u)\right](\theta_s\omega)$$

$$= \Phi^{i,(n+1)}_{0,t}(y, \theta_s\omega),$$

where the limit is taken in the L^2 sense.

$\square$

6.4 Stochastic flows, cocycle and Markov properties of SDEs

Corollary 6.4.11 Φ *is a multiplicative cocycle, i.e.*

$$\Phi_{0,s+t}(y, \omega) = \Phi_{0,t}(\Phi_{0,s}(y), \theta_s(\omega))$$

for all $s, t \geq 0$, $y \in \mathbb{R}^d$ *and almost all* $\omega \in \Omega$.

Proof By the flow property and Lemma 6.4.10, we obtain

$$\Phi_{0,s+t}(y, \omega) = \Phi_{s,s+t}(\Phi_{0,s}(y), \omega) = \Phi_{0,t}(\Phi_{0,s}(y), \theta_s \omega) \quad \text{a.s.}$$

$\square$

We can use the cocycle property to extend our two-parameter flow to a one-parameter family (as in the deterministic case) by including the action of the shift on Ω. Specifically, define $\Psi_t : \mathbb{R}^d \times \Omega \to \mathbb{R}^d \times \Omega$ by

$$\Psi_t(y, \omega) = \big(\Phi_{0,t}(y, \theta_t(\omega)), \theta_t(\omega)\big)$$

for each $t \geq 0$, $\omega \in \Omega$.

Corollary 6.4.12 *The following holds almost surely:*

$$\Psi_{0,t+s} = \Psi_{0,t} \circ \Psi_{0,s}$$

for all $s, t \geq 0$.

Proof By using the semigroup property of the shift (Exercise 6.4.8(1)) and Corollary 6.4.11, we have, for *all* $y \in \mathbb{R}^d$ and *almost all* $\omega \in \Omega$,

$$\begin{aligned}\Psi_{0,t+s}(y, \omega) &= \Phi_{0,t+s}(y, \theta_{t+s}(\omega), \theta_{t+s}(\omega)) \\ &= \Phi_{0,t}\big(\Phi_{0,s}((y, \theta_s \omega), (\theta_t \theta_s)(\omega)), (\theta_t \theta_s)(\omega)\big) \\ &= (\Psi_{0,t} \circ \Psi_{0,s})(y, \omega).\end{aligned}$$

$\square$

Of course, it would be more natural for Corollary 6.4.12 to hold for all $\omega \in \Omega$, and a sufficient condition for this is that Corollary 6.4.11 is itself valid for all $\omega \in \Omega$. Cocycles that have this property are called *perfect*, and these are also important in studying ergodic properties of stochastic flows. For conditions under which cocycles arising from Brownian flows are perfect, see Arnold and Scheutzow [12].

6.5 Interlacing for solutions of SDEs

In this section, we will apply interlacing to the solution flow $\Psi = (\Psi_{s,t}, 0 \leq s \leq t < \infty)$ associated with the solution of the modified SDE $Z = (Z(t), t \geq 0)$ in order to obtain Ψ as the (almost-sure) limit of an interlacing sequence. We assume that $\nu(B_c(0) - \{0\}) \neq 0$, where B_c is a ball of radius c, and fix a sequence $(\epsilon_n, n \in \mathbb{N})$ of positive real numbers which decrease monotonically to zero. We will give a precise form of each ϵ_n below. Let $(A_n, n \in \mathbb{N})$ be the sequence of Borel sets defined by $A_n = \{x \in B_c(0) - \{0\}; \epsilon_n < |x| < c\}$ and define a sequence of associated interlacing flows $(\Psi^n, n \in \mathbb{N})$ by

$$d\Psi^n_{s,t}(y) = b(\Psi^n_{s,t-}(y))dt + \sigma(\Psi^n_{s,t-}(y))dB(t)$$
$$+ \int_{A_n} F(\Psi^n_{s,t-}(y), x)\tilde{N}(dt, dx)$$

for each $n \in \mathbb{N}$, $0 \leq s \leq t < \infty$, $y \in \mathbb{R}^d$. In order to carry out our analysis we need to impose a stronger condition on the mapping F:

Assumption 6.5.1 We assume that for all $y \in \mathbb{R}^d$, $x \in B_c(0) - \{0\}$,

$$|F(y, x)| \leq |\rho(x)||\delta(y)|$$

where $\rho : B_c(0) - \{0\} \to \mathbb{R}$ satisfies $\int_{|x|<c} |\rho(x)|^2 \nu(dx) < \infty$ and $\delta : \mathbb{R}^d \to \mathbb{R}^d$ is Lipschitz continuous with Lipschitz constant C_δ.

Note that if Assumption 6.5.1 holds then, for each $x \in B_c(0) - \{0\}$, the mapping $y \to F(y, x)$ is continuous. Assumption 6.5.1 implies the growth condition for F in (6.17).

The following theorem generalises the result of Corollary 4.3.10 to strong solutions of SDEs. A similar result can be found in the appendix to Applebaum [10].

Theorem 6.5.2 *If Assumption 6.5.1 holds, then for each* $y \in \mathbb{R}^d$, $0 \leq s \leq t < \infty$,

$$\lim_{n \to \infty} \Psi^n_{s,t}(y) = \Psi_{s,t}(y) \quad a.s.$$

and the convergence is uniform on finite intervals of $\mathbb{R}^+$.

6.5 Interlacing for solutions of SDEs

Proof First note that for each $y \in \mathbb{R}^d$, $0 \leq s \leq t$, $n \in \mathbb{N}$,

$$\Psi_{s,t}^{n+1}(y) - \Psi_{s,t}^n(y)$$

$$= \int_s^t [b(\Psi_{s,u-}^{n+1}(y)) - b(\Psi_{s,u-}^n(y))] du$$

$$+ \int_s^t [\sigma(\Psi_{s,u-}^{n+1}(y)) - \sigma(\Psi_{s,u-}^n(y))] dB(u)$$

$$+ \int_s^t \int_{A_{n+1}} F(\Psi_{s,u-}^{n+1}(y), x) \tilde{N}(du, dx)$$

$$- \int_s^t \int_{A_n} F(\Psi_{s,u-}^n(y), x) \tilde{N}(du, dx)$$

$$= \int_s^t [b(\Psi_{s,u-}^{n+1}(y)) - b(\Psi_{s,u-}^n(y))] du$$

$$+ \int_s^t [\sigma(\Psi_{s,u-}^{n+1}(y)) - \sigma(\Psi_{s,u-}^n(y))] dB(u)$$

$$+ \int_s^t \int_{A_{n+1}-A_n} F(\Psi_{s,u-}^{n+1}(y), x) \tilde{N}(du, dx)$$

$$+ \int_s^t \int_{A_n} [F(\Psi_{s,u-}^{n+1}(y), x) - F(\Psi_{s,u-}^n(y), x)] \tilde{N}(du, dx).$$

Now take the norm of each side of this identity, apply the triangle inequality and the inequality (6.18) with $n=4$ and take expectations. Using Doob's martingale inequality, we then have that

$$\mathbb{E}\left(\sup_{s \leq u \leq t} |\Psi_{s,u}^{n+1}(y) - \Psi_{s,u}^n(y)|^2\right)$$

$$\leq 4 \left\{ \mathbb{E}\left(\left|\int_s^t [b(\Psi_{s,u-}^{n+1}(y)) - b(\Psi_{s,u-}^n(y))] du\right|^2\right) \right.$$

$$+ 4 \mathbb{E}\left(\left|\int_s^t [\sigma(\Psi_{s,u-}^{n+1}(y)) - \sigma(\Psi_{s,u-}^n(y))] dB(u)\right|^2\right)$$

$$+ 4 \mathbb{E}\left(\left|\int_s^t \int_{A_{n+1}-A_n} F(\Psi_{s,u-}^{n+1}(y), x) \tilde{N}(du, dx)\right|^2\right)$$

$$\left. + 4 \mathbb{E}\left(\left|\int_s^t \int_{A_n} [F(\Psi_{s,u-}^{n+1}(y), x) - F(\Psi_{s,u-}^n(y), x)] \tilde{N}(du, dx)\right|^2\right) \right\}.$$

Applying the Cauchy–Schwarz inequality in the first term and Itô's isometry in the other three, we obtain

$$\mathbb{E}\left(\sup_{s\leq u\leq t}|\Psi_{s,u}^{n+1}(y)-\Psi_{s,u}^n(y)|^2\right)$$

$$\leq 4\Bigg[t\int_s^t \mathbb{E}\big(|b(\Psi_{s,u}^{n+1}(y))-b(\Psi_{s,u}^n(y))|^2\big)du$$

$$+4\int_s^t \mathbb{E}\big(||a(\Psi_{s,u}^{n+1}(y),\Psi_{s,u}^{n+1}(y))-2a(\Psi_{s,u}^{n+1}(y),\Psi_{s,u}^n(y))$$
$$+a\left(\Psi_{s,u}^n(y),\Psi_{s,u}^n(y)\right)||\big)du$$

$$+4\int_s^t\int_{A_{n+1}-A_n}\mathbb{E}\big(|F(\Psi_{s,u}^{n+1}(y),x)^2|\big)v(dx)du$$

$$+4\int_s^t\int_{A_n}\mathbb{E}\big(|F\big(\Psi_{s,u}^{n+1}(y),x\big)-F\big(\Psi_{s,u}^n(y),x\big)|^2\big)v(dx)du\Bigg].$$

We can now apply the Lipschitz condition in the first, second and fourth terms. For the third term we use Assumption 6.5.1, the results of Exercise 6.1.1 and Corollary 6.2.4 to obtain

$$\int_s^t\int_{A_{n+1}-A_n}\mathbb{E}\big(|F(\Psi_{s,u}^{n+1}(y),x)^2|\big)v(dx)du$$

$$\leq \int_{A_{n+1}-A_n}|\rho(x)|^2dx\int_s^t\mathbb{E}\big(|\delta(\Psi_{s,u}^{n+1}(y))|^2\big)du$$

$$\leq C_1\int_{A_{n+1}-A_n}|\rho(x)|^2dx\int_s^t\mathbb{E}\big(|1+|\Psi_{s,u}^{n+1}(y)|^2\big)du$$

$$\leq C_2(t-s)(1+|y|^2)\int_{A_{n+1}-A_n}|\rho(x)|^2dx,$$

where $C_1, C_2 > 0$.

Now we can collect together terms to deduce that there exists $C_3(t) > 0$ such that

$$\mathbb{E}\left(\sup_{s\leq u\leq t}|\Psi_{s,u}^{n+1}(y)-\Psi_{s,u}^n(y)|^2\right)\leq C_2(t-s)(1+|y|^2)\int_{A_{n+1}-A_n}|\rho(x)|^2dx$$
$$+C_3\int_s^t\mathbb{E}\left(\sup_{s\leq v\leq u}|\Psi_{s,v}^{n+1}(y)-\Psi_{s,v}^n(y)|^2\right)du.$$

On applying Gronwall's inequality, we find that there exists $C_4 > 0$ such that

$$\mathbb{E}\left(\sup_{s \leq u \leq t} |\Psi^{n+1}_{s,u}(y) - \Psi^n_{s,u}(y)|^2\right) \leq C_4(t-s)(1+|y|^2)\int_{A_{n+1}-A_n} |\rho(x)|^2 dx.$$

Now fix each $\epsilon_n = \sup\{z \geq 0; \int_{0<|x|<z} |\rho(x)|^2 dx \leq 8^{-n}\}$ and then follow the argument of Theorem 2.5.2 to obtain the required result. □

In accordance with our usual philosophy, we can gain more insight into the structure of the paths by constructing the interlacing sequence directly.

Let $(Q_n, n \in \mathbb{N})$ be the sequence of compound Poisson processes associated with the sets $(A_n, n \in \mathbb{N})$ where, for each $t \geq 0$, $Q_n(t) = \int_0^t \int_{A_n} xN(ds,dx)$. So, for each $0 \leq s \leq t < \infty$, $Q_n(t) - Q_n(s) = \int_s^t \int_{A_n} xN(ds,dx)$. We denote the arrival times of $(Q_n(t) - Q_n(s), 0 \leq s \leq t < \infty)$ by $(S_n^m, m \in \mathbb{N})$ for each $n \in \mathbb{N}$.

We will have need of the sequence of solution flows to diffusion equations $\Gamma^n = (\Gamma^n_{s,t}, 0 \leq s \leq t < \infty)$; these are defined by

$$d\Gamma^n_{s,t}(y) = \left(b(\Gamma^n_{s,t}(y)) - \int_{A_n} F(\Gamma^n_{s,t}(y),x)\nu(dx)\right)dt + \sigma(\Gamma^n_{s,t}(y))dB(t).$$

Let $\pi_{F,x}: \mathbb{R}^d \to \mathbb{R}^d$ be defined by $\pi_{F,x}(y) = y + F(y,x)$ for each $y \in \mathbb{R}^d$, $0 \leq |x| < c$; then we can read off the following interlacing construction. For each $t \geq 0$, $y \in \mathbb{R}^d$,

$$\Psi^n_{s,t}(y) = \begin{cases} \Gamma^n_{s,t}(y) & \text{for } s \leq t < S_n^1, \\ \pi_{F,\Delta Q_n(S_n^1)} \circ \Psi^n_{0,S_n^1-}(y) & \text{for } t = S_n^1, \\ \Gamma^n_{S_n^1,t} \circ \Psi^n_{0,S_n^1}(y) & \text{for } S_n^1 < t < S_n^2, \\ \pi_{F,\Delta Q_n(S_n^2)} \circ \Psi^n_{0,S_n^2-}(y) & \text{for } t = S_n^2, \end{cases}$$

and so on, recursively.

Hence we see that Ψ is the almost-sure limit of solution flows associated to a sequence of jump-diffusion processes.

6.6 Continuity of solution flows to SDEs

Let Φ be the solution flow associated with the SDE (6.12). In this section we will investigate the continuity of the mappings from $\mathbb{R}^d$ to $\mathbb{R}^d$ given by $y \to \Phi_{s,t}(y)$ for each $0 \leq s < t < \infty$.

We will need to make an additional assumption. Fix $\gamma \in \mathbb{N}$ with $\gamma > 2$.

Assumption 6.6.1 (γ-Lipschitz condition) There exists $K_\gamma > 0$ such that, for all $y_1, y_2 \in \mathbb{R}^d$,

$$\int_{|x|<c} |F(y_1, x) - F(y_2, x)|^p v(dx) \leq K_\gamma |y_1 - y_2|^p$$

for all $2 \leq p \leq \gamma$.

Note that if Assumption 6.5.1 holds then Assumption 6.6.1 is simply the requirement that $\int_{|x|<c} |\rho(x)|^p v(dx) < \infty$ for all $p \in [2, \gamma]$. Moreover, if Assumption 6.5.1 holds with $|\rho(x)| \leq |x|$, for all $x \in B_c(0) - \{0\}$, then Assumption 6.6.1 is automatically satisfied.

We recall that the modified flow $\Psi = (\Psi_{s,t}, 0 \leq s \leq t < \infty)$ satisfies the SDE

$$d\Psi_{s,t}(y) = b(\Psi_{s,t-}(y))dt + \sigma(\Psi_{s,t-}(y))dB(t)$$
$$+ \int_{|x|<c} F(\Psi_{s,t-}(y), x)\tilde{N}(dt, dx). \quad (6.35)$$

The main result of this section depends critically on the following technical estimate for the modified flow, which is due to Fujiwara and Kunita [108], pp. 84–6.

Proposition 6.6.2 *For all $0 \leq s \leq t$, there exists $D(\gamma, t) > 0$ such that*

$$\mathbb{E}\left(\sup_{s \leq u \leq t} |\Psi_{s,t}(y_1) - \Psi_{s,t}(y_2)|^p\right) \leq D(\gamma, t)|y_1 - y_2|^p$$

for all $2 \leq p \leq \gamma$, $y_1, y_2 \in \mathbb{R}^d$.

Proof As this proof is rather detailed, we will introduce some notation to help us simplify expressions. Fix $s \geq 0$ and $y_1, y_2 \in \mathbb{R}^d$. For each $t \geq s$, $x \in B_c(0) - \{0\}$, we define:

$$\eta(t) = \Psi_{s,t}(y_1) - \Psi_{s,t}(y_2);$$
$$g_F(t, x) = F(\Psi_{s,t}(y_1), x) - F(\Psi_{s,t}(y_2), x);$$
$$A(t) = \int_s^t \left[b(\Psi_{s,u-}(y_1)) - b(\Psi_{s,u-}(y_2))\right] du;$$
$$M^C(t) = \int_s^t \left[\sigma(\Psi_{s,u-}(y_1)) - \sigma(\Psi_{s,u-}(y_2))\right] dB(u);$$
$$M^D(t) = \int_s^t g_F(u-, x)\tilde{N}(du, dx).$$

6.6 Continuity of solution flows to SDEs

From the modified-flow equation (6.35), we have

$$\eta(t) = (y_1 - y_2) + A(t) + M^C(t) + M^D(t).$$

Fix $2 \leq p \leq \gamma$. For any $a, b, c, d \in \mathbb{R}$, Jensen's inequality yields

$$|a + b + c + d|^p \leq 4^{p-1}(|a|^p + |b|^p + |c|^p + |d|^p).$$

Using the triangle inequality together with this fact, we obtain

$$\mathbb{E}\left(\sup_{s \leq u \leq t} |\eta(u)|^p\right) \leq 4^{p-1}\left[|y_1 - y_2|^p + \mathbb{E}\left(\sup_{s \leq u \leq t} |A(u)|^p\right)\right.$$
$$+ \mathbb{E}\left(\sup_{s \leq u \leq t} |M^C(u)|^p\right)$$
$$\left.+ \mathbb{E}\left(\sup_{s \leq u \leq t} |M^D(u)|^p\right)\right]. \quad (6.36)$$

By Hölder's inequality and the Lipschitz condition, we have

$$\mathbb{E}\left(\sup_{s \leq u \leq t} |A(u)|^p\right)$$
$$\leq (t-s)^{p-1} \int_s^t \mathbb{E}\left(|b(\Psi_{s,u}(y_1)) - b(\Psi_{s,u}(y_2))|^p\right) du$$
$$\leq K_1^{p/2}(t-s)^{p-1} \int_s^t \mathbb{E}\left(|\Psi_{s,u}(y_1) - \Psi_{s,u}(y_2)|^p\right) du. \quad (6.37)$$

Note that $M^C(t) = \left(M_1^C(t), \ldots, M_d^C(t)\right)$, where each

$$M_i^C(t) = \int_s^t \left[\sigma_i^j(\Psi_{s,u}(y_1)) - \sigma_i^j(\Psi_{s,u}(y_2))\right] dB_j(u).$$

By Hölder's inequality,

$$|M^C(t)|^p = \left\{\sum_{i=1}^d [M_i^C(t)]^2\right\}^{p/2} \leq d^{(p-2)/2} \sum_{i=1}^d |M_i^C(t)|^p.$$

Hence, by Doob's martingale inequality, Burkholder's inequality (Theorem

4.4.20), Hölder's inequality and bi-Lipschitz continuity, we obtain

$$\mathbb{E}\left(\sup_{s \leq u \leq t} |M^C(u)|^p\right)$$

$$\leq q^p \mathbb{E}(|M^C(t)|^p)$$

$$\leq q^p d^{(p-2)/2} \sum_{i=1}^{d} \mathbb{E}(|M_i^C(t)|^p)$$

$$\leq C(p) d^{(p-2)/2} \sum_{i=1}^{d} \mathbb{E}\left([M_i^C(t), M_i^C(t)]\right)^{p/2}$$

$$\leq C(p) d^{(p-2)/2} \sum_{i=1}^{d} \mathbb{E}\left(\int_s^t [a_i^i(\Psi_{s,u}(y_1), \Psi_{s,u}(y_1))\right.$$

$$- 2 a_i^i(\Psi_{s,u}(y_1), \Psi_{s,u}(y_2))$$

$$\left. + a_i^i(\Psi_{s,u}(y_2), \Psi_{s,u}(y_2))] du\right)^{p/2}$$

$$\leq C(p) d^{p/2} K_1^p (t-s)^{(p/2)-1} \int_s^t \mathbb{E}(|\Psi_{s,u}(y_1) - \Psi_{s,u}(y_2)|^p) du, \quad (6.38)$$

where $1/p + 1/q = 1$ and $C(p) > 0$.

The next part of the argument, which concerns the estimate of $M^D(t)$, is quite involved. To facilitate matters, we will find it convenient to introduce positive constants, $C_1, C_2 \ldots$ depending only on t and γ, at various stages.

By Itô's formula,

$$|M^D(t)|^p$$

$$= \int_s^t \int_{|x|<c} (|M^D(u-) + g_F(u-, x)|^p - |M^D(u-)|^p) \tilde{N}(du, dx)$$

$$+ \int_s^t \int_{|x|<c} (|M^D(u-) + g_F(u-, x)|^p - |M^D(u-)|^p$$

$$- p g_F^i(u-, x) M_i^D(u-) |M^D(u-)|^{p-2}) \nu(dx) du.$$

Using Doob's martingale inequality, Taylor's theorem and Jensen's inequality,

6.6 Continuity of solution flows to SDEs

we obtain, for some $0 < \theta < 1$,

$$\mathbb{E}\left(\sup_{s \leq u \leq t} |M^D(u)|^p\right)$$
$$\leq q^p \mathbb{E}(|M^D(t)|^p)$$
$$\leq \tfrac{1}{2} q^p p(p-1) \int_s^t \int_{|x|<c} \mathbb{E}(|g_F(u,x)|^2 |M^D(u) + \theta g_F(u,x)|^{p-2}) \nu(dx) du$$
$$\leq C_1 \int_s^t \int_{|x|<c} \mathbb{E}(|g_F(u,x)|^2 |M^d(u)|^{p-2} + |g_F(u,x)|^p) \nu(dx) du$$

By Assumption 6.6.1, $\int_{|x|<c} |g_F(u,x)|^p \nu(dx) \leq K_\gamma |\eta(u)|^p$. Hence, by Fubini's theorem,

$$\mathbb{E}\left(\sup_{s \leq u \leq t} |M^d(u)|^p\right) \leq C_2 \int_s^t \mathbb{E}(|\eta(u)|^2 |M^d(u)|^{p-2} + |\eta(u)|^p) du.$$

Now since, for all $s \leq u \leq t$, $M^D(u) = \eta(u) - (y_1 - y_2) - M^C(u) - A(u)$, Jensen's inequality yields

$$|M^D(u)|^{p-2} \leq 4^{p-3}\left(|\eta(u)|^{p-2} + |y_1 - y_2|^{p-2} + |M^C(u)|^{p-2} + |A(u)|^{p-2}\right).$$

We thus obtain

$$\mathbb{E}\left(\sup_{s \leq u \leq t} |M^D(u)|^p\right)$$
$$\leq C_3 \int_s^t \mathbb{E}\left(|\eta(u)|^p + |y_1 - y_2|^{p-2}|\eta(u)|^2 + |\eta(u)|^2|M^C(u)|^{p-2} + |\eta(u)|^2|A(u)|^{p-2}\right) du. \tag{6.39}$$

By Hölder's inequality and the estimate (6.38), we have

$$\mathbb{E}(|\eta(u)|^2 |M^C(u)|^{p-2}) \leq \mathbb{E}(|\eta(u)|^p)^{2/p} \mathbb{E}(|M^C(u)|^p)^{(p-2)/p}$$
$$\leq C_4 \mathbb{E}(|\eta(u)|^p)^{2/p} \mathbb{E}\left(\sup_{s \leq v \leq u} |\eta(u)|^p\right)^{(p-2)/p}$$
$$\leq C_4 \mathbb{E}\left(\sup_{s \leq v \leq u} |\eta(u)|^p\right).$$

Using (6.37), we similarly conclude that

$$\mathbb{E}\big(|\eta(u)|^2|A(u)|^{p-2}\big) \leq C_5 \, \mathbb{E}\left(\sup_{s \leq v \leq u} |\eta(u)|^p\right).$$

Substituting these into (6.39) we finally obtain the desired estimate,

$$\mathbb{E}\left(\sup_{s \leq u \leq t} |M^D(u)|^p\right)$$
$$\leq C_6 \int_s^t \mathbb{E}\left(\left(\sup_{s \leq v \leq u} |\eta(u)|^p\right) + |y_1 - y_2|^{p-2}|\eta(u)|^2\right) du. \quad (6.40)$$

Now substitute (6.37), (6.38) and (6.40) into (6.36) to find

$$\mathbb{E}\left(\sup_{s \leq u \leq t} |\eta(u)|^p\right) \leq 4^{p-1}|y_1 - y_2|^p + C_7 \int_s^t \mathbb{E}\left(\left(\sup_{s \leq v \leq u} |\eta(u)|^p\right) \right.$$
$$\left. + |y_1 - y_2|^{p-2}|\eta(u)|^2\right) du. \quad (6.41)$$

First let $p = 2$; then (6.41) becomes

$$\mathbb{E}\left(\sup_{s \leq u \leq t} |\eta(u)|^2\right) \leq 4^{p-1}|y_1 - y_2|^2 + 2C_7 \int_s^t \mathbb{E}\left(\sup_{s \leq v \leq u} |\eta(u)|^2\right) du;$$

hence, by Gronwall's inequality,

$$\mathbb{E}\left(\sup_{s \leq u \leq t} |\eta(u)|^2\right) \leq C_8|y_1 - y_2|^2.$$

Substitute this back into (6.41) to obtain

$$\mathbb{E}\left(\sup_{s \leq u \leq t} |\eta(u)|^p\right) \leq C_9|y_1 - y_2|^p + C_{10} \int_s^t \mathbb{E}\left(\sup_{s \leq v \leq u} |\eta(u)|^p\right) du,$$

and the required result follows by a further application of Gronwall's inequality. $\square$

Theorem 6.6.3 *The map* $y \to \Phi_{s,t}(y)$ *has a continuous modification for each* $0 \leq s < t < \infty$.

Proof First consider the modified flow Ψ. Take $\gamma > d \vee 2$ in Proposition 6.6.2 and appeal to the Kolmogorov continuity criterion (Theorem 1.1.17) to obtain the required continuous modification. The almost-sure continuity of $y \to \Phi_{s,t}(y)$ is then deduced from the interlacing structure in Theorem 6.2.9, using the continuity of $y \to G(y, x)$ for each $|x| \geq c$. $\square$

6.7 Solutions of SDEs as Feller processes

Note An alternative approach to proving Theorem 6.6.3 was developed in Applebaum and Tang [7]. Instead of Assumption 6.6.1 we impose Assumption 6.5.1. We first prove Proposition 6.6.2, but only in the technically simpler case of the Brownian flow

$$d\Gamma_{s,t}(y) = b(\Gamma_{s,t}(y))dt + \sigma(\Gamma_{s,t}(y))dB(t),$$

so that $y \to \Gamma_{s,t}(y)$ is continuous by the argument in the proof of Theorem 6.6.3. Now return to the interlacing theorem, Theorem 6.5.2. By Assumption 6.5.1, it follows that, for each $n \in \mathbb{N}$, $y \to \Psi^n_{s,t}(y)$ is continuous. From the proof of Theorem 6.5.2, we deduce that we have $\lim_{n\to\infty} \Psi^n_{s,t}(y) = \Psi_{s,t}(y)$ (a.s.) uniformly on compact intervals of $\mathbb{R}^d$ containing y, and the continuity of $y \to \Psi_{s,t}(y)$ follows immediately.

6.7 Solutions of SDEs as Feller processes, the Feynman–Kac formula and martingale problems

6.7.1 SDEs and Feller semigroups

Let $(T_t, t \geq 0)$ be the semigroup associated with the solution flow Φ of the SDE (6.12).

We need to make an additional assumption on the coefficients.

Assumption 6.7.1 For each $1 \leq i, j \leq d$, the mappings $y \to b^i(y)$, $y \to a^{ij}(y, y)$, $y \to F^i(y, x)$ ($|x| < c$) and $y \to G^i(y, x)$ ($|x| \geq c$) are in $C_b(\mathbb{R}^d)$.

We require Assumptions 6.6.1 and 6.7.1 both to hold in this section.

Theorem 6.7.2 For all $t \geq 0$,

$$T_t(C_0(\mathbb{R}^d)) \subseteq C_0(\mathbb{R}^d).$$

Proof First we establish continuity. Let $(y_n, n \in \mathbb{N})$ be any sequence in $\mathbb{R}^d$ that converges to $y \in \mathbb{R}^d$. Since for each $t \geq 0$, $f \in C_0(\mathbb{R}^d)$, we have $(T_t f)(y) = \mathbb{E}(f(\Phi_{0,t}(y)))$, it follows that

$$|(T_t f)(y) - (T_t f)(y_n)| = \left|\mathbb{E}\left(f(\Phi_{0,t}(y)) - f(\Phi_{0,t}(y_n))\right)\right|.$$

Since $|\mathbb{E}(f(\Phi_{0,t}(y)) - f(\Phi_{0,t}(y_n)))| \leq 2\|f\|$, we can use dominated convergence and Theorem 6.6.3 to deduce the required result.

For the limiting behaviour of $T_t f$, we first consider the modified flow Ψ. The following argument was suggested to the author by H. Kunita. From

Applebaum and Tang [7], pp. 158–162, we have the estimate

$$\mathbb{E}((1+|\Psi_{s,t}(y)|^2)^p) \leq C(p,t)(1+|y|^2)^p$$

for all $0 \leq s \leq t < \infty$, $y \in \mathbb{R}^d$, $p \in \mathbb{R}$, where $C(p,t) > 0$ (see also Fujiwara and Kunita [108], p. 92; in fact, this result can be established by using an argument similar to that in the proof of Proposition 6.6.2). If we take $p = -1$, we find that

$$\limsup_{|y|\to\infty} \mathbb{E}\big((1+|\Psi_{s,t}(y)|^2)^{-1}\big) = 0.$$

From this we deduce that

$$\lim_{|y|\to\infty} \frac{1}{1+|\Psi_{s,t}(y)|^2} = 0$$

in probability, and hence that $\lim_{|y|\to\infty} |\Psi_{s,t}(y)| = \infty$, in probability.

For each $t \geq 0$, $f \in C_0(\mathbb{R}^d)$, $y \in \mathbb{R}^d$, we define the semigroup $(S_t, t \geq 0)$ by $(S_t f)(y) = \mathbb{E}(f(\Psi_{0,t}(y)))$. We will now show that $\lim_{|y|\to\infty}(S_t f)(y) = 0$ (so that the solution of the modified SDE is a Feller process).

Since $f \in C_0(\mathbb{R}^d)$, given any $\delta > 0$ there exists $\epsilon > 0$ such that $|y| > \delta \Rightarrow |f(y)| < \epsilon/2$. Since $\lim_{|y|\to\infty}|\Psi_{0,t}(y)| = \infty$, in probability, there exists $K > 0$ such that $|y| > K \Rightarrow P(|\Psi_{0,t}(y)| < \delta) < \epsilon/(2\|f\|)$.

Now, for $|y| > \max\{\delta, K\}$ we have

$$|(S_t f)(y)| \leq \int_{\mathbb{R}^d} |f(z)| p_{\Psi_{0,t}(y)}(dz)$$
$$= \int_{B_\delta(0)} |f(z)| p_{\Psi_{0,t}(y)}(dz) + \int_{B_\delta(0)^c} |f(z)| p_{\Psi_{0,t}(y)}(dz)$$
$$\leq \sup_{z \in B_\delta(0)} |f(z)| P\big(\Psi_{0,t}(y) \in B_\delta(0)\big) + \sup_{z \in B_\delta(0)^c} |f(z)|$$
$$< \epsilon.$$

To pass to the general flow Φ, we use the interlacing structure from the proof of Theorem 6.2.9 and make use of the notation developed there. For each $t \geq 0$, define a sequence of events $(A_n(t), n \in \mathbb{N})$ by

$$A_{2n}(t) = (\tau_n = t), \qquad A_{2n-1}(t) = (\tau_{n-1} < t < \tau_n).$$

By the above discussion for the modified flow, for each $f \in C_0(\mathbb{R}^d)$, $y \in \mathbb{R}^d$, we have

$$\mathbb{E}(f(\Phi_{0,t}(y))|A_1) = \mathbb{E}(f(\Psi_{0,t}(y))) = (S_t f)(y),$$

hence $\lim_{|y|\to\infty} \mathbb{E}(f(\Phi_{0,t}(y))|A_1) = 0$.

6.7 Solutions of SDEs as Feller processes

Using dominated convergence and Assumption 6.7.1, we have

$$\mathbb{E}(f(\Phi_{0,t}(y))|A_2) = \mathbb{E}\left(f\left(\Psi_{0,\tau_1-}(y) + G(\Psi_{0,\tau_1-}(y), \Delta P(\tau_1))\right)\right) \to 0$$

as $|y| \to \infty$. Using the independence of $\Phi_{0,s}$ and $\Psi_{s,t}$, we also find that, for each $0 \leq s \leq t < \infty$,

$$\begin{aligned}
\mathbb{E}(f(\Phi_{0,t}(y))|A_3) &= \mathbb{E}\big(f(\Psi_{\tau_1,t}(\Phi_{0,\tau_1}(y)))\big) \\
&= \mathbb{E}\big(\mathbb{E}\big(f(\Psi_{s,t}(\Phi_{0,s}(y)))\big|\tau_1 = s\big)\big) \\
&= \int_0^\infty \mathbb{E}\big(f(\Psi_{s,t}(\Phi_{0,s}(y)))\big|\tau_1 = s\big) p_{\tau_1}(ds) \\
&= \int_0^\infty (T_{0,s} \circ S_{0,t-s} f)(y) p_{\tau_1}(ds) \\
&\to 0 \quad \text{as} \quad |y| \to \infty
\end{aligned}$$

by dominated convergence and the previous result. We can now use induction to establish

$$\lim_{|y|\to\infty} \mathbb{E}(f(\Phi_{0,t}(y))|A_n) = 0 \quad \text{for all } n \in \mathbb{N}.$$

Finally,

$$(T_t f)(y) = \mathbb{E}(f(\Phi_{0,t}(y))) = \sum_{n=1}^\infty \mathbb{E}(f(\Phi_{0,t}(y))|A_n) \, P(A_n)$$

and so $\lim_{|y|\to\infty}(T_t f)(y) = 0$, by dominated convergence. $\square$

Note 1 To prove Theorem 6.7.2, we only needed that part of Assumption 6.7.1 that pertains to the mapping G. The rest of the assumption is used below to ensure that the generator of $(T_t, t \geq 0)$ has some 'nice' functions in its domain.

Note 2 As an alternative to the use of Assumption 6.7.1 to prove Theorem 6.7.2, we can impose the growth condition that there exists $D > 0$ such that $\int_{|x|\geq c} |G(y,x)|^2 \nu(dx) < D(1+|y|^2)$ for all $y \in \mathbb{R}^d$. In this case, the estimate

$$\mathbb{E}\big((1 + |\Phi_{s,t}(y)|^2)^p\big) \leq C(p,t)(1 + |y|^2)^p$$

holds for all $0 \leq s \leq t < \infty$.

Note 3 To establish $T_t : C_b(\mathbb{R}^d) \to C_b(\mathbb{R}^d)$ instead of $T_t : C_0(\mathbb{R}^d) \to C_0(\mathbb{R}^d)$ in Theorem 6.7.2 is relatively trivial and requires neither Assumption 6.7.1 nor the growth condition discussed in Note 1 above.

Before we establish the second part of the Feller property, we introduce an important linear operator.

Define $\mathcal{L} : C_0^2(\mathbb{R}^d) \to C_0(\mathbb{R}^d)$ by

$$(\mathcal{L}f)(y) = b^i(y)(\partial_i f)(y) + \tfrac{1}{2}a^{ij}(y)(\partial_i \partial_j f)(y)$$
$$+ \int_{|x|<c} \left[f(y+F(y,x)) - f(y) - F^i(y,x)(\partial_i f)(y)\right]\nu(dx)$$
$$+ \int_{|x|\geq c} \left[f(y+G(y,x)) - f(y)\right]\nu(dx) \quad (6.42)$$

for each $f \in C_0^2(\mathbb{R}^d)$, $y \in \mathbb{R}^d$, and where each matrix $a(y,y)$ is written as $a(y)$.

Exercise 6.7.3 Confirm that each $\mathcal{L}f \in C_0(\mathbb{R}^d)$.

Theorem 6.7.4 $(T_t, t \geq 0)$ *is a Feller semigroup, and if $\mathcal{A}$ denotes its infinitesimal generator, then $C_0^2(\mathbb{R}^d) \subseteq \text{Dom}(\mathcal{A})$ and $\mathcal{A}(f) = \mathcal{L}(f)$ for all $f \in C_0^2(\mathbb{R}^d)$.*

Proof Let $f \in C_0^2(\mathbb{R}^d)$. By Itô's formula, for each $t \geq 0$, $y \in \mathbb{R}^d$,

$$df(\Phi_{0,t}(y))$$
$$= (\partial_i f)(\Phi_{0,t-}(y))b^i(\Phi_{0,t-}(y))dt + (\partial_i f)(\Phi_{0,t-}(y))$$
$$\times \sigma^i_j(\Phi_{0,t-}(y))dB^j(t)$$
$$+ \tfrac{1}{2}(\partial_i \partial_j f)(\Phi_{0,t-}(y))a^{ij}(\Phi_{0,t-}(y))dt$$
$$+ \int_{|x|<c} \left[f(\Phi_{0,t-}(y)+F(\Phi_{0,t-}(y),x)) - f(\Phi_{0,t-}(y))\right]\tilde{N}(ds,dx)$$
$$+ \int_{|x|\geq c} \left[f(\Phi_{0,t-}(y)+G(\Phi_{0,t-}(y),x)) - f(\Phi_{0,t-}(y))\right]N(ds,dx)$$
$$+ \int_{|x|<c} \left[f(\Phi_{0,t-}(y)+F(\Phi_{0,t-}(y),x)) - f(\Phi_{0,t-}(y))\right.$$
$$\left. - F^i(\Phi_{0,t-}(y),x)(\partial_i f)(\Phi_{0,t-}(y))\right]\nu(dx).$$

Now integrate with respect to t, take expectations and use the martingale property of stochastic integrals to obtain

$$(T_t f)(y) - f(y) = \int_0^t (T_s \mathcal{L}f)(y)ds \quad (6.43)$$

6.7 Solutions of SDEs as Feller processes

and so, using the fact that each T_t is a contraction, we obtain

$$||T_t f - f|| = \sup_{y \in \mathbb{R}^d} \left| \int_0^t (T_s \mathcal{L} f)(y) ds \right|$$

$$\leq \int_0^t ||T_s \mathcal{L} f|| ds$$

$$\leq \int_0^t ||\mathcal{L} f|| ds = t ||\mathcal{L} f|| \to 0 \quad \text{as} \quad t \to 0.$$

The fact that $\lim_{t \to 0} ||T_t f - f|| \to 0$, for all $f \in C_0(\mathbb{R}^d)$ follows by a straightforward density argument. Hence we have established the Feller property. The rest of the proof follows on applying the analysis of Section 3.2 to (6.43). □

Exercise 6.7.5 Let X be a Lévy process with infinitesimal generator $\mathcal{A}$. Show that $C_0^2(\mathbb{R}^d) \subseteq \text{Dom}\,\mathcal{A}$.

It is interesting to rewrite the generator in the Courrège form, as in Section 3.5. Define a family of Borel measures $(\mu(y, \cdot), y \in \mathbb{R}^d)$ by

$$\mu(y, A) = \begin{cases} \nu \circ [F(y, \cdot) + y]^{-1}(A) & \text{if } A \in \mathcal{B}(B_c(0) - \{0\}) \\ \nu \circ [G(y, \cdot) + y]^{-1}(A) & \text{if } A \in \mathcal{B}((B_c(0) - \{0\})^c). \end{cases}$$

Then μ is a Lévy kernel and, for all $f \in C_0^2(\mathbb{R}^d)$, $y \in \mathbb{R}^d$,

$$(\mathcal{L}f)(y) = b^i(y)(\partial_i f)(y) + \tfrac{1}{2} a^{ij}(y)(\partial_i \partial_j f)(y)$$

$$+ \int_{\mathbb{R}^d - D} [f(z) - f(y) - (z^i - y^i)(\partial_i f)(y) \phi(y, z)] \mu(y, dz),$$

$$(6.44)$$

where D is the diagonal and

$$\phi(y, \cdot) = \chi_{B_c(0) - \{0\}} \circ [F(y, \cdot) + y]^{-1} = \chi_{(F(y,\cdot)+y)(B_c(0)-\{0\})}.$$

The only difference from the Courrège form (3.20) is that ϕ is not a local unit, but this is can easily be remedied by making a minor modification to b.

Example 6.7.6 (The Ornstein–Uhlenbeck process (yet again)) We recall that the Ornstein–Uhlenbeck process $(Y(t), t \geq 0)$ is the unique solution of the Langevin equation

$$dY(t) = -\beta Y(t) dt + dB(t),$$

where $\beta > 0$ and B is a standard Brownian motion. By an immediate application of (6.42), we see that the generator has the form

$$\mathcal{L} = -\beta x \cdot \nabla + \tfrac{1}{2}\Delta$$

on $C_0^2(\mathbb{R}^d)$. However, by (4.9) we know that the process has the specific form

$$Y_x(t) = e^{-\beta t}x + \int_0^t e^{-\beta(t-s)} dB(s)$$

for each $t \geq 0$, where $Y_x(0) = x$ (a.s.). In Exercise 4.3.18, we saw that

$$Y_x(t) \sim N\left(e^{-\beta t}x, \frac{1}{2\beta}(1 - e^{-2\beta t})I\right)$$

and so, in this case, we can make the explicit calculation

$$(T_t f)(x) = \mathbb{E}(f(Y_x(t)))$$

$$= \frac{1}{(2\pi)^{d/2}} \int_{\mathbb{R}^d} f\left(e^{-\beta t}x + \sqrt{\frac{1 - e^{-2\beta t}}{2\beta}}\, y\right) e^{-|y|^2/2} dy$$

for each $t \geq 0$, $f \in C_0(\mathbb{R}^d)$, $x \in \mathbb{R}^d$. This result is known as *Mehler's formula*, and using it one can verify directly that $(T_t, t \geq 0)$ is a Feller semigroup. For an infinite-dimensional generalisation of this circle of ideas, see for example Nualart [239], pp. 49–53.

Exercise 6.7.7 Let $X = (X(t), t \geq 0)$ be a Markov process with associated semigroup $(T_t, t \geq 0)$ and transition probabilities $(p_t(x, \cdot), t \geq 0, x \in \mathbb{R}^d)$. We say that a Borel measure μ on $\mathbb{R}^d$ is an *invariant measure* for X if

$$\int_{\mathbb{R}^d} (T_t f)(x) \mu(dx) = \int_{\mathbb{R}^d} f(x) \mu(dx)$$

for all $t \geq 0$, $f \in C_0(\mathbb{R}^d)$.

(1) Show that μ is invariant for X if and only if $\int_{\mathbb{R}^d} p_t(x, A) \mu(dx) = \mu(A)$ for all $x \in \mathbb{R}^d$, $A \in \mathcal{B}(\mathbb{R}^d)$.
(2) Deduce that Lebesgue measure is an invariant measure for all Lévy processes.
(3) Show that $\mu \sim N(0, (2\beta)^{-1}I)$ is an invariant measure for the Ornstein–Uhlenbeck process.

6.7.2 The Feynman–Kac formula

If we compare (6.44) with (3.20), we see that a term is missing. In fact, if we write the Courrège generator as $\mathcal{L}_c$ then, for all $f \in C_0^2(\mathbb{R}^d)$, $y \in \mathbb{R}^d$, we have

$$(\mathcal{L}_c f)(y) = -c(y)f(y) + (\mathcal{L}f)(y).$$

Here we assume that $c \in C_b(\mathbb{R}^d)$ and that $c(y) > 0$ for all $y \in \mathbb{R}^d$.

Our aim here is to try to gain a probabilistic understanding of $\mathcal{L}_c$. Let Y be as usual the strong solution of (6.12) with associated flow Φ. We introduce a cemetery point Δ and define a process $Y_c = (Y_c(t), t \geq 0)$ on the one-point compactification $\mathbb{R}^d \cup \{\Delta\}$ by

$$Y_c(t) = \begin{cases} Y(t) & \text{for } 0 \leq t < \tau_c, \\ \Delta & \text{for } t \geq \tau_c. \end{cases}$$

Here τ_c is a stopping time for which

$$P(\tau_c > t | \mathcal{F}_t) = \exp\left(-\int_0^t c(Y(s))ds\right).$$

Note that Y_c is an adapted process. The probabilistic interpretation is that particles evolve in time according to the random dynamics Y but are 'killed' at the rate $c(Y(t))$ at time t.

We note that, by convention, $f(\Delta) = 0$ for all $f \in B_b(\mathbb{R}^d)$.

Lemma 6.7.8 *For each $0 \leq s \leq t < \infty$, $f \in B_b(\mathbb{R}^d)$, we have, almost surely,*

(1) $\mathbb{E}(f(Y_c(t))|\mathcal{F}_s) = \mathbb{E}\left(\exp\left[-\int_0^t c(Y(s))ds\right] f(Y(t)) \Big| \mathcal{F}_s\right),$

(2) $\mathbb{E}(f(Y_c(t))|Y_c(s)) = \mathbb{E}\left(\exp\left[-\int_0^t c(Y(s))ds\right] f(Y(t)) \Big| Y_c(s)\right).$

Proof (1) For all $0 \leq s \leq t < \infty$, $f \in B_b(\mathbb{R}^d)$, $A \in \mathcal{F}_s$, we have

$$\mathbb{E}(\chi_A f(Y_c(t))) = \mathbb{E}\left(\chi_A \chi_{(\tau_c > t)} f(Y_c(t))\right) + \mathbb{E}\left(\chi_A \chi_{(\tau_c \leq t)} f(Y_c(t))\right)$$
$$= \mathbb{E}\left(\mathbb{E}(\chi_A \chi_{(\tau_c > t)} f(Y_c(t))|\mathcal{F}_t)\right)$$
$$= \mathbb{E}(\chi_A f(Y_c(t)) \mathbb{E}(\chi_{(\tau_c > t)}|\mathcal{F}_t))$$
$$= \mathbb{E}\left(\chi_A \exp\left[-\int_0^t c(Y(s))ds\right] f(Y(t))\right)$$

and the required result follows. We can prove (2) similarly. □

The following is our main theorem and to simplify proofs we will work in path space $(\Omega, \mathcal{F}, P)$ with its associated shift θ. Again, we will adopt the convention of writing the solution at time t of (6.12) as $\Phi_{0,t}$, so that we can exploit the flow property.

Theorem 6.7.9 Y_c *is a sub-Feller process with associated semigroup*

$$(T_t^c f)(y) = \mathbb{E}\left(\exp\left[-\int_0^t c(\Phi_{0,s}(y))ds\right] f(\Phi_{0,t}(y))\right) \qquad (6.45)$$

for all $t \geq 0$, $f \in C_0(\mathbb{R}^d)$, $y \in \mathbb{R}^d$. *The infinitesimal generator acts as* $\mathcal{L}_c$ *on* $C_0^2(\mathbb{R}^d)$.

Proof We must first establish the Markov property. Using Lemma 6.7.8(1), the independent multiplicatives-increments property of Φ (Theorem 6.4.2) and Lemma 1.1.8 we find, for all $s, t \geq 0$, $f \in B_b(\mathbb{R}^d)$, $y \in \mathbb{R}^d$, that the following holds almost surely:

$\mathbb{E}(f(Y^c(s+t)|\mathcal{F}_s)$

$= \mathbb{E}\left(\exp\left[-\int_0^{t+s} c(\Phi_{0,r})dr\right] f(\Phi_{0,t+s})\bigg|\mathcal{F}_s\right)$

$= \mathbb{E}\left(\exp\left[-\int_0^s c(\Phi_{0,r})dr\right]\exp\left[-\int_s^{t+s} c(\Phi_{0,r})dr\right] f(\Phi_{0,t+s})\bigg|\mathcal{F}_s\right)$

$= \exp\left[-\int_0^s c(\Phi_{0,r})dr\right]\mathbb{E}\left(\exp\left[-\int_0^t c(\Phi_{0,r+s})dr\right] f(\Phi_{0,t+s})\bigg|\mathcal{F}_s\right)$

$= \exp\left[-\int_0^s c(\Phi_{0,r})dr\right]$

$\quad \times \mathbb{E}\left(\exp\left[-\int_0^t c(\Phi_{s,s+r}\Phi_{0,s})dr\right] f(\Phi_{s,s+t}\Phi_{0,s})\bigg|\mathcal{F}_s\right)$

$= \exp\left[-\int_0^s c(\Phi_{0,r})dr\right](T_{s,s+t}^c f)(\Phi_{0,s}),$

where, for each $y \in \mathbb{R}^d$,

$$(T_{s,s+t}^c f)(y) = \mathbb{E}\left(\exp\left[-\int_0^t c(\Phi_{s,s+r}(y))dr\right] f(\Phi_{s,s+t}(y))\right).$$

A similar argument using Lemma 6.7.8(2) yields

$$\mathbb{E}(f(Y^c s+t)|Y^c(s)) = \exp\left[-\int_0^s c(\Phi_{0,r})dr\right](T_{s,s+t} f)(\Phi_{0,s}) \quad \text{a.s.}$$

and we have the required result.

6.7 Solutions of SDEs as Feller processes

To see that Y^c is homogeneous Markov, we need to show that

$$(T^c_{s,s+t}f)(y) = (T^c_{0,t}f)(y)$$

for all $s, t \geq 0$, $f \in B_b(\mathbb{R}^d)$, $y \in \mathbb{R}^d$. By Lemma 6.4.10 and Exercise 6.4.8(2), we have

$(T^c_{s,s+t}f)(y)$

$= \mathbb{E}\left(\exp\left[-\int_0^t c(\Phi_{s,s+r}(y))dr\right] f(\Phi_{s,s+t}(y))\right)$

$= \int_\Omega \exp\left[-\int_0^t c(\Phi_{0,r}(y, \theta_s(\omega)))dr\right] f(\Phi_{0,t}(y, \theta_s(\omega)))dP(\omega)$

$= \int_\Omega \exp\left[-\int_0^t c(\Phi_{0,r}(y, \omega))dr\right] f(\Phi_{0,t}(y, \omega))dP(\theta_s^{-1}\omega)$

$= (T^c_{0,t}f)(y),$

as required. The fact that $(T^c_t, t \geq 0)$ is the semigroup associated with Y^c now follows easily.

To establish that each $T^c_t : C_0(\mathbb{R}^d) \to C_0(\mathbb{R}^d)$ is straightforward. To obtain strong continuity and the form of the generator, argue as in the proof of Theorem 6.7.4, using the fact that if we define

$$M_f(t) = \exp\left[-\int_0^t c(\Phi_{0,s})ds\right] f(\Phi_{0,t})$$

for each $t \geq 0$, $f \in C_0^2(\mathbb{R}^d)$, then by Itô's product formula,

$$dM_f(t) = -c(\Phi_{0,t})f(\Phi_{0,t})dt + \exp\left[\int_0^t c(\Phi_{0,s})ds\right] d(\Phi_{0,t}).$$

□

The formula (6.45) is called the *Feynman–Kac formula*. Note that the semigroup $(T^c_t, t \geq 0)$ is not conservative; indeed we have, for each $t \geq 0$,

$$T^c_t(1) = \mathbb{E}\left(\exp\left[-\int_0^t c(Y(s))ds\right]\right) = P(\tau_c > t | \mathcal{F}_t).$$

Historically, the Feynman–Kac formula can be traced back to Mark Kac's attempts to understand Richard Feynman's 'path-integral' solution to the Schrödinger equation,

$$i\frac{\partial \psi}{\partial t} = -\frac{1}{2}\frac{\partial^2 \psi}{\partial x^2} + V(x)\psi,$$

where $\psi \in L^2(\mathbb{R})$ is the wave function of a one-dimensional quantum system with potential V. Feynman [103] found a formal path-integral solution whose rigorous mathematical meaning was unclear. Kac [164] observed that if you make the time change $t \to -it$ then you obtain the diffusion equation

$$\frac{\partial \psi}{\partial t} = \frac{1}{2}\frac{\partial^2 \psi}{\partial x^2} - V(x)\psi$$

and, if $V = 0$, the solution to this is just given by

$$(T_t \psi)(y) = \mathbb{E}(\psi(y + B(t)))$$

for each $y \in \mathbb{R}$, where $B = (B(t), t \geq 0)$ is a one-dimensional Brownian motion. Moreover, Feynman's prescription when $V \neq 0$ essentially boiled down to replacing T_t by T_t^V in (6.45).

For a deeper analytic treatment of the Feynman–Kac formula in the Brownian motion case, see e.g. Durrett [85], pp. 137–42. Applications to the potential theory of Schrödinger's equation are systematically developed in Chung and Zhao [72]. For a Feynman–Kac-type approach to the problem of a relativistic particle interacting with an electromagnetic field see Ichinose and Tamura [139]. This utilises the Lévy-process approach to relativistic Schrödinger operators (see Example 3.3.9).

The problem of rigorously constructing Feynman integrals has led to a great deal of interesting mathematics. For a very attractive recent approach based on compound Poisson processes see Kolokoltsov [176]. The bibliography therein is an excellent guide to other work in this area.

The Feynman–Kac formula also has important applications to finance, where it provides a bridge between the probabilistic and PDE representations of pricing formulae; see e.g. Etheridge [98], Section 4.8, and Chapter 15 of Steele [288].

6.7.3 Weak solutions to SDEs and the martingale problem

So far, in this chapter, we have always imposed Lipschitz and growth conditions on the coefficients in order to ensure that (6.12) has a unique strong solution. In this subsection, we will drop these assumptions and briefly investigate the notion of the weak solution. Fix $x \in \mathbb{R}^d$; let D_x be the path space of all càdlàg functions ω from $\mathbb{R}^+$ to $\mathbb{R}^d$ for which $\omega(0) = x$ and let $\mathcal{G}_x$ be the σ-algebra generated by the cylinder sets. A *weak solution* to (6.12) with initial condition $Z(0) = x$ (a.s.) is a triple (Q_x, X, Z), where:

- Q_x is a probability measure on $(D_x, \mathcal{G}_x)$;
- $X = (X(t), t \geq 0)$ is a Lévy process on $(D_x, \mathcal{G}_x, Q_x)$;

6.7 Solutions of SDEs as Feller processes

- $Z = (Z(t), t \geq 0)$ is a solution of (6.12) whose Brownian motion and Poisson random measure are those associated with X through the Lévy–Itô decomposition.

A weak solution is said to be *unique in distribution* if, whenever (Q_x^1, X^1, Z^1) and (Q_x^2, X^2, Z^2) are both weak solutions,

$$Q_x^1(X^1(t) \in A) = Q_x^2(X^2(t) \in A)$$

for all $t \geq 0$, $A \in \mathcal{B}(\mathbb{R}^d)$.

Notice that, in contrast to strong solutions, where the noise is prescribed in advance, for weak solutions the construction of (a realisation of) the noise is part of the problem.

Finding weak solutions to SDEs is intimately related to martingale problems. We recall the linear operator $\mathcal{L}$ on $C_0(\mathbb{R}^d)$ as given in (6.44). We say that a probability measure Q_x on $(D_x, \mathcal{G}_x)$ solves the *martingale problem* associated with $\mathcal{L}$ if

$$f(Z(t)) - \int_0^t (\mathcal{L}f)Z(s)ds$$

is a Q_x-martingale for all $f \in C_0^2(\mathbb{R}^d)$, where $Z(t)(\omega) = \omega(t)$ for all $t \geq 0$, $\omega \in D_x$, and $Q_x(Z(0) = x) = 1$. The martingale problem is said to be *well posed* if such a measure Q_x exists and is unique. Readers for whom such ideas are new should ponder the case where $\mathcal{L}$ is the generator of the strong solution to (6.12).

The martingale-problem approach to weak solutions of SDEs has been most extensively developed in the case of the diffusion equation

$$dZ(t) = \sigma(Z(t))dB(t) + b(Z(t))dt. \tag{6.46}$$

Here the generator is the second-order elliptic differential operator

$$(\mathcal{L}_0 f)(x) = \tfrac{1}{2} a^{ij}(x)(\partial_i \partial_j f)(x) + b^i(x)(\partial_i f)(x),$$

where each $f \in C_0^2(\mathbb{R}^d)$ and $x \in \mathbb{R}^d$ and, as usual, $a(\cdot) = \sigma(\cdot)\sigma(\cdot)^T$. In this case, every solution of the martingale problem induces a weak solution of the SDE (6.46). This solution is unique in distribution if the martingale problem is well posed, and Z is then a strong Markov process (see e.g. Durrett [85], p. 189). The study of the martingale problem based on $\mathcal{L}_0$ was the subject of extensive work by D. Stroock and S.R.S. Varadhan in the late 1960s and is presented in their monograph [289]. In particular, if a and b are both bounded and measurable, with a also strictly positive definite and continuous, then the martingale problem for $\mathcal{L}_0$ is well posed. For more recent accounts and further work on weak solutions to (6.46) see Durrett [85], Sections 5.3 and 5.4 of Karatsas and Shreve [167] and Sections 5.3 and 5.4 of Rogers and Williams [262].

The well-posedness of the martingale problem for Lévy-type generators of the form (6.44) was first studied by D. Stroock [290] and the relationship to solutions of SDEs was investigated by Lepeltier and Marchal [191] and by Jacod [154], Section 14.5. An alternative approach due to Komatsu [177] exploited the fact that $\mathcal{L}$ is a pseudo-differential operator (see Courrège's second theorem, Theorem 3.5.5 in the present text) to solve the martingale problem. Recent work in this direction is due to W. Hoh [132, 133]; see also Chapter 4 of Jacob [148].

There are other approaches to weak solutions of SDEs that do not require one to solve the martingale problem. The case $dZ(t) = L(Z(t-))dX(t)$, where X is a one-dimensional α-stable Lévy process, is studied in Zanzotto [310, 311], who generalised an approach due to Engelbert and Schmidt in the Brownian motion case; see Engelbert and Schmidt [97] or Karatzas and Shreve [167], Section 5.5.

6.8 Marcus canonical equations

We recall the Marcus canonical integral from Chapter 4. Here, as promised, we will replace straight lines by curves within the context of SDEs. Let $(L_j, 1 \leq j \leq n)$ be complete C^1-vector fields so that, for each $1 \leq j \leq n$, there exist $c_j^i : \mathbb{R}^d \to \mathbb{R}$, where $1 \leq i \leq d$, such that $L_j = c_j^i \partial_i$. We will also assume that, for all $x \in \mathbb{R}^n - \{0\}$, the vector field $x^j L_j$ is complete, i.e. for each $y \in \mathbb{R}^d$ there exists an integral curve $(\xi(ux)(y), u \in \mathbb{R})$ such that

$$\frac{d\xi^i(ux)}{du} = x^j c_j^i(\xi(ux))$$

for each $1 \leq i \leq d$. Using the language of Subsection 4.4.4, we define a *generalised Marcus mapping* $\Phi : \mathbb{R}^+ \times \mathbb{R} \times \mathbb{R}^n \times \mathbb{R}^d \to \mathbb{R}^d$ by

$$\Phi(s, u, x, y) = \xi(ux)(y)$$

for each $s \geq 0, u \in \mathbb{R}, x \in \mathbb{R}^n, y \in \mathbb{R}^d$.

Note We have slightly extended the formalism of Subsection 4.4.4, in that translation in $\mathbb{R}^d$ has been replaced by the action of the deterministic flow ξ.

6.8 Marcus canonical equations

We consider the general *Marcus SDE* (or *Marcus canonical equation*)

$$dY(t) = c(Y(t-)) \diamond dX(t), \tag{6.47}$$

where $X = (X(t), t \geq 0)$ is an n-dimensional Lévy process (although we could replace this by a general semimartingale).

The meaning of this equation is given by the Marcus canonical integral of Subsection 4.4.4, which yields, for each $1 \leq i \leq d, t \geq 0$,

$$\begin{aligned}
dY^i(t) &= c^i_j(Y(t-)) \circ dX^j_c(t) + c^i_j(Y(t-))dX^j_d(s) \\
&+ \sum_{0 \leq s \leq t} \left[\xi(\Delta X(s))(Y(s-)) - Y(s-) - c^i_j(Y(s-))\Delta X^j(s) \right],
\end{aligned} \tag{6.48}$$

where X_c and X_d are the usual continuous and discontinuous parts of X and $\circ$ denotes the Stratonovitch differential. Note that when $X_d = 0$ this is just a *Stratonovitch SDE* (see e.g.Kunita [182], Section 3.4).

In order to establish the existence and uniqueness of such equations, we must write them in the form (6.12). This can be carried out by employing the Lévy–Itô decomposition,

$$X^j(t) = m^j t + \tau^j_k B^k(t) + \int_{|x|<1} x^j \tilde{N}(dt, dx) + \int_{|x| \geq 1} x^j N(dt, dx)$$

for each $t \geq 0$, $1 \leq j \leq n$. We write $a = \tau\tau^T$ as usual.

We then obtain, for each $1 \leq i \leq d, t \geq 0$,

$$\begin{aligned}
dY^i(t) &= m^j c^i_j(Y(t-))dt + \tau^j_k c^i_j(Y(t-))dB^k(t) \\
&+ \tfrac{1}{2} a^{jl} c^k_l(Y(t-))(\partial_k c^i_j)(Y(t-))dt \\
&+ \int_{|x|<1} \left[\xi^i(x)(Y(t-)) - Y(t-)^i \right] \tilde{N}(dt, dx) \\
&+ \int_{|x| \geq 1} \left[\xi^i(x)(Y(t-)) - Y(t-)^i \right] N(dt, dx) \\
&+ \int_{|x|<1} \left[\xi^i(x)(Y(t-)) - Y(t-)^i - x^j c^i_j(Y(t-)) \right] v(dx)dt.
\end{aligned} \tag{6.49}$$

We usually consider such equations with the deterministic initial condition $Y(0) = y$ (a.s.).

We can now rewrite this in the form (6.12) where, for each $1 \leq i \leq d$, $1 \leq k \leq r$, $y \in \mathbb{R}^d$,

$$b^i(y) = m^j c^i_j(y) + \tfrac{1}{2} a^{jl} c^k_l(y)(\partial_k c^i_j)(y)$$
$$+ \int_{|x|<1} \left[\xi^i(x)(y) - y^i - x^j c^i_j(y) \right] \nu(dx),$$
$$\sigma^i_k(y) = \tau^j_k c^i_j(y),$$
$$F^i(y, x) = \xi^i(x)(y) - y^i \qquad \text{for all } |x| < 1,$$
$$G^i(y, x) = \xi^i(x)(y) - y^i \qquad \text{for all } |x| \geq 1.$$

We will find it convenient to define

$$H^i(y, x) = \xi^i(x)(y) - y^i - x^j c^i_j(y)$$

for each $1 \leq i \leq d$, $|x| < 1$, $y \in \mathbb{R}^d$.

In order to prove existence and uniqueness for the Marcus equation, we will need to make some assumptions. First we introduce some notation. For each $1 \leq i \leq d$, $y \in \mathbb{R}^d$, we denote by $(\tilde{\nabla}^i c)(y)$ the $n \times n$ matrix whose (j, l)th entry is $(\partial c^i_j(y)/\partial y_k) c^k_l(y)$. We also define $(\tilde{\nabla} c)(y)$ to be the vector in $\mathbb{R}^d$ whose ith component is $\max_{1 \leq j,l \leq n} |(\tilde{\nabla}^i c(y))_{j,l}|$. For each $1 \leq j \leq n$, $c_j(y) = (c^1_j(y), \ldots, c^d_j(y))$.

Assumption 6.8.1

(1) For each $1 \leq j \leq n$ we have that c_j is globally Lipschitz, so there exists $P_1 > 0$ such that

$$\max_{1 \leq j \leq n} |c_j(y_1) - c_j(y_2)| \leq P_1 |y_1 - y_2|$$

for all $y_1, y_2 \in \mathbb{R}^d$.

(2) For each $1 \leq j \leq n$ we have that $(\tilde{\nabla} c)_j$ is globally Lipschitz, so there exists $P_2 > 0$ such that

$$|(\tilde{\nabla} c)(y_1) - (\tilde{\nabla} c)(y_2)| \leq P_2 |y_1 - y_2|$$

for all $y_1, y_2 \in \mathbb{R}^d$.

Exercise 6.8.2 Show that for Assumption 6.8.1(1), (2) to hold, it is sufficient that each $c^i_j \in C^2_b(\mathbb{R}^d)$.

Note that Assumption 6.8.1(1) is enough to ensure that each $x^j L_j$ is complete, by Theorem 6.1.3.

We will have need of the following technical lemma.

Lemma 6.8.3 *For each $y, y_1, y_2 \in \mathbb{R}^d$, $|x| < 1$, there exist $M_1, M_2, M_3 \geq 0$ such that:*

(1) $|F(y, x)| \leq M_1 |x|(1 + |y|)$;
(2) $|F(y_1, x) - F(y_2, x)| \leq M_2 |x| |y_1 - y_2|$;
(3) $|H(y_1, x) - H(y_2, x)| \leq M_3 |x|^2 |y_1 - y_2|$.

Proof We follow Tang [294], pp. 34–6.

(1) Using the Cauchy–Schwarz inequality and Exercise 6.1.1, there exists $Q > 0$ such that, for each $u \in \mathbb{R}$,

$$|F(y, ux)| = \left| \int_0^u x^j c_j(\xi(ax)(y)) da \right|$$

$$\leq d^{1/2} |x| \int_0^u \max_{1 \leq j \leq n} |c_j(\xi(ax)(y))| \, da$$

$$\leq Q d^{1/2} |x| \int_0^u [1 + |\xi(ax)(y)|] da$$

$$\leq Q d^{1/2} |x| \int_0^u [(1 + |y|) + |F(y, ax)|] da$$

$$\leq Q d^{1/2} |x| u (1 + |y|) + Q d^{1/2} |x| \int_0^u (|F(y, ax)|) da,$$

and the required result follows by Gronwall's inequality.

(2) For each $u \in \mathbb{R}$, using the Cauchy–Schwarz inequality we have

$$|F(ux, y_1) - F(ux, y_2)| = \left| [\xi(ux)(y_1) - y_1] - [\xi(ux)(y_2) - y_2] \right|$$

$$= \left| \int_0^u x^j [c_j(\xi(ax)(y_1)) - c_j(\xi(ax)(y_2))] da \right|$$

$$\leq P_1 |x| \int_0^u |\xi(ax)(y_1) - \xi(ax)(y_2)| da.$$

(6.50)

From this we deduce that

$$|(\xi(ux)(y_1) - (\xi(ux)(y_2)| \leq |y_1 - y_2|$$
$$+ P_1 u |x| \int_0^u |\xi(ax)(y_1) - \xi(ax)(y_2)| da$$

and hence, by Gronwall's inequality,

$$|(\xi(ux)(y_1) - (\xi(ux)(y_2)| \leq e^{P_1 u |x|}|y_2 - y_1|. \qquad (6.51)$$

The required result is obtained on substituting (6.51) into (6.50).

(3) For each $1 \leq i \leq d$,

$$H^i(y_1, x) - H^i(y_2, x)$$
$$= [\xi^i(x)(y_1) - y_1] - [\xi^i(x)(y_2) - y_2] - x^j[c_j^i(y_1) - c_j^i(y_2)]$$
$$= \int_0^1 x^j \{[c_j^i(\xi(ax)(y_1)) - c_j^i(y_1)] - [c_j^i(\xi(ax)(y_2)) - c_j^i(y_2)]\} da$$
$$= \int_0^1 \int_0^a x^j \left(\frac{\partial c_j^i(\xi(bx)(y_1))}{\partial y_k} \frac{\partial \xi^k(bx)(y_1)}{\partial b} \right.$$
$$\left. - \frac{\partial c_j^i(\xi(bx)(y_2))}{\partial y_k} \frac{\partial \xi^k(bx)(y_2)}{\partial b} \right) db\, da$$
$$= \int_0^1 \int_0^a x^j x^l \left(\frac{\partial c_j^i(\xi(bx)(y_1))}{\partial y_k} c_l^k(\xi(bx)(y_1)) \right.$$
$$\left. - \frac{\partial c_j^i(\xi(bx)(y_2))}{\partial y_k} c_l^k(\xi(bx)(y_2)) \right) db\, da$$
$$= \int_0^1 \int_0^a x^j [\tilde{\nabla} c^i(\xi(bx)(y_1))_{jl} - \tilde{\nabla} c^i(\xi(bx)(y_2))_{jl}] x^l db\, da$$

Now use the Cauchy–Schwarz inequality and Assumption 6.8.1 to obtain

$$|H^i(y_1, x) - H^i(y_2, x)|$$
$$\leq |x|^2 \int_0^1 \int_0^a |(\tilde{\nabla} c)(\xi(bx)(y_1))^i - (\tilde{\nabla} c)(\xi(bx)(y_2))^i| db\, da$$
$$\leq |x|^2 \int_0^1 \int_0^a |\xi(bx)^i(y_1) - \xi(bx)^i(y_1)| db\, da.$$

The result follows on substituting (6.51) into the right-hand side. $\square$

Exercise 6.8.4 Deduce that $y \to G(y, x)$ is globally Lipschitz for each $|x| > 1$.

Theorem 6.8.5 *There exists a unique strong solution to the Marcus equation (6.49). Furthermore the associated solution flow has an almost surely continuous modification.*

Proof A straightforward application of Lemma 6.8.3 ensures that the Lipschitz and growth conditions on b, σ and F are satisfied; these ensure that the modified equation has a unique solution. We can then extend this to the whole equation by interlacing. Lemma 6.8.3(1) also ensures that Assumptions 6.5.1 and 6.6.1 hold, and so we can apply Theorem 6.6.3 to deduce the required continuity. □

Note that, since Assumption 6.5.1 holds, we also have the interlacing construction of Theorem 6.5.2. This gives a strikingly beautiful interpretation of the solution of the Marcus canonical equation: as a diffusion process controlled by the random vector fields $L_j X(t)^j$, with jumps at the 'fictional time' $u = 1$ along the integral curves $(\xi(u\Delta X(t)^j L_j), u \in \mathbb{R})$.

If each $c_j^i \in C_b^1(\mathbb{R}^d)$, we may introduce the linear operator $\mathcal{N} : C_b^2(\mathbb{R}^d) \to C_b(\mathbb{R}^d)$ where, for each $f \in C_b^2(\mathbb{R}^d)$, $y \in \mathbb{R}^d$,

$$(\mathcal{N} f)(y) = m^j (L_j f)(y) + \tfrac{1}{2} a^{jm}(L_j L_m f)(y)$$
$$+ \int_{\mathbb{R}^n - \{0\}} \left[f(\xi(x)y) - f(y) - x^j (L_j f)(y) \chi_{\hat{B}}(x) \right] v(dx)$$

and each

$$(L_j L_m f)(y) = (L_j (L_m f))(y)$$
$$= c_j^i(y) \left[\partial_i c_m^k(y) \right] (\partial_k f)(y) + c_j^i(y) c_m^k(y) (\partial_i \partial_k f)(y).$$

Now apply Itô's formula to the solution flow $\Phi = (\Phi_{s,t}, 0 \leq s \leq t < \infty)$ of equation (6.49), to find that, for all $f \in C_b^2(\mathbb{R}^d)$, $y \in \mathbb{R}^d$, $0 \leq s \leq t < \infty$,

$$f(\Phi_{s,t}(y)) = f(y) + (\mathcal{N} f)(\Phi_{s,t-}(y))dt + \tau_k^j (L_j f)(\Phi_{s,t-}(y)) dB^k(t)$$
$$+ \int_{|x|<1} \left[f(\xi^i(x)(\Phi_{s,t-}(y))) - f(\Phi_{s,t-}(y)) \right] \tilde{N}(dt, dx)$$
$$+ \int_{|x|\geq 1} \left[f(\xi^i(x)(\Phi_{s,t-}(y))) - f(\Phi_{s,t-}(y)) \right] N(dt, dx).$$

If c is such that $\mathcal{N} : C_0^2(\mathbb{R}^d) \to C_0(\mathbb{R}^d)$ then $(T_t, t \geq 0)$ is a Feller semigroup by Theorem 6.7.4. The following exercise gives a sufficient condition on c for this to hold.

Exercise 6.8.6 Show that Assumption 6.7.1 is satisfied (and so the solution to (6.49) is a Feller process) if $c_j^i \in C_0(\mathbb{R}^d)$ for each $1 \leq j \leq n$, $1 \leq i \leq d$.

(Hint: You need to show that $\lim_{|y| \to \infty} [\xi^i(x)(y) - y^i] = 0$ for each $x \neq 0$, $1 \leq i \leq d$. First use induction to prove this for each member of the sequence of Picard iterates and then approximate.)

The structure of $\mathcal{N}$ can be thought of as a higher level Lévy–Khintchine type-formula in which the usual translation operators are replaced by integral curves of the vector fields $x^j L_j$. This is the key to further generalisations of stochastic flows and Lévy processes to differentiable manifolds (see e.g. the article by the author in [20]).

We now discuss briefly the homeomorphism and diffeomorphism properties of solution flows to Marcus canonical equations. As the arguments are very lengthy and technical we will not give full proofs but be content with providing an outline. We follow the account in Kunita [183]. The full story can be found in Fujiwara and Kunita [109]. For alternative approaches, see Kurtz, Pardoux and Protter [187] or Applebaum and Tang [7].

Theorem 6.8.7 Φ *is a stochastic flow of homeomorphisms.*

Proof First consider the modified flow Ψ. For each $0 \leq s < t < \infty$, $y_1, y_2 \in \mathbb{R}^d$, with $y_1 \neq y_2$, we define

$$\chi_{s,t}(y_1, y_2) = \frac{1}{\Psi_{s,t}(y_1) - \Psi_{s,t}(y_2)}.$$

The key to establishing the theorem for Ψ is the following pair of technical estimates. For each $0 \leq s < t < \infty$, $p > 2$, there exist $K_1, K_2 > 0$ such that, for all $y_1, y_2, z_1, z_2, y \in \mathbb{R}^d$, $y_1 \neq y_2$, $z_1 \neq z_2$,

$$\mathbb{E}\left(|\chi_{s,t}(y_1, y_2) - \chi_{s,t}(z_1, z_2)|^{-p}\right) \leq K_1(|y_1 - z_1|^{-p} + |y_2 - z_2|^{-p}), \tag{6.52}$$

$$\mathbb{E}\left(\sup_{s \leq r \leq t} |1 + \Psi_{s,r}(y)|^{-p}\right) \leq K_2 |1 + y|^{-p}. \tag{6.53}$$

By Kolmogorov's continuity criterion (Theorem 1.1.17) applied to (6.52), we see that the random field on $\mathbb{R}^{2d} - D$ given by $(y_1, y_2) \to \chi_{s,t}(y_1, y_2)$ is almost surely continuous. From this we deduce easily that $y_1 \neq y_2 \Rightarrow \Psi_{s,t}(y_1) \neq \Psi_{s,t}(y_2)$ (a.s.) and hence that each $\Psi_{s,t}$ is almost surely injective.

To prove surjectivity, suppose that $\liminf_{|y| \to \infty} \inf_{r \in [s,t]} |\Psi_{r,t}(y)| = a \in [0, \infty)$ (a.s.). Applying the reverse Fatou lemma (see e.g. Williams [304], p. 53), we get

$$\limsup_{|y| \to \infty} \mathbb{E}\left(\sup_{s \leq r \leq t} |1 + \Psi_{s,r}(y)|^{-p}\right) = \frac{1}{a},$$

6.8 Marcus canonical equations

contradicting (6.53). So we must have each $\liminf_{|y|\to\infty} |\Psi_{s,t}(y)| = \infty$ (a.s.). But we know that each $\Psi_{s,t}$ is continuous and injective. In the case $d = 1$, a straightforward analytic argument then shows that $\Psi_{s,t}$ is surjective and has a continuous inverse. For $d > 1$, more sophisticated topological arguments are necessary (see e.g. Kunita [182], p. 161). This proves that the modified flow Ψ comprises homeomorphisms almost surely.

For the general result, we apply the interlacing technique of Theorem 6.2.9. Let $(\tau_n, n \in \mathbb{N})$ be the arrival times for the jumps of $P(s,t) = \int_{s,t} \int_{|x|\geq c} x N(t, dx)$ and suppose that $\tau_{n-1} < t < \tau_n$; then

$$\Phi_{s,t} = \Psi_{\tau_{n-1},t} \circ \xi(\Delta P(s, \tau_{n-1})) \circ \Psi_{\tau_{n-2},\tau_{n-1}-} \circ \cdots \circ \xi(\Delta P(s, \tau_1)) \circ \Psi_{s,\tau_1-}.$$

Recalling Exercise 6.1.11, we see that $\Phi_{s,t}$ is the composition of a finite number of almost-sure homeomorphisms and so is itself a homeomorphism (a.s.). □

Exercise 6.8.8 Let Φ be a stochastic flow of homeomorphisms. Show that, for each $0 \leq s < t < \infty$, $\Phi_{s,t} = \Phi_{0,t} \circ \Phi_{0,s}^{-1}$ (a.s).

In order to establish the diffeomorphism property, we need to make an additional assumption on the driving vector fields. Fix $m \in \mathbb{N} \cup \{\infty\}$.

Assumption 6.8.9 $c_j^i \in C_b^{m+2}(\mathbb{R}^d)$ for $1 \leq j \leq d, 1 \leq j \leq n$.

Theorem 6.8.10 *If Assumption 6.8.9 holds, then Φ is a stochastic flow of C^m-diffeomorphisms.*

Proof (Sketch) We fix $m = 1$ and again deal with the modified flow Ψ. Let $\{e_1, \ldots, e_d\}$ be the natural basis for $\mathbb{R}^d$, so that each

$$e_j = (0, \ldots, 0, \overset{(j)}{1}, 0, \ldots 0).$$

For $h \in \mathbb{R}, h \neq 0, 1 \leq j \leq d, 0 \leq s \leq t < \infty, y \in \mathbb{R}^d$, define

$$(\Delta_j \Psi_{s,t})(y, h) = \frac{\Psi_{s,t}(y + h e_j) - \Psi_{s,t}(y)}{h};$$

then, for all $0 \leq s \leq t < \infty$, $p > 2$, there exists $K > 0$ such that

$$\mathbb{E}\left(\sup_{s \leq r \leq t} |(\Delta_j \Psi_{s,t})(y_1, h_1) - (\Delta_j \Psi_{s,t})(y_2, h_2)|^p\right)$$
$$\leq K(|y_1 - y_2|^p + |h_1 - h_2|^p)$$

for all $y_1, y_2 \in \mathbb{R}^d$ and $h_1, h_2 \in \mathbb{R} - \{0\}$.

By Kolmogorov's continuity criterion (Theorem 1.1.17), we see that

$$(y, h) \to \sup_{s \leq r \leq t} \Delta_j \Psi_{s,t}(y, h)$$

is a continuous random field on $\mathbb{R}^d \times (\mathbb{R} - \{0\})$. In fact it is uniformly continuous and so it has a continuous extension to $\mathbb{R}^d \times \mathbb{R}$. Hence Ψ is differentiable.

To show that $\Psi_{s,t}^{-1}$ is differentiable we first must show that the Jacobian matrix of $\Psi_{s,t}$ is non-singular, and the result then follows by the implicit function theorem.

To see that $\Phi_{s,t}$ is a diffeomorphism, use the interlacing argument from the end of Theorem 6.8.7 together with the result of Theorem 6.1.7.

The result for general m is established by induction. □

In the second paper of Fujiwara and Kunita [109], it was shown that for each $t > 0$, the inverse flow $(\Phi_{s,t}^{-1}, 0 \leq s \leq t)$ satisfies a backwards Marcus SDE

$$d\Phi_{s,t}^{-1} = -c(\Phi_{s,t}^{-1}) \diamond_b dX(s)$$

(with final condition $\Phi_{t,t}^{-1}(y) = y$ (a.s.) for all $y \in \mathbb{R}^d$).

We conjecture that, just as in the Brownian case ([182], Chapter 4), every Lévy flow on $\mathbb{R}^d$ with reasonably well-behaved characteristics can be obtained as the solution of an SDE driven by an infinite-dimensional Lévy process. This problem was solved for a class of Lévy flows with stationary multiplicative increments in Fujiwara and Kunita [108].

Just as in the case of ODEs, stochastic flows driven by Marcus canonical equations make good sense on smooth manifolds. Investigations of such Lévy flows can be found in Fujiwara [111] in the case where M is compact and in Kurtz, Pardoux and Protter [187], Applebaum and Kunita [5] or Applebaum and Tang [8] for more general M.

Stochastic flows of diffeomorphisms of $\mathbb{R}^d$ have also been investigated as solutions of the Itô SDE

$$d\Phi_{s,t} = c(\Phi_{s,t})dX(t);$$

see e.g. Section 5.7 of Protter [255], or Meyer [225] or Léandre [189]. However, as is pointed out in Kunita [183], in general these will not even be homeomorphisms unless, as is shown by the interlacing structure, the maps from $\mathbb{R}^d$ to $\mathbb{R}^d$ given by $y \to y + x^j c_j(y)$ are also homeomorphisms for each $x \in \mathbb{R}^d$, and this is a very strong and quite unnatural constraint.

6.9 Notes and further reading

Like so many of the discoveries described in this book, the concept of a stochastic differential equation is due to Itô [145]. In fact, he established the existence and uniqueness of strong solutions of SDEs driven by Lévy processes that are essentially of the form (6.12). His treatment of SDEs is standard fare in textbooks on stochastic calculus (see for example the list at the end of Chapter 4), but the majority of these omit jump integrals. Of course, during the 1960s, 1970s and 1980s, when most of these text books were written, stochastic analysts were absorbed in exploring the rich world of diffusion processes, where Brownian motion reigns supreme. The natural tendency of mathematicians towards greater abstraction and generality led to interest in SDEs driven by semimartingales with jumps, and the first systematic accounts of these were due to C. Doléans-Dade [82] and J. Jacod [154]. This is also one of the main themes of Protter [255]. Recent developments in flows driven by SDEs with jumps were greatly stimulated by the important paper of Fujiwara and Kunita [108]. They worked in a context more general than that described above, by employing an infinite-dimensional Lévy process taking values in $C(\mathbb{R}^d, \mathbb{R}^d)$ to drive SDEs. The non-linear stochastic integration theory of Carmona and Nualart [65] was a further generalisation of this approach.

The study of SDEs and associated flows is much more advanced in the Brownian case than in the general Lévy case. In particular, there are some interesting results giving weaker conditions than the global Lipschitz condition for solutions to exist for all time. Some of these are described in Section 5.3 of Durrett [85]. For a powerful recent result, see Li [194]. For the study of an SDE driven by a Poisson random measure with non-Lipschitz coefficients, see Fournier [105]. An important resource for Brownian flows on manifolds and stochastic differential geometry is Elworthy [93].

Simulation of the paths of SDEs can be very important in applications. A valuable guide to this for SDEs driven by α-stable Lévy processes is given by Janicki and Weron [157]. For a discussion of the simulation of more general SDEs driven by Lévy processes, which is based on the Euler approximation scheme, see Protter and Talay [254]. For a nice introduction to applications of SDEs to filtering and control, respectively, in the Brownian case, see Chapters 6 and 11 of Øksendal [241]. A fuller account of filtering, which includes the use of jump processes, can be found in Chapters 15 to 20 of Liptser and Shiryaev [201]. A forthcoming monograph by Øksendal and Sulem [240] is devoted to the stochastic control of jump diffusions. A wide range of applications of SDEs, including some in material science and ecology, can be found in Grigoriu [122].

SDEs driven by Lévy processes and more general semimartingales are sure to see much further development in the future. Here are some future directions in which work is either ongoing or not yet started.

- The study of Lévy flows as random dynamical systems is far more greatly developed in the Brownian case than for general Lévy processes, and a systematic account of this is given in Arnold [13]. Some asymptotic and ergodic properties of Lévy flows have been investigated by Kunita and Oh [179] and Applebaum and Kunita [6]. So far there has been little work on the behaviour of Lyapunov exponents (see Baxendale [31] and references therein for the Brownian case, and Mao and Rodkina [211] for general semimartingales with jumps). Liao [195] has studied these in the special case of flows on certain homogeneous spaces induced by Lévy processes in Lie groups. An applications-oriented approach to Lyapunov exponents of SDEs, which includes the case of compound Poisson noise, can be found in Section 8.7 of Grigoriu [122].
- One of the most important developments of stochastic calculus within the last thirty years is the Malliavin calculus. This has been applied to find conditions for solutions of SDEs driven by Brownian motion to have smooth densities; see e.g. Subsection 2.3.2 of Nualart [239] or Subsection 3.1.4 of Huang and Yan [135]. Recent work of Kunita and Oh [180] and Kunita [184] (see also Picard [251]) has established conditions for the solutions of Marcus canonical SDEs to have smooth densities. So far, the smoothness results require strong conditions to be imposed on the Lévy measure.
- Another important application of Malliavin calculus is in establishing the existence and uniqueness of solutions to SDEs driven by Brownian motion when the standard initial condition is weakened in such a way that Z_0 is no longer independent of the driving noise; see e.g. Chapter 3 of Nualart [239]. As yet, to the author's knowledge there has been no work in this direction for SDEs with jumps.
- A new approach to stochastic calculus has recently been pioneered by T.J. Lyons [202, 203], in which SDEs driven by Brownian motion are solved pathwise as deterministic differential equations with an additional driving term given by the Lévy area. This has been extended to SDEs driven by Lévy processes in Williams [305]. Related ideas are explored in Mikosch and Norvaiša [227]. It will be interesting to see whether this approach generates new insights about SDEs with jumps in the future.
- One area of investigation that has been neglected until very recently is the extent to which the solution of an SDE can inherit interesting probabilistic

6.9 Notes and further reading

properties from its driving noise. For example, Samorodnitsky and Grigoriu [270] recently studied the SDE

$$dZ(t) = -f(Z(t-))dt + dX(t),$$

where X has heavy tails (e.g. X might be an α-stable Lévy process) and $d = 1$. Under certain restrictions on f, they were able to show that Z also has heavy tails.

- An important area related to SDEs is the study of stochastic partial differential equations (SPDEs). These are partial differential equations perturbed by a random term, so that the solution, if it exists, is a space–time random field. The case based on Brownian noise has been extensively studied, see e.g. the lecture notes by Walsh [299] and the survey by Pardoux [245], but so far there has been little work on the case of Lévy-type noise.

In Applebaum and Wu [9], existence and uniqueness were established for an SPDE written formally as

$$\frac{\partial u(t,x)}{\partial t} = \frac{\partial^2 u(t,x)}{\partial x^2} + a(t,x,u(t,x)) + b(t,x,u(t,x))\dot{F}_{t,x}$$

on the region $[0, \infty) \times [0, L]$, where $L > 0$, with initial and Dirichlet boundary conditions. Here $F_{t,x}$ is a Lévy space–time white noise. An extensive study of the case where F is a Poisson random measure is due to Saint Loubert Bié [269]; see also Albeverio et al. [4].

In a separate development, Mueller [233] established the short-time existence for solutions to the equation

$$\frac{\partial u(t,x)}{\partial t} = -(-\Delta)^{p/2} u(t,x) + u(t,x)^\gamma \dot{M}_{t,x}$$

on $\mathbb{R}^+ \times D$, where D is a domain in $\mathbb{R}^d$, with given initial condition and u vanishing on the complement of D. Here $0 \leq \alpha < 1$, $p \in (0, 2]$, $\gamma > 0$ and $(M(t,x), t \geq 0, x \in D)$ is an α-stable space–time white noise. There seems to be an intriguing relationship between this equation and stable measure-valued branching processes. Weak solutions of this equation in the case where $1 < \alpha < 2$ and $p = 2$ have recently been found by Mytnik [234].

Solutions of certain SPDEs driven by Poisson noise generate interesting examples of quantum field theories. For recent work in this area, see Gielerak and Lugiewicz [117] and references therein.

References

[1] R. Abraham, J. E. Marsden, T. Ratiu, *Manifolds, Tensor Analysis and Applications* (second edition), Springer-Verlag (1988); first edition Addison-Wesley (1983).

[2] V. Akgiray, G. G. Booth, The stable-law model of stock returns, *J. Business and Econ. Statis.* **6**, 51–7 (1988).

[3] S. Albeverio, S. Song, Closability and resolvent of Dirichlet forms perturbed by jumps, *Pot. Anal.* **2**, 115–30 (1993).

[4] S. Albeverio, J.-L. Wu, T.-S. Zhang, Parabolic SPDEs driven by Poisson white noise, *Stoch. Proc. Appl.* **74**, 21–36 (1998).

[5] D. Applebaum, H. Kunita, Lévy flows on manifolds and Lévy processes on Lie groups, *J. Math. Kyoto Univ.* **33**, 1103–23 (1993).

[6] D. Applebaum, H. Kunita, Invariant measures for Lévy flows of diffeomorphisms, *Roy. Soc. Edin. Proc. A (Math.)* **130**, 925–46 (2000).

[7] D. Applebaum, F. Tang, The interlacing construction for stochastic flows of diffeomorphisms on Euclidean spaces, *Sankhyā, Series A*, **63**, 139–78 (2001).

[8] D. Applebaum, F. Tang, Stochastic flows of diffeomorphisms on manifolds driven by infinite-dimensional semimartingales with jumps, *Stoch. Proc. Appl.* **92**, 219–36 (1992).

[9] D. Applebaum, J.-L. Wu, Stochastic partial differential equations driven by Lévy space–time white noise, *Random Ops. and Stoch. Eqs.* **8**, 245–61 (2000).

[10] D. Applebaum, Compound Poisson processes and Lévy processes in groups and symmetric spaces, *J. Theoret. Prob.* **13**, 383–425 (2000).

[11] A. Araujo, E. Giné, *The Central Limit Theorem for Real and Banach Valued Random Variables*, Wiley (1980).

[12] L. Arnold, M. Scheutzow, Perfect cocycles through stochastic differential equations, *Prob. Theory Rel. Fields* **101**, 65–88 (1995).

[13] L. Arnold, *Random Dynamical Systems*, Springer-Verlag (1998).

[14] R. B. Ash, C. A. Doléans-Dade, *Probability and Measure Theory*, Academic Press (2000).

[15] F. Avram, T. Chan, M. Usabel, On the valuation of constant barrier options under spectrally one-sided exponential Lévy models and Carr's approximation for American puts, *Stoch. Proc. Appl.* **100**, 75–107 (2002).

[16] L. Bachelier, Théorie de la spéculation, *Ann. Ecole Norm. Sup.* **17** (1900). Reprinted in *The Random Character of Stock Market Prices*, ed. P. H. Cootner, MIT Press, pp. 17–78 (1967).
[17] M. T. Barlow, J. Hawkes, Application de l'entropie métrique à la continuité des temps locaux des processus de Lévy, *C. R. Acad. Sci. Paris* **301**, 237–9.
[18] M. T. Barlow, Necessary and sufficient conditions for the continuity of local times of Lévy processes, *Ann. Prob.* **15**, 241–62.
[19] O. E. Barndorff-Nielsen, S. E. G. Raverson, T. Mikosch (eds.), *Mini-Proceedings of Conference on Lévy Processes: Theory and Applications, MaPhySto Miscelleanea* **11** (1999) (http://www.maphysto.dk).
[20] O. E. Barndorff-Nielsen, C. Halgreen, Infinite divisibility of the hyperbolic and generalised inverse Gaussian distributions, *Z. Wahrsch. verw. Geb.* **38**, 309–12 (1977).
[21] O. E. Barndorff-Nielsen, J. L. Jensen, M. Sørensen, Some stationary processes in discrete and continuous time, *Adv. Appl. Prob.* **30**, 989–1007 (1998).
[22] O. E. Barndorff-Nielsen, S. Z. Levendorskiĭ, Feller processes of normal inverse Gaussian type, *Quant. Finance* **1**, 318–31 (2001).
[23] O. E. Barndorff-Nielsen, T. Mikosch, S. Resnick (eds.), *Lévy Processes: Theory and Applications*, Birkhäuser (2001).
[24] O. E. Barndorff-Nielsen, K. Prause, Apparent scaling, *Finance Stochast.* **5**, 103–13 (2001).
[25] O. E. Barndorff-Nielsen, N. Shephard, Non-Gaussian Ornstein–Uhlenbeck-based models and some of their uses in financial economics, *J.R. Statis. Soc. B* **63**, 167–241 (2001).
[26] O. E. Barndorff-Nielsen, Exponentially decreasing distributions for the logarithm of particle size, *Proc. Roy. Soc. London Ser. A* **353**, 401–19 (1977).
[27] O. E. Barndorff-Nielsen, Normal inverse Gaussian distributions and stochastic volatility modelling, *Scand. J. Statis.* **24**, 1–13 (1997).
[28] O. E. Barndorff-Nielsen, Processes of normal inverse Gaussian type, *Finance Stochast.* **2**, 41–68 (1998).
[29] R. F. Bass, Uniqueness in law for pure jump Markov processes, *Prob. Theory Rel. Fields* **79**, 271–87 (1988).
[30] P. Baxendale, Brownian motions in the diffeomorphism group I, *Comp. Math.* **53**, 19–50 (1984).
[31] P. Baxendale, Stability and equilibrium properties of stochastic flows of diffeomorphisms. In *Diffusions and Related Problems in Analysis*, vol. 2, *Stochastic Flows*, eds. M. Pinsky, V. Wihstutz, *Progress in Probability* **27**, Birkhäuser, pp. 3–35 (1992).
[32] M. Baxter, A. Rennie, *Financial Calculus*, Cambridge University Press (1996).
[33] E. Benhamou, Option pricing with Lévy processes, LSE preprint (2000).
[34] C. Berg, G. Forst, Non-symmetric translation invariant Dirichlet forms, *Invent. Math.* **21**, 199–212 (1973).
[35] C. Berg, G. Forst, *Potential Theory on Locally Compact Abelian Groups*, Springer-Verlag (1975).
[36] J. Bertoin, *Lévy Processes*, Cambridge University Press (1996).

[37] J. Bertoin, Subordinators: examples and applications. In *Ecole d'Eté de Probabilités de St Flour XXVII*, ed. P. Bernard, Lecture Notes in Mathematics 1717, Springer-Verlag, pp. 4–79 (1999).
[38] J. Bertoin, Some elements of Lévy processes. In *Handbook of Statistics*, vol. 19, eds. D. N. Shanbhag, C. R. Rao, Elsevier Science, pp. 117–44 (2001).
[39] A. Beurling, J. Deny, Espaces de Dirichlet I, le case élémentaire, *Acta. Math.* **99**, 203–24 (1958).
[40] A. Beurling, J. Deny, Dirichlet spaces, *Proc. Nat. Acad. Sci. USA* **45**, 208–15 (1959).
[41] P. Biane, J. Pitman, M. Yor, Probability laws related to the Jacobi theta and Riemann zeta functions, and Brownian excursions, *Bull. Amer. Math. Soc.* **38**, 435–67 (2001).
[42] M. Bibby, M. Sørensen, A hyperbolic diffusion model for stock prices, *Finance Stochast.* **1**, 25–41 (1997).
[43] K. Bichteler, *Stochastic Integration With Jumps*, Cambridge University Press (2002).
[44] P. Billingsley, *Probability and Measure* (second edition), Wiley (1986).
[45] P. Billingsley, *Convergence of Probability Measures* (second edition), Wiley (1999).
[46] N. H. Bingham, C. M. Goldie, J. L. Teugels, *Regular Variation*, Cambridge University Press (1987).
[47] N. H. Bingham, R. Kiesel, *Risk-Neutral Valuation, Pricing and Hedging of Financial Derivatives*, Springer-Verlag (1998).
[48] N. H. Bingham, R. Kiesel, Modelling asset returns with hyperbolic distributions. In *Return Distributions in Finance*, eds. J. Knight, S. Satchell, Butterworth-Heinemann, pp. 1–20 (2001).
[49] N. H. Bingham, R. Kiesel, Semiparametric modelling in finance: theoretical foundations, *Quantitative Finance* **2**, 241–50 (2002).
[50] F. Black, M. Scholes, The pricing of options and corporate liabilities, *J. Political Economy* **81**, 637–59 (1973).
[51] R. M. Blumenthal, R. K. Getoor, *Markov Processes and Potential Theory*, Academic Press (1968).
[52] W. R. Bloom, H. Heyer, *Harmonic Analysis of Probability Measures on Hypergroups*, de Gruyter (1995).
[53] S. Bochner, *Harmonic Analysis and the Theory of Probability*, University of California Press (1955).
[54] M. Born, *Einstein's Theory of Relativity*, Methuen (1924); Dover (1962).
[55] N. Bouleau, F. Hirsch, *Dirichlet Forms and Analysis on Wiener Space*, de Gruyter (1991).
[56] S. I. Boyarchenko, S. Z. Levendorskiĭ, Barrier options and touch-and-out options under regular Lévy processes of exponential type, *Annals Appl. Prob.* **12**, 1261–99 (2002).
[57] S. I. Boyarchenko, S. Z. Levendorskiĭ, *Non-Gaussian Merton–Black–Scholes Theory*, World Scientific (2002).
[58] L. Breiman, *Probability*, Addison-Wesley (1968); republished in *SIAM* (1992).
[59] P. Brémaud, *Point Processes and Queues: Martingale Dynamics*, Springer-Verlag (1981).

[60] J. L. Bretagnolle, p-variation de fonctions aléatoires, 2ème partie: processus à accroissements indépendants. In *Séminaire de Probabilités VI*, Lecture Notes in Mathematics 258, Springer-Verlag, pp. 64–71 (1972).
[61] J. L. Bretagnolle, Processus à accroissements indépendants. In *Ecole d'Eté de Probabilités*, Lecture Notes in Mathematics 307, Springer-Verlag, pp. 1–26 (1973).
[62] D. L. Burkholder, B. J. Davis, R. F. Gundy, Integral inequalities for convex functions of operators on martingales. In *Proc. 6th Berkeley Symp. on Math. Stat. and Prob*, vol. 2, University of California Press, pp. 223–40 (1972).
[63] R. H. Cameron, W. T. Martin, Transformation of Wiener integrals under translation, *Ann. Math.* **45**, 386–96 (1944).
[64] R. Carmona, W. C. Masters, B. Simon, Relativistic Schrödinger operators: asymptotic behaviour of the eigenvalues, *J. Funct. Anal.* **91**, 117–42 (1990).
[65] R. Carmona, D. Nualart, *Non-Linear Stochastic Integrators, Equations and Flows*, Gordon and Breach (1990).
[66] P. Carr, E. C. Chang, D. B. Madan, The variance gamma process and option pricing, *Eur. Finan. Rev.* **2**, 79–105 (1998).
[67] P. Carr, H. Geman, D. B. Madan, M. Yor, Stochastic volatility for Lévy processes, *Mathematical Finance*, **13**, 345–82 (2003).
[68] P. Carr, H. Geman, D. B. Madan, M. Yor, The fine structure of asset returns: an empirical investigation, *J. Business* **75**, 305–33 (2002).
[69] T. Chan, Pricing contingent claims on stocks driven by Lévy processes, *Annals Appl. Prob.* **9**, 504–28 (1999).
[70] S. Chandrasekhar, Stochastic problems in physics and astronomy, *Rev. Mod. Phys.* **15**, 1–89 (1943); reprinted in N. Wax (ed.), *Selected Papers on Noise and Stochastic Processes*, Dover (1954).
[71] K. L. Chung, R. J. Williams, *Introduction to Stochastic Integration* (second edition), Birkhäuser (1990).
[72] K. L. Chung, Z. Zhao, *From Brownian Motion to Schrödinger's Equation*, Springer-Verlag (1995).
[73] D. L. Cohn, *Measure Theory*, Birkhäuser (1980).
[74] P. Courrège, Génerateur infinitésimal d'un semi-group de convolution sur $\mathbb{R}^n$ et formule de Lévy–Khintchine, *Bull. Sci. Math. 2e Série* **88**, 3–30 (1964).
[75] P. Courrège, Sur la forme intégro-différentielle des opérateurs de C_k^∞ dans C satifaisant au principe du maximum, *Sém. Théorie du Potential* exposé **2**, (1965/66) 38pp.
[76] E. B. Davies, *One-Parameter Semigroups*, Academic Press (1980).
[77] F. Delbaen, W. Schachermeyer, A general version of the fundamental theorem of asset pricing, *Math. Ann.* **300**, 463–520 (1994).
[78] C. Dellacherie, P. A. Meyer, *Probabilités et Potential: Theorie des Martingales*, Hermann (1980).
[79] J. Deny, Méthodes Hilbertiennes et théorie du potentiel. In *Potential Theory*, ed. M. Brelot, Centro Internazionale Matematico Estivo, Edizioni Cremonese, pp. 123–201 (1970).
[80] N. Dinculeanu, *Vector Integration and Stochastic Integration in Banach Space*, Wiley (2000).

[81] C. Doléans-Dade, Quelques applications de la formule de changement de variables pour les martingales, *Z. Wahrsch. verv. Geb.* **16**, 181–94 (1970).
[82] C. Doléans-Dade, On the existence and unicity of solutions of stochastic differential equations, *Z. Wahrsch. verv. Geb.* **36**, 93–101 (1976).
[83] J. L. Doob, *Stochastic Processes*, Wiley (1953).
[84] R. M. Dudley, *Real Analysis and Probability*, Wadsworth and Brooks/Cole Advanced Books and Software (1989); republished, Cambridge University Press (2002).
[85] R. Durrett, *Stochastic Calculus: A Practical Introduction*, CRC Press (1996).
[86] E. B. Dynkin, *Markov Processes* (two volumes), Springer-Verlag (1965).
[87] E. B. Dynkin, *Markov Processes and Related Problems of Analysis*, London Math. Soc. Lecture Notes Series 54, Cambridge University Press (1982).
[88] E. Eberlein, U. Keller, K. Prause, New insights into smile, mispricing and value at risk: the hyperbolic model, *J. Business* **71**, 371–405 (1998).
[89] E. Eberlein, U. Keller, Hyperbolic distributions in finance, *Bernoulli* **1**, 281–99 (1995).
[90] E. Eberlein, F. Özkan, The defaultable Lévy term structure: ratings and restructuring, *Mathematical Finance*, **13**, 277–300 (2003).
[91] E. Eberlein, S. Raible, Term structure models driven by general Lévy processes, *Mathematical Finance* **9**, 31–53 (1999).
[92] A. Einstein, *Investigations on the Theory of the Brownian Movement*, Dover (1956).
[93] K. D. Elworthy, Geometric aspects of diffusions on manifolds. In *Ecole d'Eté de Probabilités de St Flour XVII*, ed. P. L. Hennequin, Lecture Notes in Mathematics 1362, Springer-Verlag, pp. 276-425 (1988).
[94] P. Embrechts, C. Klüppelberg, T. Mikosch, *Modelling Extremal Events for Insurance and Finance*, Springer-Verlag (1997).
[95] P. Embrechts, M. Maejima, *Selfsimilar Processes*, Princeton University Press (2002).
[96] M. Emery, M. Yor (eds.) *Séminaires de Probabilités, 1967–1980: A Selection in Martingale Theory*, Lecture Notes in Mathematics 1771, Springer-Verlag (2002).
[97] H. J. Engelbert, W. Schmidt, On solutions of stochastic differential equations without drift, *Z. Wahrsch. verw. Geb.* **68**, 287–317 (1985).
[98] A. Etheridge, *A Course in Financial Calculus*, Cambridge University Press (2002).
[99] S. N. Ethier, T. G. Kurtz, *Markov Processes, Characterisation and Convergence*, Wiley (1986).
[100] W. Farkas, N. Jacob, R. L. Schilling, Feller semigroups, L^p-sub-Markovian semigroups and applications to pseudo-differential operators with negative definite symbols, *Forum Math.* **13**, 51–90 (2001).
[101] W. Feller, *An Introduction to Probability Theory and its Applications*, vol. 1 (second edition), Wiley (1957).
[102] W. Feller, *An Introduction to Probability Theory and its Applications*, vol. 2 (second edition), Wiley (1971).
[103] R. P. Feynman, Space–time approach to non-relativistic quantum mechanics, *Rev. Mod. Phys.* **20**, 367–87 (1948).

[104] H. Föllmer, M. Schweizer, Hedging of contingent claims under incomplete information. In *Applied Stochastic Analysis*, eds. M. H. A. Davis, R. J. Elliot, Gordon and Breach, pp. 389–414 (1991).
[105] N. Fournier, Jumping SDEs: absolute continuity using monotonicity, *Stoch. Proc. Appl.* **98**, 317–30 (2002).
[106] B. Fristedt, L. Gray, *A Modern Approach to Probability Theory*, Birkhaüser (1997).
[107] B. Fristedt, Sample functions of stochastic processes with stationary, independent increments. In *Advances in Probability*, vol. 3, eds. P. Ney and S. Port, Marcel Dekker, pp. 241–396 (1974).
[108] T. Fujiwara, H. Kunita, Stochastic differential equations of jump type and Lévy processes in diffeomorphisms group, *J. Math. Kyoto Univ.* **25**, 71–106 (1985).
[109] T. Fujiwara, H. Kunita, Canonical SDEs based on semimartingales with spatial parameters. Part 1: Stochastic flows of diffeomorphisms; Part II: Inverse flows and backwards SDEs, *Kyushu J. Math.* **53**, 265–300; 301–31 (1999).
[110] T. Fujiwara, Y. Miyahara, The minimal entropy martingale measures for geometric Lévy processes, *Finance Stochast.* **7**, 509–31 (2003).
[111] T. Fujiwara, Stochastic differential equations of jump type on manifolds and Lévy flows, *J. Math Kyoto Univ.* **31**, 99–119 (1991).
[112] M. Fukushima, Y. Oshima, M. Takeda, *Dirichlet Forms and Symmetric Markov Processes*, de Gruyter (1994).
[113] I. M. Gelfand, N. Y. Vilenkin, *Generalised Functions*, vol. 4, Academic Press (1964).
[114] H. Geman, D. B. Madan, M. Yor, Asset prices are Brownian motion: only in business time. In *Quantitative Analysis in Financial Markets: Collected Papers of the New York University Mathematical Finance Seminar*, vol. 2, World Scientific, pp. 103–46 (2001).
[115] H. Geman, D. B. Madan, M. Yor, Time changes for Lévy processes, *Math. Finance* **11**, 79–86 (2001).
[116] H. U. Gerber, E. S. W. Shiu, Option pricing by Esscher transforms, *Trans. Soc. Actuaries* **46**, 99–191 (1991).
[117] R. Gielerak, P. Lugiewicz, From stochastic differential equation to quantum field theory, *Rep. Math. Phys.* **44**, 101–10 (1999).
[118] I. I. Gihman, A. V. Skorohod, *Stochastic Differential Equations*, Springer-Verlag (1972).
[119] E. Giné, M. B. Marcus, The central limit theorem for stochastic integrals with respect to Lévy processes, *Ann. Prob.* **11**, 58–77 (1983).
[120] I. V. Girsanov, On transforming a certain class of stochastic processes by absolutely continuous substitution of measures, *Theor. Prob. Appl.* **5**, 285–301 (1960).
[121] B. V. Gnedenko, A. N. Kolmogorov, *Limit Distributions for Sums of Independent Random Variables* (second edition), Addison-Wesley (1968).
[122] M. Grigoriu, *Stochastic Calculus: Applications in Science and Engineering*, Birkhaüser (2002).
[123] G. R. Grimmett, D. R. Stirzaker, *Probability and Random Processes* (third edition), Clarendon Press (2001).
[124] E. Grosswald, The Student t-distribution of any degree of freedom is infinitely divisible, *Z. Wahrsch. verw. Geb.* **36**, 103–9 (1976).

[125] C. Halgreen, Self-decomposability of the generalised inverse Gaussian and hyperbolic distributions, *Z. Wahrsch. verw. Geb.* **47**, 13–17 (1979).
[126] P. Hall, A comedy of errors: the canonical form for a stable characteristic function, *Bull. London Math. Soc.* **13**, 23–7 (1981).
[127] J. M. Harrison, S. R. Pliska, Martingales and stochastic integrals in the theory of continuous trading, *Stoch. Proc. Appl.* **11**, 215–60 (1981).
[128] H. Heyer, *Probability Measures on Locally Compact Groups*, Springer-Verlag (1977).
[129] E. Hille, R. S. Phillips, *Functional Analysis and Semigroups* (revised edition), American Mathematical Society Colloquium Publ. 31 (1957).
[130] E. Hille, Notes on linear transformations I, *Trans. Amer. Math. Soc.* **39**, 131–53 (1936).
[131] E. W. Hobson, *The Theory of Functions of a Real Variable*, Cambridge University Press (1921).
[132] W. Hoh, The martingale problem for a class of pseudo-differential operators, *Math. Ann.* **300**, 121–47 (1994).
[133] W. Hoh, Pseudo-differential operators with negative definite symbols and the martingale problem, *Stoch. and Stoch. Rep.* **55**, 225–52 (1995).
[134] E. Hsu, Analysis on path and loop spaces. In *Probability Theory and Applications*, eds. E. Hsu, S. R. S. Varadhan, IAS/Park City Math. Series 6, American Mathematical Society, pp. 279–347 (1999).
[135] Z. Huang, J. Yan, *Introduction to Infinite-Dimensional Stochastic Analysis*, Kluwer (2000).
[136] B. W. Huff, The strict subordination of differential processes, *Sankhyā Series A* **31**, 403–12 (1969).
[137] J. C. Hull, *Options, Futures and Other Derivatives* (fourth edition), Prentice Hall (2000).
[138] G. A. Hunt, Markoff processes and potentials I, II and III, *Illinois J. Math.* **1**, 44–93; 316–69 (1957); **2**, 151–213 (1958).
[139] T. Ichinose, H. Tamura, Imaginary-time path integral for a relativistic spinless particle in an electromagnetic field, *Commun. Math. Phys.* **105**, 239–57 (1986).
[140] N. Ikeda, S. Watanabe, *Stochastic Differential Equations and Diffusion Processes* (second edition), North Holland-Kodansha (1989).
[141] K. Itô, H. P. McKean, *Diffusion Processes and Their Sample Paths*, Springer-Verlag (1965); second edition (1974).
[142] K. Itô, On stochastic processes: I, Infinitely divisible laws of probability, *Japan. J. Math.* **18**, 261–301 (1942). Reprinted in *Kiyosi Itô Selected Papers*, eds. D. W. Stroock, S. R. S. Varadhan, Springer-Verlag, pp. 1–42 (1987).
[143] K. Itô, Stochastic integral, *Proc. Imp. Acad. Tokyo* **20**, 519–24 (1944). Reprinted in *Kiyosi Itô Selected Papers*, eds. D. W. Stroock, S. R. S. Varadhan, Springer-Verlag, pp. 85–92 (1987).
[144] K. Itô, On a formula concerning stochastic differentials, *Nagoya Math. J.* **3**, 55–65 (1951). Rreprinted in *Kiyosi Itô Selected Papers*, eds. D. W. Stroock, S. R. S. Varadhan, Springer-Verlag, pp. 169–181 (1987).
[145] K. Itô, *On Stochastic Differential Equations*, Mem. Amer. Math. Soc 4 (1951). Reprinted in *Kiyosi Itô Selected Papers*, eds. D. W. Stroock, S. R. S. Varadhan, Springer-Verlag, pp. 117–169 (1987).

[146] K. Itô, *Introduction to Probability Theory*, Cambridge University Press (1984).
[147] N. Jacob, R. L. Schilling, An analytic proof of the Lévy–Khintchine formula on $\mathbb{R}^n$, *Publ. Math. Debrecen* **53**(1–2), 69–89 (1998).
[148] N. Jacob, *Pseudo-Differential Operators and Markov Processes*, Akademie-Verlag, Mathematical Research 94 (1996).
[149] N. Jacob, *Pseudo-Differential Operators and Markov Processes: 1, Fourier Analysis and Semigroups*, World Scientific (2001).
[150] N. Jacob, *Pseudo-Differential Operators and Markov Processes: 2, Generators and Their Potential Theory*, World Scientific (2002).
[151] J. Jacod, A. N. Shiryaev, *Limit Theorems for Stochastic Processes*, Springer-Verlag (1987); second edition (2003).
[152] J. Jacod, Multivariate point processes: predictable representation, Radon–Nikodým derivatives, representation of martingales, *Z. Wahrsch. verw. Geb.* **31**, 235–53 (1975).
[153] J. Jacod, A general theorem for representation of martingales, *Proc. Symp. Pure Math.* **31**, 37–53 (1977).
[154] J. Jacod, *Calcul Stochastique et Problèmes de Martingales*, Lecture Notes in Mathematics 714, Springer-Verlag (1979).
[155] J. Jacod, Grossissement de filtration et processus d'Ornstein–Uhlenbeck généralisé, in *Séminaire de Calcul Stochastiques*, eds. Th. Jeulin, M. Yor, Lecture Notes in Mathematics 1118, Springer-Verlag, pp. 36–44 (1985).
[156] A. Janicki, Z. Michna, A. Weron, Approximation of stochastic differential equations driven by α-stable Lévy motion, *Appl. Math.* **24**, 149–68 (1996).
[157] A. Janicki, A. Weron, *Simulation and Chaotic Behaviour of α-Stable Stochastic Processes*, Marcel Dekker (1994).
[158] M. Jeanblanc, J. Pitman, M. Yor, Self-similar processes with independent increments associated with Lévy and Bessel processes, *Stoch. Proc. Appl.* **100**, 223–31 (2002).
[159] J. Jessen, A. Wintner, Distribution functions and the Riemann zeta function, *Trans. Amer. Math. Soc.* **38**, 48–88 (1935).
[160] S. Johansen, On application of distributions, *Z. Wahrsch. verw. Geb.* **5**, 304–16 (1966).
[161] Z. J. Zurek, J. D. Mason, *Operator-Limit Distributions in Probability Theory*, Wiley (1993).
[162] Z. J. Jurek, W. Vervaat, An integral representation for selfdecomposable Banach space-valued random variables, *Z. Wahrsch. verw. Geb.* **62**, 247–62 (1983).
[163] Z. J. Jurek, A note on gamma random variables and Dirichlet series, *Statis. Prob. Lett.* **49**, 387–92 (2000).
[164] M. Kac, On some connections between probability theory and differential and integral equations. In *Proc. 2nd Berkeley Symp. on Math. Stat. and Prob.*, University of California Press, pp. 189–215 (1951).
[165] O. Kallenberg, *Random Measures* (fourth edition), Akademie-Verlag and Academic Press (1976).
[166] O. Kallenberg, *Foundations of Modern Probability*, Springer-Verlag, (1997).
[167] I. Karatzas, S. Shreve, *Brownian Motion and Stochastic Calculus* (second edition), Springer-Verlag (1991).

[168] T. Kato, *Perturbation Theory for Linear Operators* (second edition), Springer-Verlag (1995).
[169] J. F. C. Kingman, *Poisson Processes*, Oxford University Press (1993).
[170] F. C. Klebaner, *Introduction to Stochastic Calculus with Applications*, Imperial College Press (1998).
[171] F. B. Knight, *Essentials of Brownian Motion and Diffusion*, American Mathematical Society (1981).
[172] A. N. Kolmogorov, Uber die analytischen Methoden in Wahrscheinlichkeitsrechnung, *Math. Ann.* **104**, 415–58 (1931).
[173] A. N. Kolmogorov, The Wiener helix and other interesting curves in Hilbert space, *Dokl. Acad. Nauk.* **26**, 115–8 (in Russian) (1940).
[174] V. N. Kolokoltsov, Symmetric stable laws and stable-like jump diffusions, *Proc. London Math. Soc.* **80**, 725–68 (2000).
[175] V. N. Kolokoltsov, *Semiclassical Analysis for Diffusions and Stochastic Processes*, Lecture Notes in Mathematics 1724, Springer-Verlag (2000).
[176] V. N. Kolokoltsov, A new path integral representation for the solutions of Schrödinger, heat and stochastic Schrödinger equations, *Math. Proc. Camb. Phil. Soc.*, **132**, 353–75 (2002).
[177] T. Komatsu, Markov processes associated with certain integro-differential operators, *Osaka J. Math.* **10**, 271–305 (1973).
[178] N. V. Krylov, *Introduction to the Theory of Diffusion Processes*, American Mathematical Society (1995).
[179] H. Kunita, J.-P. Oh, Asymptotic properties of Lévy flows, *J. Korean Math. Soc.* **27**, 255–80 (1990).
[180] H. Kunita, J.-P. Oh, Malliavin calculus on the Wiener–Poisson space and its application to canonical SDE with jumps, preprint (2001).
[181] H. Kunita, S. Watanabe, On square-integrable martingales, *Nagoya Math. J.* **30**, 209–45 (1967).
[182] H. Kunita, *Stochastic Flows and Stochastic Differential Equations*, Cambridge University Press (1990).
[183] H. Kunita, Stochastic differential equations with jumps and stochastic flows of diffeomorphisms. In *Itô's Stochastic Calculus and Probability Theory*, eds. N. Ikeda *et al.*, Springer-Verlag, pp. 197–212 (1996).
[184] H. Kunita, Smooth density of canonical stochastic differential equation with jumps, preprint (2001).
[185] H. Kunita, Mathematical finance for price processes with jumps, preprint (2002).
[186] H. Kunita, Representation of martingales with jumps and applications to mathematical finance. To appear in *Stochastic Analysis and Related Topics in Kyoto*, Advanced Studies in Pure Mathematics, eds. H. Kunita, Y. Takahashi, S. Watanabe, Mathematical Society of Japan.
[187] T. G. Kurtz, E. Pardoux, P. Protter, Stratonovitch stochastic differential equations driven by general semimartingales, *Ann. Inst. Henri Poincaré Prob. Stat.* **31**, 351–77 (1995).
[188] D. Lamberton, B. Lapeyre, *Introduction to Stochastic Calculus Applied to Finance*, translated by N. Rabeau and F. Mantion, Chapman and Hall (1996). Original French language edition, *Introduction au Calcul Stochastique Appliqué à la Finance*, Edition Marketing (1991).

[189] R. Léandre, Flot d'une équation differentielle stochastique avec semimartingale directrice discontinue. In *Séminaire de Probabilités XIX*, Lecture Notes in Mathematics 1123, Springer-Verlag, pp. 271–4 (1985).
[190] P. M. Lee, Infinitely divisible stochastic processes, *Z. Wahrsch. verw. Geb.* **7**, 147–60 (1967).
[191] J.-P. Lepeltier, B. Marchal, Problème des martingales et équations différentielles stochastiques associées à un opérateur intégro-différentiel, *Ann. Inst. Henri Poincaré B: Prob. Stat.* **12**, 43–103 (1976).
[192] P. Lévy, *Théorie de l'Addition des Variables Aléatoire* (second edition), Gauthier-Villars (1953).
[193] P. Lévy, *Processus Stochastiques et Mouvement Brownien* (second edition) Gauthier-Villars (1954).
[194] X.-M. Li, Strong p-completeness of stochastic differential equations and the existence of smooth flows on noncompact manifolds, *Prob. Theory Rel. Fields* **100**, 485–511 (1994).
[195] M. Liao, Lévy processes in semisimple Lie groups and stability of stochastic flows, *Trans. Amer. Math. Soc.* **350**, 501–22 (1998).
[196] E. H. Lieb, M. Loss, *Analysis*, American Mathematical Society (1997).
[197] G. W. Lin, C.-Y. Hu, The Riemann zeta distribution, *Bernoulli* **7**, 817–28 (2001).
[198] W. Linde, *Probability in Banach Spaces – Stable and Infinitely Divisible Distributions*, Wiley (1986).
[199] Ju. V. Linnik, I. V. Ostrovskii, *Decomposition of Random Variables and Vectors*, Translations of Mathematical Monographs 48, American Mathematical Society (1977).
[200] R. S. Liptser, A. N. Shiryaev, *Statistics of Random Processes: 1. General Theory*, Springer-Verlag (2001); first published in Russian (1974).
[201] R. S. Liptser, A. N. Shiryaev, *Statistics of Random Processes: 2. Applications*, Springer-Verlag (2001); first published in Russian (1974).
[202] T. J. Lyons, Differential equations driven by rough signals, *Revista Math. Ibero-America* **14**, 215–310 (1998).
[203] T. J. Lyons, Z. Qian, *System Control and Rough Paths*, Oxford University Press (2002).
[204] Z.-M. Ma, M. Röckner, *Introduction to the Theory of Non-Symmetric Dirichlet Forms*, Springer-Verlag (1992).
[205] J. H. McCulloch, Financial applications of stable distributions. In *Handbook of Statistics*, vol. 14, eds. G. S. Maddala, C. R. Rao, Elsevier Science (1996).
[206] D. B. Madan, E. Seneta, The variance gamma (V. G.) model for share market returns, *J. Business* **63**, 511–24 (1990).
[207] P. Malliavin (with H. Aurault, L. Kay, G. Letac), *Integration and Probability*, Springer-Verlag (1995).
[208] P. Malliavin, *Stochastic Analysis*, Springer-Verlag, (1997).
[209] B. Mandelbrot, The variation of certain speculative prices, *J. Business* **36**, 394–419 (1963).
[210] B. Mandelbrot, *Fractals and Scaling in Finance: Discontinuity, Concentration, Risk*, Springer-Verlag (2000).
[211] X. Mao, A. E. Rodkina, Exponential stability of stochastic differential equations driven by discontinuous semimartingales, *Stoch. and Stoch. Rep.* **55**, 207–24 (1995).

[212] M. B. Marcus, J. Rosen, Sample path properties of the local times of strongly symmetric Markov processes via Gaussian processes, *Ann. Prob.* **20**, 1603–84 (1992).

[213] S. I. Marcus, Modelling and analysis of stochastic differential equations driven by point processes, *IEEE Trans. Inform. Theory* **IT-24**, 164–72 (1978).

[214] S. I. Marcus, Modelling and approximation of stochastic differential equations driven by semimartingales, *Stochastics* **4**, 223–45 (1981).

[215] G. Maruyama, Infinitely divisible processes, *Theory of Prob. and Appl.* **15**, 1–22 (1970).

[216] M. M. Meerschaert, H.-P. Scheffler, *Limit Distributions for Sums of Independent Random Vectors*, Wiley (2001).

[217] M. M. Meerschaert, H.-P. Scheffler, Portfolio modelling with heavy tailed random vectors. In *Handbook of Heavy Tailed Distributions in Finance*, ed. S. T. Rachev, Elsevier, pp. 595–640 (2003).

[218] A. V. Mel'nikov, *Financial Markets, Stochastic Analysis and the Pricing of Derivative Securities*, Translations of Mathematical Monographs 184, American Mathematical Society (1999).

[219] R. C. Merton, Theory of Rational Option Pricing, *Bell J. Econ. and Man. Sci.* **4**, 141–83 (1973).

[220] R. C. Merton, Option pricing when underlying stock returns are discontinuous, *J. Financial Econ.* **3**, 125–44 (1976).

[221] M. Métivier, *Semimartingales*, de Gruyter (1982).

[222] P. A. Meyer, A decomposition theorem for supermartingales, *Ill. J. Math.* **6**, 193–205 (1962).

[223] P. A. Meyer, Decompositions of supermartingales: the uniqueness theorem, *Ill. J. Math.* **6**, 1–17 (1963).

[224] P. A. Meyer, Un cours sur les intégrales stochastiques. In *Séminaire de Prob. X*, Lecture Notes in Mathematics 511, Springer-Verlag, pp. 246–400 (1976).

[225] P. A. Meyer, Flot d'une équation différentielle stochastique. In *Séminaire de Prob. XV*, Lecture Notes in Mathematics 850, Springer-Verlag, pp. 103–17 (1981).

[226] P. A. Meyer, *Quantum Probability for Probabilists* (second edition), Lecture Notes in Mathematics 1538, Springer-Verlag (1995).

[227] T. Mikosch, R. Norvaiša, Stochastic integral equations without probability, *Bernoulli* **6**, 401–35 (2000).

[228] T. Mikosch, *Elementary Stochastic Calculus with Finance in View*, World Scientific (1998).

[229] P. W. Millar, Stochastic integrals and processes with stationary and independent increments. In *Proc. 6th Berkeley Symp. on Math. Stat. and Prob.*, University of California Press, pp. 307–31 (1972).

[230] P. W. Millar, Exit properties of stochastic processes with stationary independent increments, *Trans. Amer. Math. Soc.* **178**, 459–79 (1973).

[231] I. Monroe, Processes that can be embedded in Brownian motion, *Ann. Prob.* **6**, 42–56 (1978).

[232] M. Motoo, S. Watanabe, On a class of additive functionals of Markov processes, *J. Math. Kyoto Univ.* **4**, 429–69 (1965).

[233] C. Mueller, The heat equation with Lévy noise, *Stoch. Proc. Appl.* **74** 67–82 (1998).

[234] L. Mytnik, Stochastic partial differential equation driven by stable noise, *Prob. Theory Rel. Fields* **123**, 157–201 (2002).
[235] A. Negoro, Stable-like processes: construction of the transition density and the behaviour of sample paths near $t = 0$, *Osaka J. Math.* **31**, 189–214.
[236] E. Nelson, *Dynamical Theories of Brownian Motion*, Princeton University Press (1967).
[237] D. Nualart, W. Schoutens, Chaotic and predictable properties of Lévy processes. *Stoch. Proc. Appl.* **90**, 109–122 (2000).
[238] D. Nualart, W. Schoutens, Backwards stochastic differential equations and Feynman–Kac formula for Lévy processes, with applications in finance, *Bernoulli*, **7**, 761–76 (2001).
[239] D. Nualart, *The Malliavin Calculus and Related Topics*, Springer-Verlag (1995).
[240] B. Øksendal, A. Sulem, *Applied Stochastic Control of Jump Diffusions*, Springer-Verlag (2004).
[241] B. Øksendal, *Stochastic Differential Equations* (fifth edition), Springer-Verlag (1998).
[242] L. S. Ornstein, G. E. Uhlenbeck, On the theory of Brownian motion, *Phys. Rev.* **36**, 823–41 (1930).
[243] Y. Ouknine, Reflected backwards stochastic differential equations with jumps, *Stoch. and Stoch. Rep.* **65**, 111–25 (1998).
[244] R. E. A. C. Paley, N. Wiener, *Fourier Transforms in the Complex Domain*, American Mathematical Society (1934).
[245] E. Pardoux, Stochastic partial differential equations, a review, *Bull. Sc. Math. 2e Série* **117**, 29–47 (1993).
[246] K. R. Parthasarathy, K. Schmidt, *Positive Definite Kernels, Continuous Tensor Products and Central Limit Theorems of Probability Theory*, Lecture Notes in Mathematics 272, Springer-Verlag (1972).
[247] K. R. Parthasarathy, *Probability Measures on Metric Spaces*, Academic Press (1967).
[248] K. R. Parthasarathy, Square integrable martingales orthogonal to every stochastic integral, *Stoch. Proc. Appl.* **7**, 1–7 (1978).
[249] K. R. Parthasarathy, *An Introduction to Quantum Stochastic Calculus*, Birkhäuser (1992).
[250] S. J. Patterson, *An Introduction to the Theory of the Riemann Zeta-Function*, Cambridge University Press (1988).
[251] J. Picard, On the existence of smooth densities for jump processes, *Prob. Theory Rel. Fields* **105**, 481–511 (1996).
[252] S. C. Port, C. J. Stone, Infinitely divisible processes and their potential theory I and II, *Ann. Inst. Fourier* **21** (2), 157–275, (4), 179–265 (1971).
[253] N. U. Prabhu, *Stochastic Storage Processes, Queues, Insurance Risks and Dams*, Springer-Verlag (1980).
[254] P. Protter, D. Talay, The Euler scheme for Lévy driven stochastic differential equations, *Ann. Prob.* **25**, 393–423 (1997).
[255] P. Protter, *Stochastic Integration and Differential Equations*, Springer-Verlag (1992).
[256] B. S. Rajput, J. Rosiński, Spectral representations of infinitely divisible processes, *Prob. Theory Rel. Fields* **82**, 451–87 (1989).

[257] M. Reed, B. Simon, *Methods of Modern Mathematical Physics*, vol. 1, *Functional Analysis* (revised and enlarged edition), Academic Press (1980).
[258] M. Reed, B. Simon, *Methods of Modern Mathematical Physics*, vol. 4. *Analysis of Operators*, Academic Press (1978).
[259] S. Resnick, *Extreme Values, Regular Variation and Point Processes*, Springer-Verlag (1987).
[260] D. Revuz, M. Yor *Continuous Martingales and Brownian Motion*, Springer-Verlag (1990); third edition (1999).
[261] L. C. G. Rogers, D. Williams, *Diffusions, Markov Processes and Martingales*, vol. 1, *Foundations*, Wiley (1979, 1994); Cambridge University Press (2000).
[262] L. C. G. Rogers, D. Williams, *Diffusions, Markov Processes and Martingales*, vol. 2, *Itô Calculus*, Wiley (1994); Cambridge University Press (2000).
[263] L. C. G. Rogers, Arbitrage with fractional Brownian motion, *Math. Finance* **7**, 95–105 (1997).
[264] J. S. Rosenthal, *A First Look at Rigorous Probability Theory*, World Scientific (2000).
[265] J. Rosiński, W. A. Woyczyński, On Itô stochastic integration with respect to p-stable motion: inner clock, integrability of sample paths, double and multiple integrals, *Ann. Prob.* **14**, 271–86 (1986).
[266] H. L. Royden, *Real Analysis* (third edition), Macmillan (1988).
[267] W. Rudin *Functional Analysis* (second edition), Mc-Graw Hill (1991).
[268] M. Ryznar, Estimates of Green function for relativistic α-stable process, *Pot. Anal.* **17**, 1–23 (2002).
[269] E. Saint Loubert Bié, Etude d'une EDPS conduite par un bruit poissonnien, *Prob. Theory Rel. Fields* **111**, 287–321 (1998).
[270] G. Samorodnitsky, M. Grigoriu, Tails of solutions of stochastic differential equations driven by heavy tailed Lévy motions, *Stoch. Proc. Appl.* **105**, 690–97 (2003).
[271] G. Samorodnitsky, M. S. Taqqu, *Stable Non-Gaussian Random Processes*, Chapman and Hall (1994).
[272] J. de Sam Lazaro, P. A. Meyer, Questions des théorie des flots. In *Séminaire de Prob. IX*, Lecture Notes in Mathematics 465, Springer, pp. 1–153 (1975).
[273] K.-I. Sato, M. Yamazoto, Stationary processes of Ornstein–Uhlenbeck type. In *Probability and Mathematical Statistics*, eds. K. Itô, J. V. Prohorov, Lecture Notes in Mathematics 1021, Springer-Verlag, pp. 541–5 (1982).
[274] K.-I. Sato, *Lévy Processes and Infinite Divisibility*, Cambridge University Press (1999).
[275] R. L. Schilling, Subordination in the sense of Bochner and a related functional calculus, *J. Austral. Math. Soc (Series A)* **64**, 368–96 (1998).
[276] R. L. Schilling, Conservativeness and extensions of Feller semigroups, *Positivity* **2**, 239–56 (1998).
[277] R. L. Schilling, Growth and Hölder conditions for the sample paths of Feller processes, *Prob. Theory Rel. Fields* **112**, 565–611 (1998).
[278] R. L. Schilling, Feller processes generated by pseudo-differential operators: on the Hausdorff dimension of their sample paths, *J. Theor. Prob.* **11**, 303–30 (1998).
[279] R. L. Schilling, Dirichlet operators and the positive maximum principle, *Integral Eqs. and Op. Theory* **41**, 74–92 (2001).

[280] W. Schoutens, *Lévy Processes in Finance: Pricing Financial Derivatives*, Wiley (2003).
[281] M. Schürmann, *White Noise on Bialgebras*, Lecture Notes in Mathematics 1544, Springer-Verlag (1991).
[282] V. Seshadri, *The Inverse Gaussian Distribution: A Case Study in Exponential Families*, Clarendon Press, Oxford (1993).
[283] D. C. Shimko, *Finance in Continuous Time: A Primer*, Kolb Publishing (1992).
[284] A. N. Shiryaev, *Essentials of Stochastic Finance: Facts, Models, Theory*, World Scientific (1999).
[285] M. F. Shlesinger, G. M. Zaslavsky, U. Frisch (eds.), *Lévy Flights and Related Topics in Physics*, Springer-Verlag (1995).
[286] R. Situ, On solutions of backwards stochastic differential equations with jumps and applications, *Stoch. Proc. Appl.* **66**, 209–36 (1997).
[287] A. V. Skorohod, *Random Processes with Independent Increments*, Kluwer (1991).
[288] J. M. Steele, *Stochastic Calculus and Financial Applications*, Springer-Verlag (2001).
[289] D. Stroock, S. R. S. Varadhan, *Multidimensional Diffusion Processes*, Springer-Verlag (1979).
[290] D. Stroock, Diffusion processes associated with Lévy generators, *Z. Wahrsch. verw. Geb.* **32**, 209–44 (1975).
[291] D. Stroock, *Probability Theory: An Analytic View*, Cambridge University Press (1993).
[292] D. Stroock, *An Introduction to the Analysis of Paths on a Riemannian Manifold*, American Mathematical Society (2000).
[293] F. Tang, The interlacing construction for stochastic flows of diffeomorphisms, Nottingham Trent University Ph.D. thesis (1999).
[294] M. S. Taqqu, A bibliographic guide to self-similar processes and long-range dependence. In *Dependence in Probability and Statistics*, eds. E. Eberlein, M. S. Taqqu, Birkhaüser, pp. 137–62 (1986).
[295] M. E. Taylor, *Partial Differential Equations II: Qualitative Studies of Linear Equations*, Springer-Verlag (1996).
[296] M. Tsuchiya, Lévy measure with generalised polar decomposition and the associated SDE with jumps, *Stoch. and Stoch. Rep.* **38**, 95–117 (1992).
[297] V. V. Uchaikin, V. M. Zolotarev, *Chance and Stability: Stable Distributions and their Applications*, VSP (1999).
[298] T. Uemura, On some path properties of symmetric stable-like processes for one dimension, *Pot. Anal.* **16**, 79–91 (2002).
[299] J. B. Walsh, An introduction to stochastic partial differential equations. In *Ecole d'Eté de Probabilités de St Flour XIV*, Lecture Notes in Mathematics 1180, Springer-Verlag, pp. 266–439 (1986).
[300] G. N. Watson, *A Treatise on the Theory of Bessel Functions*, Cambridge University Press (1922).
[301] N. Wiener, A. Siegel, B. Rankin, W. T. Martin, *Differential Space, Quantum Systems and Prediction*, MIT Press (1966).
[302] N. Wiener, Differential Space, *J. Math. and Physics* **58**, 131–74 (1923).

[303] D. Williams, To begin at the beginning ... In *Stochastic Integrals*, *Proc. LMS Durham Symp. 1980*, Lecture Notes in Mathematics 851, Springer-Verlag, pp. 1–55 (1981).
[304] D. Williams, *Probability with Martingales*, Cambridge University Press (1991).
[305] D. R. E. Williams, Path-wise solutions of stochastic differential equations driven by Lévy processes, *Revista Ibero-Americana* **17**, 295–329 (2001).
[306] S. J. Wolfe, On a continuous analogue of the stochastic difference equation $X_n = \rho X_{n-1} + B_n$, *Stoch. Proc. Appl.* **12**, 301–12 (1982).
[307] M. Yor, *Some Aspects of Brownian Motion*, part 1, Birkhäuser (1992).
[308] M. Yor, *Some Aspects of Brownian Motion*, part 2, Birkhäuser (1997).
[309] K. Yosida, *Functional Analysis* (sixth edition), Springer-Verlag, (1980).
[310] P. A. Zanzotto, On solutions of one-dimensional stochastic differential equations driven by stable Lévy motion, *Stoch. Proc. Appl.* **68**, 209–28 (1997).
[311] P. A. Zanzotto, On stochastic differential equations driven by Cauchy process and the other stable Lévy motions, *Ann. Prob.* **30**, 802–26 (2002).
[312] V. M. Zolotarev, *One-Dimensional Stable Distributions*, American Mathematical Society (1986).

Index of notation

$\cdot_b$	backwards stochastic integral	241
$\cong$	Hilbert-space isomorphism	193
$\otimes$	Hilbert-space tensor product	193
$\langle \cdot, \cdot \rangle_{T,\rho}$	inner product in $\mathcal{H}_2(T, E)$	193
$\circ$	Itô's circle (Stratonovitch integral)	236
$\circ_b$	backwards Stratonovitch integral	242
$\diamond$	Marcus canonical integral	238
$\diamond_b$	backwards Marcus canonical integral	243
α	index of stability	32
$\Gamma(\alpha)$	gamma function $\int_0^\infty x^{\alpha-1} e^{-x} dx$	50
δ	mesh of a partition	95
η	Lévy symbol or characteristic exponent	30
η_X	Lévy symbol of a Lévy process X	42
δ_x	Dirac measure at $x \in \mathbb{R}^n$	22
$\Delta X(t) = X(t) - X(t-)$	jump process	85
η_Z	Lévy symbol of the subordinated process Z	55
μ	intensity measure	87
$\mu_1 * \mu_2$	convolution of probability measures	20
ν	Lévy measure	27
$\xi_t, t \in \mathbb{R}$	solution flow to an ODE	296
$\rho(A)$	resolvent set of A	134
$\rho_{s,t}$	transition density	122
$\sigma(T)$	spectrum of an operator T	183
$(\tau_a, a \in \mathbb{R}^d)$	translation group of $\mathbb{R}^d$	137
ϕ	local unit	157
χ_A	indicator function of the set A	5
ψ	Laplace exponent	50

Index of notation

$\Psi = (\Psi_{s,t}, 0 \leq s \leq t < \infty)$	solution flow to an SDE	328
$(\Omega, \mathcal{F}, P)$	probability space	3
A	infinitesimal generator of a semigroup	131
A^X	infinitesimal generator of a Lévy process	145
A^Z	infinitesimal generator of a subordinated Lévy process	145
(b, A, ν)	characteristics of an infinitely divisible distribution	42
(b, λ)	characteristics of a subordinator	49
$\hat{B}$	$\{x \in \mathbb{R}^d; |x| < 1\}$	22
$B = (B(t), t \geq 0)$	standard Brownian motion	43
$B_A(t)$	Brownian motion with covariance A	46
$\mathcal{B}(S)$	Borel σ-algebra of $S \subseteq \mathbb{R}^d$	2
$B_b(S)$	bounded Borel measurable functions from S to $\mathbb{R}$	5
$\mathcal{C}(I)$	cylinder functions over $I = [0, T]$	259
$C_c(S)$	continuous functions with compact support on S	5
$C_0(S)$	continuous functions from S to $\mathbb{R}$ that vanish at ∞	5
$C^n(\mathbb{R}^d)$	n-times differentiable functions from $\mathbb{R}^d$ to $\mathbb{R}$	23
$\text{Cov}(X, Y)$	covariance of X and Y	7
dY	stochastic differential of a semimartingale Y	209
D	diagonal, $\{(x, x); x \in \mathbb{R}^d\}$	157
D_A	domain of a linear operator A	131
$D_\phi F$	directional derivative of Wiener functional F in direction ϕ	261
$\mathbb{E}$	expectation	6
$\mathbb{E}_\mathcal{G}$	conditional expectation mapping	9
$\mathbb{E}_s$	$\mathbb{E}(\cdot|\mathcal{F}_s)$ conditional expectation given $\mathcal{F}_s$	72
$\mathbb{E}(X; A)$	$\mathbb{E}(X\chi_A)$	6
$\mathbb{E}(X|\mathcal{G})$	conditional expectation of X given $\mathcal{G}$	9
$\mathcal{E}$	closed form, Dirichlet form	165
$\mathcal{E}_Y$	stochastic (Doléans-Dade) exponential of Y	247
$\hat{f}$	Fourier transform of f	138
$f^+(x)$	$\max\{f(x), 0\}$	4
$f^-(x)$	$-\min\{f(x), 0\}$	4
f_X	probability density function (pdf) of a random variable X	9
$\mathcal{F}$	σ-algebra	2
$(\mathcal{F}_t, t \geq 0)$	filtration	70
$(\mathcal{F}_t^X, t \geq 0)$	natural filtration of the process X	71
$\mathcal{F}_\infty$	$\bigvee_{t \geq 0} \mathcal{F}_t$	71
$\mathcal{F}_{t+}$	$\bigcap_{\epsilon > 0} \mathcal{F}_{t+\epsilon}$	72
G_T	graph of the linear operator T	179

Index of notation

$(\mathcal{G}_t, t \geq 0)$	augmented filtration	72
$\mathcal{G}^X, t \geq 0$	augmented natural filtration of the process X	72
$\mathbb{H}(I)$	Cameron–Martin space over $I = [0, T]$	258
$\mathcal{H}_2(T, E)$	Hilbert space of square-integrable, predictable mappings on $[0, T] \times E \times \Omega$	192
$\mathcal{H}_2^-(s, E)$	Hilbert space of square-integrable, backwards predictable mappings on $[0, s] \times E \times \Omega$	241
I	identity matrix	22
I	identity operator	121
$IG(\delta, \gamma)$	inverse Gaussian random variable	51
$I_T(F)$	Itô stochastic integral of F	199
$\hat{I}_T(F)$	extended Itô stochastic integral of F	203
K_ν	Bessel function of the third kind	282
l_X	lifetime of a sub-Markov process X	128
$L(B)$	space of bounded linear operators in a Banach space B	129
$L^p(S, \mathcal{F}, \mu; \mathbb{R}^d)$	L^p-space of equivalence classes of mappings from S to $\mathbb{R}^d$	7
$L(x, t)$	local time at x in $[0, t]$	66
$M = (M(t), t \geq 0)$	martingale	65
$\mathcal{M}$	martingale space	78
$M(\cdot)$	random measure	89
$\langle M, N \rangle$	Meyer angle bracket	81
$[M, N]$	quadratic variation of M and N	220
$\mathcal{M}_1(\mathbb{R}^d)$	set of all Borel probability measures on $\mathbb{R}^d$	20
$N = (N(t), t \geq 0)$	Poisson process	46
$\tilde{N} = (\tilde{N}(t), t \geq 0)$	compensated Poisson process	46
$N(t, A)$	$\#\{0 \leq s \leq t; \Delta X(s) \in A\}$	17
$\tilde{N}(t, A)$	compensated Poisson random measure	90
p	Poisson point process	90
$p_{s,t}(x, A)$	transition probabilities	122
$(p_t, t \geq 0)$	convolution semigroup of probability measures	60
$p_{t_1, t_2, \ldots, t_n}$	finite-dimensional distributions of a stochastic process	18
$p_t(x, A)$	homogeneous transition probabilities	125
p_X	probability law (distribution) of a random variable X	4
$\mathcal{P}$	partition	95
$\mathcal{P}$	predictable σ-algebra	192
$\mathcal{P}^-$	backwards predictable σ-algebra	241
$\mathcal{P}_2(T, E)$	predictable processes whose squares are a.s. integrable on $[0, T] \times E$	201

$\mathcal{P}_2^-(s,E)$ backwards predictable processes whose squares are a.s. integrable on $[s,T] \times E$	241
$P(\cdot)$ projection-valued measure	152
$P(A\|\mathcal{G})$ conditional probability of the set A given $\mathcal{G}$	10
$P_{Y\|\mathcal{G}}$ conditional distribution of a random variable Y, given $\mathcal{G}$	10
q_t probability law of a Lévy process at time t	122
$R_\lambda(A)$ resolvent of A at the point λ	134
$(S, \mathcal{F}, \mu)$ measure space	2
$S(\mathbb{R}^d)$ Schwartz space	138
$S(T,E)$ simple processes on $[0,T]$	193
$S^-(s,E)$ backwards simple processes on $[s,T]$	241
$(S(t), t \geq 0)$ stock price process	271
$(\tilde{S}(t), t \geq 0)$ discounted stock price process	271
T stopping time	78
T_A first hitting time to a set A	79
$\overline{T}$ closure of a closable operator T	179
T^c dual operator to T	181
T^* adjoint operator to T	181
$T = (T(t), t \geq 0)$ subordinator	49
$(T_{s,t}, 0 \leq s \leq t < \infty)$ Markov evolution	121
$(T_t, t \geq 0)$ semigroup of linear operators	128
$(T_t^X, t \geq 0)$ semigroup associated with a Lévy process X	145
$(T_t^Z, t \geq 0)$ semigroup associated with a subordinated Lévy process Z	145
$V(g)$ variation of a function g	95
$\mathrm{Var}(X)$ variance of X	7
$\mathcal{W}_0(I)$ Wiener space over $I = [0,T]$	258
$X \sim N(m,A)$ X is Gaussian with mean m and covariance A	25
$X \sim \pi(c)$ X is Poisson with intensity c	25
$X \sim \pi(c, \mu_Z)$ X is compound Poisson with Poisson intensity c and Lévy measure $c\mu(\cdot)$	26
$X \sim S\alpha S$ X is symmetric α-stable	34
$X(t-)$ left limit of X at t	85

Subject index

σ-additivity, 89
adjoint operator, 181
σ-algebra, 2
almost all, 3
almost everywhere (a.e.), 3
almost surely (a.s.), 3
American call option, 269
anomalous diffusion, 319
arbitrage, 269
asset pricing
   fundamental theorem of, 270, 271

Bachelier, L., 44, 271
background-driving Lévy process, 217
backwards adapted, 240
backwards filtration, 240
   natural, 240
backwards Marcus canonical integral, 243
backwards martingale, 240
backwards martingale-valued measure, 240
backwards predictable mapping, 241
backwards predictable σ-algebra, 241
backwards simple process, 241
backwards stochastic differential
   equations, 313
Bernstein function, 52
Bessel equation of order ν, 289
   modified, 289
Bessel function
   modified, 290
Bessel function
   of the first kind, 289
   of the second kind, 289
   of the third kind, 282, 290

Beurling–Deny formula, 167
bi-Lipschitz continuity, 303
Black–Scholes papers on option
   pricing, 267
Black–Scholes portfolio, 277
Black–Scholes pricing formula, 278
Bochner integral, 131
Bochner's theorem, 16
Borel σ-algebra of S, 2
Borel measurable mapping, 4
Borel measure, 2
Borel set, 2
bounded below, 87
bounded jumps, 101
bounded operator, 177
Brownian flow, 321
Brownian motion
   standard, xvi, 43
   with covariance A, 46
   with drift, xvi, 45
Brownian motions
   one-dimensional, 44
Brownian sheet, 192
Burkholder's inequality, 234

càdlàg functions, 117
càdlàg paths, 70, 76
càglàd functions, 117
Cameron–Martin space, 258
Cameron–Martin–Maruyama theorem, 259
canonical Markov process, 125
Cauchy process, 56
Cauchy–Schwarz inequality, 8
cemetery point, 128
Chapman–Kolmogorov equations, 123

characteristic exponent, 30
characteristic function, 15
Chebyshev–Markov inequality, 7
Chung–Fuchs criterion, 64
class D, 80
closable linear operator, 179
closed subspace, 8
closed symmetric form, 184
closure, 185
cocycle, 325
  perfect, 327
coercive bilinear form, 173
compensated Poisson process, 46
compensated Poisson random
    measure, 90
completely monotone, 52
compound Poisson process, 46
conditional distribution, 10
conditional expectation, 9
conditional probability, 10
conditionally positive definite, 16
conservative, 148
contingent claim, 269
continuity of measure, 3
contraction, 178
convergence
  in distribution, 13
  in mean square, 13
  in probability, 13
  weak, 14
convolution, 186
convolution nth root, 23
convolution of probability
    measures, 20
convolution semigroup, 60
co-ordinate process, 18
core, 167, 180
countably generated, 8
counting measure, 3
Courrège's first theorem, 158
Courrège's second theorem, 159
covariance, 7
creep, 111
cylinder function, 259
cylinder sets, 18

degenerate Gaussian, 25
densely defined linear operator, 177
diffusion, anomalous, 319
diffusion measure, 167
diffusion operator, 160
diffusion process, 160, 317

Dirac mass, xvi
Dirichlet class, 80
Dirichlet form, 165, 171, 173
  local, 167
  non-local, 167
Dirichlet integral, 166
Dirichlet operator, 165
discounted process, 270
discounting, 270
distribution, 4
distribution function, 5
Doléans-Dade exponential, 247
Doob's martingale inequality, 74
Doob's optional stopping theorem, 79
drift, xvi
dual operator, 181
dual space, 181

energy integral, 166
Esscher transform, 280
European call option, 269
event, 3
exercise price, 269
expectation, 6
expiration time, 269
explosion time, 312
exponential map, 300
exponential martingale, 254
exponential type of Lévy process, 288
extended stochastic integral, 203
extension (of an operator), 177

Fatou's lemma, 7
Feller process, 126
  sub-Feller process, 129
Feller semigroup, 126
  strong, 126
  sub-Feller semigroup, 129
Feynman–Kac formula, 345
filtered probability space, 71
filtration, 70
  augmented, 72
  augmented natural, 72
  natural, 71
financial derivative, 268
first hitting time of a process to a
    set, 79
flow, 298
  Brownian, 321
  Lévy, 321
  stochastic, 320
Fokker–Planck equation, 164

Subject index

form
   closed symmetric, 184
   Markovian, 168
   quasi-regular, 174
Fourier transform, 138, 185
Fourier transformation, 185
Friedrichs extension, 185
fundamental theorem of asset pricing, 270, 271
Föllmer–Schweizer minimal measure, 279

G-invariant measure, 3
gamma process, 52
Gaussian
   degenerate, 25
   non-degenerate, 24
Gaussian space–time white noise, 191
generalised inverse Gaussian, 282
generalised Marcus mapping, 348
generator, 131
geometric Brownian motion, 272
Girsanov's theorem, 257
global solution, 312
graph, 179
graph norm, 179
Gronwall's inequality, 295

hedge, 270
helices, 326
hermitian, 16
Hille–Yosida theorem, 135
Hille–Yosida–Ray theorem, 156
Holtsmark distribution, 34
homogeneous Markov process, 125
Hunt process, 171
Hurst index, 48
hyperbolic distribution, 282
Hölder's inequality, 8
Hörmander class, 188

identity matrix, xxiii
i.i.d., 11
implied volatility, 286
independent increments, 39
independently scattered property, 89
index of stability, 32
indicator function, 5
infinite-dimensional analysis, 260
infinitely divisible, 24
infinitesimal covariance, 317

infinitesimal mean, 317
inhomogeneous Markov process, 125
integrable function, 6
integrable process, 80
integral, 5
integral curve, 300
intensity measure, 87
interest rate, 269
interlacing, 48
invariant measure, 342
   G-invariant measure, 3
inverse Gaussian, 51
inverse Gaussian subordinator, 51, 83
isometric isomorphisms, 178
isometry, 178
Itô backwards stochastic integral, 241
Itô calculus, 230
Itô correction, 231
Itô differentials, 209
Itô diffusion, 317
Itô formula, xxi
Itô stochastic integral, 199
Itô's circle, 236
Itô's isometry, 199
Itô's product formula, 231

Jacobi identity, 300
Jacod's martingale representation theorem, 266
Jensen's inequality, 6
joint distribution, 5
jump, 111
jump-diffusion process, 319
jump measure, 167
jump process, 85

killed subordinator, 53
killing measure, 167
Kolmogorov backward equation, 164
Kolmogorov forward equation, 164
Kolmogorov's consistency criteria, 18
Kolmogorov's continuity criterion, 20
Ky Fan metric, 13

L^p-Markov semigroup, 148
L^p-positivity-preserving semigroup, 148
Landau notation, xxiv
Langevin equation, 316, 341
Laplace exponent, 50
law, 4
Lebesgue measurable sets, 4

Subject index

Lebesgue measure, 2
Lebesgue symmetric, 152
Lebesgue's dominated-convergence
  theorem, 7
lifetime, 128
Lipschitz, 293
Lipschitz condition, 293
Lipschitz constant, 293
local Dirichlet form, 167
local martingale, 80
local solution, 312
local unit, 157
localisation, 80
localised martingale-valued measure, 243
Lévy exponent, 30
Lévy flights, 69
Lévy flow, 321
Lévy kernel, 157
Lévy measure, 27
Lévy process, 39, 171
  background-driving, 217
  canonical, 60
  of exponential type, 288
  recurrent, 63
  stable, 48
  transient, 64
Lévy stochastic integral, 210
Lévy symbol, 30
Lévy type, 158
Lévy walk, 69
Lévy–Itô decomposition, xviii, 108
Lévy–Khintchine formula, xv, 28
Lévy-type backwards stochastic
  integral, 242
Lévy-type stochastic integral, 208

Marcus canonical equation, 323, 349
Marcus canonical integral, 238
  backwards, 243
Marcus mapping, 238
  generalised, 348
Marcus stochastic differential equation, 349
marginal distribution, 5
market
  complete, 271
  incomplete, 271
Markov process, 71
  canonical, 125
  homogeneous, 125
  inhomogeneous, 125

normal, 123
sub-, 128
Markovian form, 168
martingale, 72
  centred, 73
  closed, 73
  continuous, 73
  L^2, 73
  local, 80
martingale measure, 271
martingale problem, 347
  well-posed, 347
martingale representation theorem, 265
martingale space, 78
mean, 6
measurable function, simple, 5
measurable partition, 3
measurable space, 2
measure, 2
  absolutely continuous, 8
  completion of, 4
  counting, 3
  diffusion, 167
  equivalent, 8
  σ-finite, 2
  G-invariant, 3
  jump, 167
  killing, 167
  Lévy, 27
  of type $(2, \rho)$, 191
  probability, 3
  product, 11
  projection-valued, 183
measure space, 2
  product, 11
Meyer's angle-bracket process, 81
minimal entropy martingale measures, 281
mixing measure, 282
modification, 62
moment, 7
moment-generating function, 15
monotone-convergence theorem, 7
moving-average process, 214
multi-index, 187

n-step transition probabilities, 160
natural backwards filtration, 240
non-degenerate Gaussian vector, 24
non-local Dirichlet form, 167
norm, 178
normal Markov process, 123

Subject index

normal vector, 24
Novikov criterion, 255

Ornstein–Uhlenbeck process, 216, 316, 341
  generalised, 317
  integrated, 218
  orthogonal, 8

pdf, 9
phase multiplication, 185
Picard iteration, 293
Poisson integral, 91
  compensated, 94
Poisson point process, 90
Poisson process, 46
  compensated, 46
  compound, 46
Poisson random measure, 90
Poisson random variable, 25
  compound, 26
polarisation identity, xxiv
portfolio, 270
  replicating, 271
  self-financing, 271
positive definite, xxiii
  strictly, xxiii
positive maximum principle, 156
predictable, 192
predictable σ-algebra, 192
predictable process, 80
probability density function, 9
probability measure, 3
probability mixture, 282
probability space, 3
product measure, 11
product of measure spaces, 11
projection, 182
projection-valued measure, 183
pseudo-differential operator, 188
pseudo-Poisson process, 160
put option, 269

quadratic variation process, 220
quantisation, 142
quasi-left-continuous process, 171
quasi-regular form, 174

Radon–Nikodým derivative, 9
Radon–Nikodým theorem, 9
random field, 19
random measure, 89

Poisson, 90
random variable, 4
  characteristics of, 29
  complex, 4
  inverse Gaussian, 51
  stable, 32
  strictly stable, 32
  symmetric, 4
random variables
  identically distributed, 4
recurrent Lévy process, 64
regular, 167
relativistic Schrödinger operator, 142
replicating portfolio, 271
restriction, 177
return on investment, 272
Riemann hypothesis, 37
Riemann zeta function, 36
risk-neutral measure, 271

sample path, 19
Schoenberg correspondence, 16
Schwartz space, 187
self-adjoint semigroup, 152, 182
  essentially, 182
self-decomposable, 36
semigroup, 126
  C_0-, 130
  conservative, 148
  L^p-Markov, 148
  L^p-positivity-preserving, 148
  self-adjoint, 152
  sub-Markovian, 148
  weakly continuous, 60
separable, 8
separable stochastic process, 19
sequence of random variables
  i.i.d, 11
  independent, 11
shift, 325
spectrum, 180
Spitzer criterion, 64
stable Lévy process, 48
stable stochastic process, 48
standard normal random vector, 25
stationary increments, 39
stochastic calculus, 230
stochastic differentials, 209
stochastic exponential, 247, 315
stochastic flow, 320
stochastic integral
  extended, 203

stochastic process, 17
  adapted, 71
  finite-dimensional distributions of, 18
  G-invariant, 19
  Lebesgue symmetric, 152
  rotationally invariant, 19
  sample path of, 19
  separable, 19
  stable, 48
  stochastically continuous, 39
  strictly stationary, 215
  symmetric, 19
  μ-symmetric, 152
stochastic processes, independent, 17
stochastically continuous stochastic process, 39
stock drift, 272
stopped σ-algebra, 79
stopped random variable, 79
stopping time, 78
Stratonovitch integral, 236
Stratonovitch stochastic differential equation, 349
strictly stationary stochastic process, 215
strike price, 269
strong Markov property, 83, 171
sub-σ-algebra, 9
subdiffusion, 319
sub-Feller semigroup, 129
sub-Feller process, 129, 171
subfiltration, 71
sub-Markov process, 128
sub-Markovian semigroup, 148
submartingale, 73
subordinator, 49
  characteristics of, 49
  inverse Gaussian, 51
  killed, 53
superdiffusion, 319
supermartingale, 73
symbol, 188
symmetric operator, 182

Tanaka's formula, 244
tempered distributions, 187
Teugels martingales, 111
time change, 287
total mass, 2
  finite, 2
total set of vectors, 125
trace, xxiii
trading strategy, 270
transient Lévy process, 64
transition function, 161
transition operator, 160
transition probabilities, n-step, 161
translation, 185
translation-invariant semigroup, 137
translation semigroup, 130
transpose, xxiii
two-parameter filtration, 243

underlying, 268
uniformly integrable family of random variables, 80
unique in distribution, 347
upper semicontinuous mapping, 157

vaguely convergent to δ_0, 62
value, 269
variance, 7
variation, 95
  finite, 95
volatility, 272
  implied, 286
volatility smile, 287

weak convergence, 14
weak-sector condition, 173
weak solution, 346
weakly convergent to δ_0, 58
Wiener integral, 213
Wiener measure, 258
Wiener space, 258
  integration by parts in, 261
Wiener–Hopf factorisation, 65
Wiener–Lévy integral, 213